INTRODUCTION TO ELEMENTARY PARTICLE PHYSICS

The Standard Model is the theory of the elementary building blocks of matter and of their forces. It is the most comprehensive physical theory ever developed, and has been experimentally tested with high accuracy.

This textbook conveys the basic elements of the Standard Model using elementary concepts, without theoretical rigour. While most texts on this subject emphasise theoretical aspects, this textbook contains examples of basic experiments, before going into the theory. This allows readers to see how measurements and theory interplay in the development of physics. The author examines leptons, hadrons and quarks, before presenting the dynamics and surprising properties of the charges of the different forces. The textbook concludes with a brief discussion on the recent discoveries in physics beyond the Standard Model, and its connections with cosmology.

Quantitative examples are given throughout the book, and the reader is guided through the necessary calculations. Each chapter ends in exercises so readers can test their understanding of the material. Solutions to some problems are included in the book, and complete solutions are available to instructors at www.cambridge.org/9780521880213. This textbook is suitable for advanced undergraduate students and graduate students.

ALESSANDRO BETTINI is Professor of Physics at the University of Padua, Italy, former Director of the Gran Sasso Laboratory, Italy, and Director of the Canfranc Underground Laboratory, Spain. His research includes measurements of hadron quantum numbers, the lifetimes of charmed mesons, intermediate bosons and neutrino physics, and the development of detectors for particle physics.

INTRODUCTION TO ELEMENTARY PARTICLE PHYSICS

ALESSANDRO BETTINI
University of Padua, Italy

CAMBRIDGE
UNIVERSITY PRESS

CAMBRIDGE UNIVERSITY PRESS
Cambridge, New York, Melbourne, Madrid, Cape Town, Singapore, São Paulo, Delhi

Cambridge University Press
The Edinburgh Building, Cambridge CB2 8RU, UK

Published in the United States of America by Cambridge University Press, New York

www.cambridge.org
Information on this title: www.cambridge.org/9780521880213

First published 2008

Printed in the United Kingdom at the University Press, Cambridge

A catalogue record for this publication is available from the British Library

ISBN 978-0-521-88021-3 hardback

Contents

Preface

This book is mainly intended to be a presentation of subnuclear physics, at an introductory level, for undergraduate physics students, not necessarily for those specialising in the field. The reader is assumed to have already taken, at an introductory level, nuclear physics, special relativity and quantum mechanics, including the Dirac equation. Knowledge of angular momentum, its composition rules and the underlying group theoretical concepts is also assumed at a working level. No prior knowledge of elementary particles or of quantum field theories is assumed.

The Standard Model is the theory of the fundamental constituents of matter and of the fundamental interactions (excluding gravitation). A deep understanding of the 'gauge' quantum field theories that are the theoretical building blocks of this model requires skills that the readers are not assumed to have. However, I believe it to be possible to convey the basic physics elements and their beauty even at an elementary level. 'Elementary' means that only knowledge of elementary concepts (in relativistic quantum mechanics) is assumed. However it does not mean a superficial discussion. In particular, I have tried not to cut corners and I have avoided hiding difficulties, whenever was the case. I have included only well-established elements with the exception of the final chapter, in which I survey the main challenges of the present experimental frontier.

The text is designed to contain the material that may be accommodated in a typical undergraduate course. This condition forces the author to hard, and sometimes difficult, choices. The chapters are ordered in logical sequence. However, for a short course, a number of sections, or even chapters, can be left out. This is achieved at the price of a few repetitions. In particular, the treatments of oscillation and of the CP violation phenomena are given in an increasingly advanced way, first for the K mesons, then for the B mesons and finally for neutrinos.

The majority of the texts on elementary particles place special emphasis on theoretical aspects. However, physics is an experimental science and only experiment can decide which of the possible theoretical schemes has been chosen by Nature. Moreover, the progress of our understanding is often due to the discovery of unexpected phenomena. I have tried to select examples of basic experiments first, and then to go on to the theoretical picture.

A direct approach to the subject would start from leptons and quarks and their interactions and explain the properties of hadrons as consequences. A historical approach would also discuss the development of ideas. The former is shorter, but is lacking in depth. I tried to arrive at a balance between the two views.

The necessary experimental and theoretical tools are presented in the first chapter. From my experience, students have a sufficient knowledge of special relativity, but need practical exercise in the use of relativistic invariants and Lorentz transformations. In the first chapter I also include a summary of the artificial and natural sources of high-energy particles and of detectors. This survey is far from being complete and is limited to what is needed for the understanding of the experiments described in the following chapters.

The elementary fermions fall into two categories: the leptons, which can be found free, and the quarks, which always live inside the hadrons. Hadrons are non-elementary, compound structures, rather like nuclei. Three chapters are dedicated to the ground-level hadrons (the S wave nonets of pseudoscalar and vector mesons and the S wave octet and decimet of baryons), to their symmetries and to the measurement of their quantum numbers (over a few examples). The approach is partly historical.

There is a fundamental difference between hadrons on the one hand and atoms and nuclei on the other. While the electrons in atoms and nucleons in nuclei move at non-relativistic speeds, the quarks in the nucleons move almost at the speed of light. Actually, their rest energies are much smaller than their total energies. Subnuclear physics is fundamentally relativistic quantum mechanics.

The mass of a system can be measured if it is free from external interaction. Since the quarks are never free, for them the concept of mass must be extended. This can be done in a logically consistent way only within quantum chromodynamics (QCD).

The discoveries of an ever-increasing number of hadrons led to a confused situation at the beginning of the 1960s. The development of the quark model suddenly put hadronic spectroscopy in order in 1964. An attempt was subsequently made to develop the model further to explain the hadron mass spectrum. In this programme the largest fraction of the hadron mass was assumed to be due to the quark masses. Quarks were supposed to move slowly, at non-relativistic speeds inside the hadrons. This model, which was historically important in the development of the correct description of hadronic dynamics, is not satisfactory however. Consequently, I will limit the use of the quark model to classification.

The second part of the book is dedicated to the fundamental interactions and the Standard Model. The approach is substantially more direct. The most important experiments that prove the crucial aspects of the theory are discussed in some detail. I try to explain at an elementary level the space-time and gauge structure of the different types of 'charge'. I have included a discussion of the colour factors, giving examples of their attractive or repulsive character. I try to give some hint of the origin of hadron masses and of the nature of vacuum. In the weak interaction chapters the chiralities of the fermions and their weak couplings are discussed. The Higgs mechanism, the theoretical mechanism that gives rise to the masses of the particles, has not yet been tested experimentally. This will be done at the new high-energy large-hadron collider, LHC, now becoming operational at CERN. I shall only give a few hints about this frontier challenge.

In the final chapter I touch on the physics that has been discovered beyond the Standard Model. Actually, neutrino mixing, masses, oscillations and flavour transitions in matter make a beautiful set of phenomena that can be properly described at an elementary level, namely using only the basic concepts of quantum mechanics. Other clues to the physics beyond the Standard Model are already before our eyes. They are due mainly to the increasing interplay between particle physics and cosmology, astrophysics and nuclear physics. The cross fertilisation between these sectors will certainly be one of the main elements of fundamental research over the next few years. I limit the discussion to a few glimpses to give a flavour of this frontier research.

Problems

Numbers in physics are important; the ability to calculate a theoretical prediction on an observable or an experimental resolution is a fundamental characteristic of any physicist. More than 200 numerical examples and problems are presented. The simplest ones are included in the main text in the form of questions. Other problems covering a range of difficulty are given at the end of each chapter (except the last one). In every case the student can arrive at the solution without studying further theoretical material. Physics rather than mathematics is emphasised.

The physical constants and the principal characteristics of the particles are not given explicitly in the text of the problems. The student is expected to look for them in the tables given in the appendices. Solutions for about half of the problems are given at the end of the book.

Appendices

One appendix contains the dates of the main discoveries in particle physics, both experimental and theoretical. It is intended to give a bird's-eye view of the history of the field. However, keep in mind that the choice of the issues is partially arbitrary and that history is always a complex, non-linear phenomenon. Discoveries are seldom due to a single person and never happen instantaneously.

Tables of the Clebsch–Gordan coefficients, the spherical harmonics and the rotation functions in the simplest cases are included in the appendices. Other tables give the main properties of gauge bosons, of leptons, of quarks and of the ground levels of the hadronic spectrum.

The principal source of the data in the tables is the 'Review of Particle Properties' (Yao *et al.* 2006). This 'Review', with its website http://pdg.lbl.gov/, may be very useful to the reader too. It includes not only the complete data on elementary particles, but also short reviews of topics such as tests of the Standard Model, searches for hypothetical particles, particle detectors, probability and statistical methods, etc. However, it should be kept in mind that these 'mini-reviews' are meant to be summaries for the expert and that a different literature is required for a deeper understanding.

Reference material on the Internet

There are several URLs present on the Internet that contain useful material for further reading and data on elementary particles, which are systematically adjourned. The URLs cited in this work were correct at the time of going to press, but the publisher and the author make no undertaking that the citations remain live or accurate or appropriate.

Acknowledgments

It is a pleasure to thank G. Carugno, E. Conti, A. Garfagnini, S. Limentani, G. Puglierin, F. Simonetto and F. Toigo for helpful discussions and critical comments during the preparation of this work and Dr Christine Pennison for her precious help with the English language.

The author would very much appreciate any comments and corrections to mistakes or misprints (including the trivial ones). Please address them to alessandro. bettini@pd.infn.it.

Every effort has been made to secure necessary permissions to reproduce copyright material in this work. If any omissions are brought to our notice, we will be happy to include appropriate acknowledgments on reprinting.

I am indebted to the following authors, institutions and laboratories for the permission to reproduce or adapt the following photographs and diagrams:

BABAR Collaboration and M. Giorgi for kind agreement on the reproduction of Fig. 8.11 before publication

Brookhaven National Laboratory for Fig. 4.19

CERN for Figs. 7.27, 9.6 and 9.25

Fermilab for Fig. 4.31

INFN for Fig. 1.14

Kamioka Observatory, Institute for Cosmic Ray Research, University of Tokyo and Y. Suzuki for Fig. 1.16

Lawrence Berkeley Laboratory for Fig. 1.18

Derek Leinweber, CSSM, University of Adelaide for Fig. 6.32

Salvatore Mele, CERN for Fig. 6.25

The Nobel Foundation for: Fig. 2.7 from F. Reines, Nobel Lecture 1995, Fig. 5; Fig. 2.8 from M. Schwartz, Nobel Lecture 1988, Fig. 1; Fig. 4.14 from l. Alvarez, Nobel Lecture 1968, Fig. 10; Figs. 4.22 and 4.23 from S. Ting, Nobel Lecture 1976, Fig. 3 and Fig. 12; Figs. 4.24, 4.25 and 4.26 from B. Richter, Nobel Lecture 1976, Figs. 5, 6 and 18; Fig. 4.28 from L. Lederman, Nobel Lecture 1988, Fig. 12;

Fig. 6.10(a) from R. E. Taylor, Nobel Lecture 1990, Fig. 14; Fig. 6.12 from J. Friedman, Nobel Lecture 1990, Fig. 1; Figs. 8.4 and 8.5 from M. Fitch, Nobel Lecture 1980, Figs. 1 and 3; Figs. 9.14(a), 9.14(b), 9.19(a) and 9.19(b) from C. Rubbia, Nobel Lecture 1984, Figs. 16(a), (b), 25 and 26

Particle Data Group and the Institute of Physics for Figs. 1.6, 1.8, 1.9, 4.3, 5.26, 6.3, 6.14, 9.24, 9.30, 9.33, 10.10

Stanford Linear Accelerator Center for Fig. 6.10

Super-Kamiokande Collaboration and Y. Suzuki for Fig. 1.17

I acknowledge the permission of the following publishers and authors for reprinting or adapting the following figures:

Elsevier for: Figs. 5.34(a) and (b) from P. Achard *et al.*, *Phys. Lett.* **B623** (2005) 26; Figs. 6.2 and 6.6 from B. Naroska, *Phys. Rep.* **148** (1987); Fig. 6.7 from S. L. Wu, *Phys. Rep.* **107** (1984) 59; Fig. 6.4 from H. J. Beherend *et al.*, *Phys. Lett.* **B183** (1987) 400; Fig. 7.16 from F. Koks and J. van Klinken, *Nucl. Phys.* **A272** (1976) 61; Fig. 8.2 from S. Gjesdal *et al.*, *Phys. Lett.* **B52** (1974) 113; Fig. 9.9 from D. Geiregat *et al.*, *Phys. Lett.* **B259** (1991) 499; Fig. 9.12 from C. Albajar *et al.*, *Z. Phys.* **C44** (1989) 15; Fig. 9.15 adapted from G. Arnison *et al.*, *Phys. Lett.* **B122** (1983) 103; Figs. 9.17 and 9.18(b) from C. Albajar *et al.*, *Z. Phys.* **C44** (1989) 15; Fig. 9.20 adapted from G. Arnison *et al.*, *Phys. Lett.* **B126** (1983) 398; Fig. 9.21 from C. Albajar *et al.*, *Z. Phys.* **C44** (1989) 15; Fig. 9.22 from C. Albajar *et al.*, *Z. Phys.* **C36** (1987) 33 and 18.21

Springer, D. Plane and the OPAL Collaboration for Fig. 5.31 from G. Abbiendi *et al.*, *Eur. Phys. J.* **C33** (2004) 173

Springer, E. Gallo and the ZEUS Collaboration for Fig. 6.15 from S. Chekanov *et al.*, *Eur. Phys. J.* **C21** (2001) 443

Progress of Theoretical Physics and Prof. K. Niu for Fig. 4.27 from K. Niu *et al.*, *Progr. Theor. Phys.* **46** (1971) 1644

John Wiley & Sons, Inc. and the author J. W. Rohlf for Figs. 6.3, 9.15 and 9.20 adapted from Figs. 18.3, 18.17 and 18.21 of *Modern Physics from a to Z^0*, 1994

The American Physical Society http://publish.aps.org/linkfaq.html and D. Nygren, S. Vojcicki, P. Schlein, A. Pevsner, R. Plano, G. Moneti, M. Yamauchi, Y. Suzuki and K. Inoue for Fig. 1.7 from H. Aihara *et al.*, *Phys. Rev. Lett.* **61** (1988) 1263; for Fig. 4.4 from L. Alvarez *et al.*, *Phys. Rev. Lett.* **10** (1963) 184; for Fig. 4.5 from P. Schlein *et al.*, *Phys. Rev. Lett.* **11** (1963) 167; for Fig. 4.13(a) from A. Pevsner *et al.*, *Phys. Rev. Lett.* **7** (1961) 421; for Fig. 4.13(b) and (c) and Fig. 4.14 (b) from C. Alff *et al.*, *Phys. Rev. Lett.* **9** (1962) 325; for Fig. 4.29 from A. Andrews *et al.*, *Phys. Rev. Lett.* **44** (1980) 1108; for Fig. 8.8 from K. Abe *et al.*, *Phys. Rev.* **D71** (2005) 072003; for Fig. 10.7 from Y. Ashie *et al.*, *Phys. Rev.* **D71** (2005) 112005; for Fig. 10.11 from T. Araki *et al.*, *Phys. Rev. Lett.* **94** (2005) 081801.

1

Preliminary notions

1.1 Mass, energy, linear momentum

Elementary particles have generally very high speeds, close to that of light. Therefore, we recall a few simple properties of relativistic kinematics and dynamics in this section and in the next three.

Let us consider two reference frames in rectilinear uniform relative motion $S(t,x,y,z)$ and $S'(t',x',y',z')$. We choose the axes as represented in Fig. 1.1. At a certain moment, which we take as $t' = t = 0$, the origins and the axes coincide. The frame S' moves relative to S with speed $\mathbf{V}$, in the direction of the x-axis.

We introduce the following two dimensionless quantities relative to the motion in S of the origin of S'

$$\boldsymbol{\beta} \equiv \frac{\mathbf{V}}{c} \tag{1.1}$$

and

$$\gamma = \frac{1}{\sqrt{1 - \beta^2}} \tag{1.2}$$

called the 'Lorentz factor'. An event is defined by the four-vector of the coordinates $(ct,\mathbf{r})$. Its components in the two frames (t,x,y,z) and (t',x',y',z') are linked by the Lorentz transformations (Lorentz 1904, Poincaré 1905)

$$
\begin{aligned}
x' &= \gamma(x - \beta ct) \\
y' &= y \\
z' &= z \\
ct' &= \gamma(ct - \beta x).
\end{aligned}
\tag{1.3}
$$

The Lorentz transformations form a group that H. Poincaré, who first recognised this property in 1905, called the Lorentz group. The group contains the parameter c,

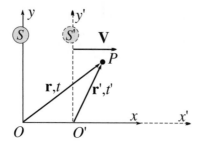

Fig. 1.1.   Two reference frames in rectilinear relative motion.

a constant with the dimensions of the velocity. A physical entity moving at speed c in a reference frame moves with the same speed in any other frame. In other words, c is invariant under Lorentz transformations. It is the propagation speed of all the fundamental perturbations: light and gravitational waves (Poincaré 1905).

The same relationships are valid for any four-vector. Of special importance is the energy-momentum vector (E/c, $\mathbf{p}$) of a free particle

$$p_{x'} = \gamma \left(p_x - \beta \frac{E}{c} \right)$$
$$p_{y'} = p_y$$
$$p_{z'} = p_z \tag{1.4}$$
$$\frac{E'}{c} = \gamma \left(\frac{E}{c} - \beta p_x \right).$$

The transformations that give the components in S as functions of those in S', the inverse of (1.3) and (1.4), can be most simply obtained by changing the sign of the speed $\mathbf{V}$.

The norm of the energy-momentum vector is, as for all the four-vectors, an invariant; the square of the mass of the system multiplied by the invariant factor c^4

$$m^2 c^4 = E^2 - p^2 c^2. \tag{1.5}$$

This is a fundamental expression: it is the definition of the mass. It is, we repeat, valid only for a free body but is completely general: for point-like bodies, such as elementary particles, and for composite systems, such as nuclei or atoms, even in the presence of internal forces.

The most general relationship between the linear momentum (we shall call it simply momentum) $\mathbf{p}$, the energy E and the speed $\mathbf{v}$ is

$$\mathbf{p} = \frac{E}{c^2} \mathbf{v} \tag{1.6}$$

which is valid both for bodies with zero and non-zero mass.

For massless particles (1.5) can be written as

$$pc = E. \tag{1.7}$$

The photon mass is exactly zero. Neutrinos have non-zero but extremely small masses in comparison to the other particles. In the kinematic expressions involving neutrinos, their mass can usually be neglected.

If $m \neq 0$ the energy can be written as

$$E = m\gamma c^2 \tag{1.8}$$

and (1.6) takes the equivalent form

$$\mathbf{p} = m\gamma\mathbf{v}. \tag{1.9}$$

We call the reader's attention to the fact that one can find in the literature, and not only in that addressed to the general public, concepts that arose when the theory was not yet well understood and that are useless and misleading. One of these is the 'relativistic mass' that is the product $m\gamma$, and the dependence of mass on velocity. The mass is a Lorentz invariant, independent of the speed; the 'relativistic mass' is simply the energy divided by c^2 and as such the fourth component of a four-vector; this of course if $m \neq 0$, while for $m = 0$ relativistic mass has no meaning at all. Another related term to be avoided is the 'rest mass', namely the 'relativistic mass' at rest, which is simply the mass.

The concept of mass applies, to be precise, only to the stationary states, i.e. to the eigenstates of the free Hamiltonian, just as only monochromatic waves have a well-defined frequency. Even the barely more complicated wave, the dichromatic wave, does not have a well-defined frequency. We shall see that there are two-state quantum systems, such as K^0 and B^0, which are naturally produced in states different from stationary states. For the former states it is not proper to speak of mass and of lifetime. As we shall see, the nucleons, as protons and neutrons are collectively called, are made up of quarks. The quarks are never free and consequently the definition of quark mass presents difficulties, which we shall discuss later.

Example 1.1 Consider a source emitting a photon with energy E_0 in the frame of the source. Take the x-axis along the direction of the photon. What is the energy E of the photon in a frame in which the source moves in the x direction at the speed $v = \beta c$? Compare with the Doppler effect.

Call S' the frame of the source. Remembering that photon energy and momentum are proportional, we have $p'_x = p' = E_0/c$. The inverse of the last equation in (1.4) gives

$$\frac{E}{c} = \gamma\left(\frac{E_0}{c} + \beta p'_x\right) = \gamma\frac{E_0}{c}(1 + \beta)$$

and we have $\dfrac{E}{E_0} = \gamma(1 + \beta) = \sqrt{\dfrac{1 + \beta}{1 - \beta}}.$

Doppler effect theory tells us that, if a source emits a light wave of frequency ν_0, an observer who sees the source approaching at speed $v = \beta c$ measures the frequency ν, such that $\dfrac{\nu}{\nu_0} = \sqrt{\dfrac{1 + \beta}{1 - \beta}}.$ This is no wonder, in fact quantum mechanics tells us that $E = h\nu$.

1.2 The law of motion of a particle

The 'relativistic' law of motion of a particle was found by Planck in 1906 (Planck 1906). As in Newtonian mechanics, a force $\mathbf{F}$ acting on a particle of mass $m \neq 0$ results in a variation in time of its momentum. Newton's law in the form $\mathbf{F} = d\mathbf{p}/dt$ (the form used by Newton himself) is also valid at high speed, provided the momentum is expressed by Eq. (1.9). The expression $\mathbf{F} = m\mathbf{a}$, used by Einstein in 1905, on the contrary, is wrong. It is convenient to write explicitly

$$\mathbf{F} = \frac{d\mathbf{p}}{dt} = m\gamma\mathbf{a} + m\frac{d\gamma}{dt}\mathbf{v}. \tag{1.10}$$

Taking the derivative, we obtain

$$m\frac{d\gamma}{dt}\mathbf{v} = m\frac{d\left(1 - \frac{v^2}{c^2}\right)^{-1/2}}{dt}\mathbf{v} = -m\frac{1}{2}\left(1 - \frac{v^2}{c^2}\right)^{-3/2}\left(-2\frac{v}{c^2}a_t\right)\mathbf{v}$$
$$= m\gamma^3(\mathbf{a}\cdot\boldsymbol{\beta})\boldsymbol{\beta}.$$

Hence

$$\mathbf{F} = m\gamma\mathbf{a} + m\gamma^3(\mathbf{a}\cdot\boldsymbol{\beta})\boldsymbol{\beta}. \tag{1.11}$$

We see that the force is the sum of two terms, one parallel to the acceleration and one parallel to the velocity. Therefore, we cannot define any 'mass' as the ratio between acceleration and force. At high speeds, the mass is *not* the inertia to motion.

To solve for the acceleration we take the scalar product of the two members of Eq. (1.11) with $\boldsymbol{\beta}$. We obtain

$$\mathbf{F}\cdot\boldsymbol{\beta} = m\gamma\mathbf{a}\cdot\boldsymbol{\beta} + m\gamma^3\beta^2\mathbf{a}\cdot\boldsymbol{\beta} = m\gamma(1 + \gamma^2\beta^2)\mathbf{a}\cdot\boldsymbol{\beta} = m\gamma^3\mathbf{a}\cdot\boldsymbol{\beta}.$$

Hence

$$\mathbf{a} \cdot \boldsymbol{\beta} = \frac{\mathbf{F} \cdot \boldsymbol{\beta}}{m\gamma^3}$$

and, by substitution into (1.11)

$$\mathbf{F} - (\mathbf{F} \cdot \boldsymbol{\beta})\boldsymbol{\beta} = m\gamma\mathbf{a}.$$

The acceleration is the sum of two terms, one parallel to the force, and one parallel to the speed.

Force and acceleration have the same direction in two cases only: (1) force and velocity are parallel: $\mathbf{F} = m\gamma^3\mathbf{a}$; (2) force and velocity are perpendicular: $\mathbf{F} = m\gamma\mathbf{a}$. Notice that the proportionality constants are different.

In order to have simpler expressions in subnuclear physics the so-called 'natural units' are used. We shall discuss them in Section 1.5, but we anticipate here one definition: without changing the SI unit of time, we define the unit of length in such a way that $c = 1$. In other words, the unit length is the distance the light travels in a second in vacuum, namely 299 792 458 m, a very long distance. With this choice, in particular, mass, energy and momentum have the same physical dimensions. We shall often use as their unit the electronvolt (eV) and its multiples.

1.3 The mass of a system of particles, kinematic invariants

The mass of a system of particles is often called 'invariant mass', but the adjective is useless; the mass is always invariant.

The expression is simple only if the particles of the system do not interact amongst themselves. In this case, for n particles of energies E_i and momenta $\mathbf{p}_i$, the mass is

$$m = \sqrt{E^2 - P^2} = \sqrt{\left(\sum_{i=1}^{n} E_i\right)^2 - \left(\sum_{i=1}^{n} \mathbf{p}_i\right)^2}. \qquad (1.12)$$

Consider the square of the mass which we shall indicate by s, obviously an invariant quantity

$$s = E^2 - P^2 = \left(\sum_{i=1}^{n} E_i\right)^2 - \left(\sum_{i=1}^{n} \mathbf{p}_i\right)^2. \qquad (1.13)$$

Notice that s cannot be negative

$$s \geq 0. \qquad (1.14)$$

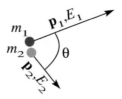

Fig. 1.2.   System of two non-interacting particles.

Let us see its expression in the 'centre of mass' (CM) frame that is defined as the reference in which the total momentum is zero. We see immediately that

$$s = \left(\sum_{i=1}^{n} E_i^* \right)^2 \tag{1.15}$$

where E_i^* are the energies in the centre of mass frame. In words, the mass of a system of non-interacting particles is also its energy in the centre of mass frame.

Consider now a system made up of two non-interacting particles. It is the simplest system and also a very important one. Figure 1.2 defines the kinematic variables.

The expression of s is

$$s = (E_1 + E_2)^2 - (\mathbf{p}_1 + \mathbf{p}_2)^2 = m_1^2 + m_2^2 + 2E_1E_2 - 2\mathbf{p}_1 \cdot \mathbf{p}_2 \tag{1.16}$$

and, in terms of the velocity, $\boldsymbol{\beta} = \mathbf{p}/E$

$$s = m_1^2 + m_2^2 + 2E_1E_2(1 - \boldsymbol{\beta}_1 \cdot \boldsymbol{\beta}_2). \tag{1.17}$$

Clearly in this case, and as is also true in general, the mass of a system is not the sum of the masses of its constituents, even if these do not interact. It is also clear from Eq. (1.12) that energy and momentum conservation implies that the mass is a conserved quantity: in a reaction such as a collision or decay, the mass of the initial system is always equal to that of the final system. For the same reason the sum of the masses of the bodies present in the initial state is generally different from the sum of the masses of the final bodies.

Example 1.2   We find the expressions for the mass of the system of two photons of the same energy E, if they move in equal or in different directions.

The energy and the momentum of the photon are equal, because its mass is zero, $p = E$. The total energy $E_{\text{tot}} = 2E$.

If the photons have the same direction then the total momentum is $p_{\text{tot}} = 2E$ and therefore the mass is $m = 0$.

If the velocities of the photons are opposite, $E_{tot} = 2E$, $p_{tot} = 0$, and hence $m = 2E$.

In general, if θ is the angle between the velocities, $p_{tot}^2 = 2p^2 + 2p^2 \cos \theta = 2E^2(1 + \cos \theta)$ and hence $m^2 = 2E^2(1 - \cos \theta)$.

Notice that the system does not contain any matter, but only energy. Contrary to intuition, mass is *not* a measure of the quantity of matter in a body.

Now consider one of the basic processes of subnuclear physics, collisions. In the initial state two particles, a and b, are present, in the final state we may have two particles (not necessarily a and b) or more. Call these c, d, e, ... The process is

$$a + b \rightarrow c + d + e + \cdots. \tag{1.18}$$

If the final state contains the initial particles, and only them, then the collision is said to be elastic.

$$a + b \rightarrow a + b. \tag{1.19}$$

We specify that the excited state of a particle must be considered as a different particle.

The time spent by the particles in the interaction, the collision time, is extremely short and we shall think of it as instantaneous. Therefore, the particles in both the initial and final states can be considered as free.

We shall consider two reference frames, the centre of mass frame already defined above and the laboratory frame (L). The latter is the frame in which, before the collision, one of the particles called the target is at rest, while the other, called the beam, moves against it. Let a be the beam particle, m_a its mass, $\mathbf{p}_a$ its momentum and E_a its energy; let b be the target particle and m_b its mass. Figure 1.3 shows the system in the initial state.

In the laboratory frame, s is given by

$$s = (E_a + m_b)^2 - p_a^2 = m_a^2 + m_b^2 + 2m_b E_a. \tag{1.20}$$

In practice, the energy of the projectile is often, but not always, much larger than both the projectile and the target masses. If this is the case, we can approximate Eq. (1.20) by

$$s \approx 2m_b E_a \qquad (E_a, E_b \gg m_a, m_b). \tag{1.21}$$

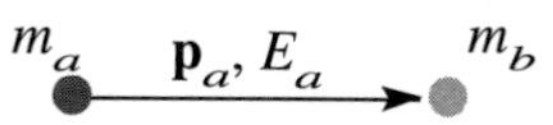

Fig. 1.3.   The laboratory frame (L).

We are often interested in producing new types of particles in the collision, and therefore in the energy available for such a process. This is obviously the total energy in the centre of mass, which, as seen in (1.21), grows proportionally to the square root of the beam energy.

Let us now consider the centre of mass frame, in which the two momenta are equal and opposite, as in Figure 1.4. If the energies are much larger than the masses, $E_a^* \gg m_a$ and $E_b^* \gg m_b$, the energies are approximately equal to the momenta: $E_a^* \approx p_a^*$ and $E_b^* \approx p_b^*$, hence equal to each other, and we call them simply E^*. The total energy squared is

$$s \approx (2E^*)^2 \quad (E^* \gg m_a, m_b). \tag{1.22}$$

We see that the total centre of mass energy is proportional to the energy of the colliding particles. In the centre of mass frame, all the energy is available for the production of new particles, in the laboratory frame only part of it is available, because momentum must be conserved.

Now let us consider a collision with two particles in the final state: the two-body scattering

$$a + b \rightarrow c + d. \tag{1.23}$$

Figure 1.5 shows the initial and final kinematics in the laboratory and in the centre of mass frames. Notice in particular that in the centre of mass frame the final momentum is in general different from the initial momentum; they are equal *only* if the scattering is elastic.

Since s is an invariant it is equal in the two frames; since it is conserved it is equal in the initial and final states. We have generically in any reference frame

$$s = (E_a + E_b)^2 - (\mathbf{p}_a + \mathbf{p}_b)^2 = (E_c + E_d)^2 - (\mathbf{p}_c + \mathbf{p}_d)^2. \tag{1.24}$$

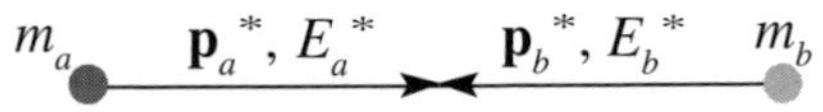

$$m_a \quad \mathbf{p}_a^*, E_a^* \quad \mathbf{p}_b^*, E_b^* \quad m_b$$

Fig. 1.4.   The centre of mass reference frame (CM).

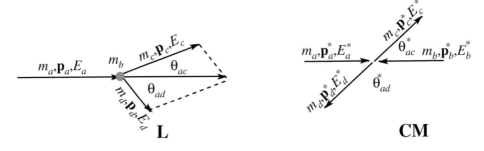

Fig. 1.5.   Two-body scattering in the L and CM frames.

These properties are useful to solve a number of kinematic problems, as we shall see later in the 'Problems' section.

In a two-body scattering, there are two other important kinematic variables that have the dimensions of the square of an energy: the a–c four-momentum transfer t, and the a–d four-momentum transfer u. The first is defined as

$$t \equiv \left(E_c - E_a\right)^2 - \left(\mathbf{p}_c - \mathbf{p}_a\right)^2. \tag{1.25}$$

It is easy to see that the energy and momentum conservation implies

$$t = \left(E_c - E_a\right)^2 - \left(\mathbf{p}_c - \mathbf{p}_a\right)^2 = \left(E_d - E_b\right)^2 - \left(\mathbf{p}_d - \mathbf{p}_b\right)^2. \tag{1.26}$$

In a similar way

$$u \equiv \left(E_d - E_a\right)^2 - \left(\mathbf{p}_d - \mathbf{p}_a\right)^2 - \left(E_c - E_b\right)^2 - \left(\mathbf{p}_c - \mathbf{p}_b\right)^2. \tag{1.27}$$

The three variables are not independent. It is easy to show (see Problems) that

$$s + t + u = m_a^2 + m_b^2 + m_c^2 + m_d^2. \tag{1.28}$$

Notice finally that

$$t \leq 0 \qquad u \leq 0. \tag{1.29}$$

1.4 Systems of interacting particles

Let us now consider a system of interacting particles. We immediately stress that its total energy is not in general the sum of the energies of the single particles, $E \neq \sum_{i=1}^{n} E_i$, because the field responsible for the interaction itself contains energy. Similarly, the total momentum is not the sum of the momenta of the particles, $\mathbf{P} \neq \sum_{i=1}^{n} \mathbf{p}_i$, because the field contains momentum. In conclusion, Eq. (1.12) does not in general give the mass of the system. We shall restrict ourselves to a few important examples in which the calculation is simple.

Let us first consider a particle moving in an external, given field. This means that we can consider the field independent of the motion of the particle.

Let us start with an atomic electron of charge q_e at the distance r from a nucleus of charge Zq_e. The nucleus has a mass $M_N \gg m_e$, hence it is not disturbed by the electron motion. The electron then moves in a constant potential $\phi = -\frac{1}{4\pi\varepsilon_0}\frac{Zq_e}{r}$. The electron energy (in SI units) is

$$E = \sqrt{m_e^2 c^4 + p^2 c^2} - \frac{1}{4\pi\varepsilon_0}\frac{Zq_e^2}{r} \approx m_e c^2 + \frac{p^2}{2m_e} - \frac{1}{4\pi\varepsilon_0}\frac{Zq_e^2}{r}$$

where, in the last member, we have taken into account that the atomic electron speeds are much smaller than c. The final expression is valid in non-relativistic situations, as is the case in an atom, and it is the non-relativistic expression of the energy, apart from the irrelevant constant $m_e c^2$.

Let us now consider a system composed of an electron and a positron. The positron, as we shall see, is the antiparticle of the electron. It has the same mass and opposite charge. The difference from the hydrogen atom is that there is no longer a fixed centre of force. We must consider not only the two particles but also the electromagnetic field in which they move, which, in turn, depends on their motion. If the energies are high enough, quantum processes happen at an appreciable frequency: the electron and the positron can annihilate each other, producing photons; inversely, a photon of the field can 'materialise' in a positron–electron pair. In these circumstances, we can no longer speak of a potential.

In conclusion, the concept of potential is non-relativistic: we can use it if the speeds are small in comparison to c or, in other words, if energies are much smaller than the masses. It is correct for the electrons in the atoms, to first approximation, but not for the quarks in the nucleons.

Example 1.3   Consider the fundamental level of the hydrogen atom. The energy needed to separate the electron from the proton is $\Delta E = 13.6$ eV. The mass of the atom is smaller than the sum of the masses of its constituents by this quantity, $m_H + \Delta E = m_p + m_e$. The relative mass difference is

$$\frac{m_H - m_p - m_e}{m_H} = \frac{13.6}{9.388 \times 10^8} = 1.4 \times 10^{-8}.$$

This quantity is extremely small, justifying the non-relativistic approximation.

Example 1.4   The processes we have mentioned above, of electron–positron annihilation and pair production, can take place only in the presence of another body. Otherwise, energy and momentum cannot be conserved simultaneously. Let us now consider the following processes:

- $\gamma \rightarrow e^+ + e^-$. Let E_+ be the energy and $\mathbf{p}_+$ the momentum of e^+, E_- and $\mathbf{p}_-$ those of e^-. In the initial state $s = 0$; in the final state $s = (E_+ + E_-)^2 - (\mathbf{p}_+ + \mathbf{p}_-)^2 = 2m_e^2 + 2(E_+ E_- - p_+ p_- \cos\theta) > 2\,m_e^2 > 0$. This reaction cannot occur.
- $e^+ + e^- \rightarrow \gamma$. This is just the inverse reaction, it cannot occur either.
- $\gamma + e^- \rightarrow e^-$. Let the initial electron be at rest, let E_γ be the energy of the photon, E_f, $\mathbf{p}_f$ the energy and the momentum of the final electron. Initially $s = (E_\gamma + m_e)^2 - p_\gamma^2 = 2m_e E_\gamma + m_e^2$, in the final state $s = E_f^2 - p_f^2 = m_e^2$. Setting

the two expressions equal we obtain $2m_eE_\gamma = 0$, which is false. The same is true for the inverse process $e^- \rightarrow e^- + \gamma$. This process happens in the Coulomb field of the nucleus, in which the electron accelerates and radiates a photon. The process is known by the German word bremsstrahlung.

Example 1.5 Macroscopically inelastic collision. Consider two bodies of the same mass m moving initially one against the other with the same speed v (for example two wax spheres). The two collide and remain attached in a single body of mass M.

The total energy does not vary, but the initial kinetic energy has disappeared. Actually, the rest energy has increased by the same amount. The energy conservation is expressed as $2\gamma mc^2 = Mc^2$. The mass of the composite body is $M > 2m$, but by just a little.

Let us see by how much, as a percentage, for a speed of $v = 300$ m/s. This is rather high by macroscopic standards, but small compared to c, $\beta = v/c = 10^{-6}$.

Expanding in series: $M = 2\gamma m = \dfrac{2m}{\sqrt{1 - \beta^2}} \approx 2m(1 + \tfrac{1}{2}\beta^2)$. The relative mass

difference is: $\dfrac{M - 2m}{2m} \approx \tfrac{1}{2}\beta^2 \approx 10^{-12}$.

It is so small that we cannot measure it directly; we do it indirectly by measuring the increase in temperature with a thermometer.

Example 1.6 Nuclear masses. Let us consider a ^{4}He nucleus, which has a mass of $m_{\mathrm{He}} = 3727.41$ MeV. Recalling that $m_p = 938.27$ MeV and $m_n = 939.57$ MeV, the mass defect is $\Delta E = (2m_p + 2m_n) - m_{\mathrm{He}} = 28.3$ MeV, or, in relative

terms, $\dfrac{\Delta E}{m_{\mathrm{He}}} = \dfrac{28.3}{3727.41} = 0.8\%$.

In general, the mass defects in the nuclei are much larger than in the atoms; indeed, they are bound by a much stronger interaction.

1.5 Natural units

In the following, we shall normally use the so-called 'natural units' (NU). Actually, we have already started to do so. We shall also use the electronvolt instead of the joule as the unit of energy.

Let us start by giving $\hbar$ and c in useful units:

$$\hbar = 6.58 \times 10^{-16} \, \mathrm{eV \, s}. \tag{1.30}$$

$$c = 3 \times 10^{23} \, \mathrm{fm \, s}^{-1}. \tag{1.31}$$

$$\hbar c = 197 \, \text{MeV fm (or GeV am)}. \tag{1.32}$$

As we have already done, we keep the second as unit of time and define the unit of length such that $c = 1$. Therefore, in dimensional equations we shall have $[L] = [T]$.

We now define the unit of mass in such a way as to have $\hbar = 1$. Mass, energy and momentum have the same dimensions: $[M] = [E] = [P] = [L^{-1}]$.

For unit conversions the following relationships are useful:

$$1 \, \text{MeV} = 1.52 \times 10^{21} \, \text{s}^{-1} \qquad 1 \, \text{MeV}^{-1} = 197 \, \text{fm}$$

$$1 \, \text{s} = 3 \times 10^{23} \, \text{fm} \qquad 1 \, \text{s}^{-1} = 6.5 \times 10^{-16} \, \text{eV} \qquad 1 \, \text{ps}^{-1} = 0.65 \, \text{meV}$$

$$1 \, \text{m} = 5.07 \times 10^{6} \, \text{eV}^{-1} \qquad 1 \, \text{m}^{-1} = 1.97 \times 10^{-7} \, \text{eV}.$$

The square of the electron charge is related to the fine structure constant a by the relation

$$\frac{q_e^2}{4\pi\varepsilon_0} = a\hbar c \approx 2.3 \times 10^{-28} \, \text{J m}. \tag{1.33}$$

Being dimensionless, a has the same value in all unit systems (note that, unfortunately, one can still find in the literature the Heaviside–Lorentz units, in which $\varepsilon_0 = \mu_0 = 1$),

$$a = \frac{q_e^2}{4\pi\varepsilon_0 \hbar c} \approx \frac{1}{137}. \tag{1.34}$$

Notice that the symbol m can mean both the mass and the rest energy mc^2, but remember that the first is Lorentz-invariant, the second is the fourth component of a four-vector. To be complete, the same symbol may also mean the reciprocal of the Compton length times 2π, $\dfrac{2\pi\hbar}{mc}$.

Example 1.7 Measuring the lifetime of the π^0 meson one obtains $\tau_{\pi^0} = 8.4 \times 10^{-17}$ s; what is its width? Measuring the width of the η meson one obtains $\Gamma_\eta = 1.3 \, \text{keV}$; what is its lifetime? We simply use the uncertainty principle:

$$\Gamma_{\pi^0} = \hbar/\tau_{\pi^0} = \left(6.6 \times 10^{-16} \, \text{eV s}\right)/\left(8.4 \times 10^{-17} \, \text{s}\right) = 8 \, \text{eV}$$

$$\tau_\eta = \hbar/\tau_\eta = \left(6.6 \times 10^{-16} \, \text{eV s}\right)/(1300 \, \text{eV}) = 5 \times 10^{-19} \, \text{s}.$$

In conclusion, lifetime and width are completely correlated. It is sufficient to measure one of the two. The width of the π^0 particle is too small to be measured, and so we measure its lifetime; vice versa in the case of the η particle.

Example 1.8 Evaluate the Compton wavelength of the proton.

$$\lambda_p = 2\pi/m = (6.28/938)\,\text{MeV}^{-1} = 6.7 \times 10^{-3}\,\text{MeV}^{-1}$$
$$= 6.7 \times 10^{-3} \times 197\,\text{fm} = 1.32\,\text{fm}.$$

1.6 Collisions and decays

As we have already stated, subnuclear physics deals with two types of processes: collisions and decays. In both cases the transition amplitude is given by the matrix element of the interaction Hamiltonian between final $|f\rangle$ and initial $|i\rangle$ states

$$M_{fi} = \langle f|H_{\text{int}}|i\rangle. \tag{1.35}$$

We shall now recall the basic concepts and relations.

Collisions Consider the collision $a + b \rightarrow c + d$. Depending on what we measure, we can define the final state with more or fewer details: we can specify or not specify the directions of c and d, we can specify or not specify their polarisations, we can say that particle c moves in a given solid angle around a certain direction without specifying the rest, etc. In each case, when computing the **cross section** of the observed process we must integrate on the non-observed variables.

Given the two initial particles a and b, we can have different particles in the final state. Each of these processes is called a 'channel' and its cross section is called the 'partial cross section' of that channel. The sum of all the partial cross sections is the total cross section.

Decays Consider, for example, the three-body decay $a \rightarrow b + c + d$: again, the final state can be defined with more or fewer details, depending on what is measured. Here the quantity to compute is the decay rate in the measured final state. Integrating over all the possible kinematic configurations, one obtains the partial decay rate Γ_{bcd}, or partial width, of a into the $b\,c\,d$ channel. The sum of all the partial decay rates is the **total width** of a. The latter, as we have anticipated in Example 1.7, is the reciprocal of the lifetime: $\Gamma = 1/\tau$.

The **branching ratio** of a into $b\,c\,d$ is the ratio $R_{bcd} = \Gamma_{bcd}/\Gamma$.

For both collisions and decays, one calculates the number of interactions per unit time, normalising in the first case to one target particle and one beam particle, in the second case to one decaying particle.

Let us start with the collisions, more specifically with 'fixed target' collisions. There are two elements:

1. The beam, which contains particles of a definite type moving, approximately, in the same direction and with a certain energy spectrum. The beam intensity I_b is the number of incident particles per unit time, the beam flux Φ_b is the intensity per unit normal section.

2. The target, which is a piece of matter. It contains the scattering centres of interest to us, which may be the nuclei, the nucleons, the quarks or the electrons, depending on the case. Let n_t be the number of scattering centres per unit volume and N_t be their total number (if the beam section is smaller than that of the target, N_t is the number of centres in the beam section).

The interaction rate R_i is the number of interactions per unit time (the quantity that we measure). By definition of the cross section σ of the process, we have

$$R_i = \sigma N_t \Phi_b = W N_t \qquad (1.36)$$

where W is the rate per particle in the target. To be rigorous, one should consider that the incident flux diminishes with increasing penetration depth in the target, due to the interactions of the beam particles. We shall consider this issue soon. We find N_t by recalling that the number of nucleons in a gram of matter is in all cases, with sufficient accuracy, the Avogadro number N_A. Consequently, if M is the target mass in kg we must multiply by 10^3, obtaining

$$N_{nucleons} = M(\text{kg})(10^3\text{kg/g})N_A. \qquad (1.37)$$

If the targets are nuclei of mass number A

$$N_{nuclei} = \frac{M(\text{kg})(10^3\text{kg/g})N_A}{A(\text{mol/g})}. \qquad (1.38)$$

The cross section has the dimensions of a surface. In nuclear physics one uses as a unit the barn $= 10^{-28}$ m². Its order of magnitude is the geometrical section of a nucleus with $A \approx 100$. In subnuclear physics the cross sections are smaller and submultiples are used: mb, μb, pb, etc.

In NU, the following relationships are useful

$$1 \text{ mb} = 2.5 \text{ GeV}^{-2}, \qquad 1 \text{ GeV}^{-2} = 389 \,\mu\text{b}. \qquad (1.39)$$

Consider a beam of initial intensity I_0 entering a long target of density ρ (kg/m³). Let z be the distance travelled by the beam in the target, measured from its entrance point. We want to find the beam intensity $I(z)$ as a function of this distance. Consider a generic infinitesimal layer between z and $z + dz$. If dR_i is the total number of interactions per unit time in the layer, the variation of the intensity in crossing the layer is $dI(z) = -dR_i$. If Σ is the normal section of the target, $\Phi_b(z) = I(z)/\Sigma$ is the flux and σ_{tot} is the total cross section, we have

$$dI(z) = -dR_i = -\sigma_{tot}\Phi_b(z)\,dN_t = -\sigma_{tot}\frac{I(z)}{\Sigma}n_t\Sigma\,dz$$

or

$$\frac{dI(z)}{I(z)} = -\sigma_{tot}n_t\,dz.$$

In conclusion, we have

$$I(z) = I_0 e^{-n_t\sigma_{tot}z}. \tag{1.40}$$

The 'absorption length', defined as the distance at which the beam intensity is reduced by the factor $1/e$, is

$$L_{abs} = 1/(n_t\sigma_{tot}). \tag{1.41}$$

Another related quantity is the 'luminosity' $\mathcal{L}$ $[\mathrm{m^{-2}\,s^{-1}}]$, often given in $[\mathrm{cm^{-2}\,s^{-1}}]$, defined as the number of collisions per unit time and unit cross section

$$\mathcal{L} = R_i/\sigma. \tag{1.42}$$

Let N_b be the number of incident particles per unit time and Σ the beam section; then $N_b = \Phi_b\Sigma$. Equation (1.36) gives

$$\mathcal{L} = \frac{R_i}{\sigma} = \Phi_b N_t = \frac{N_b N_t}{\Sigma}. \tag{1.43}$$

We see that the luminosity is given by the product of the number of incident particles in a second times the number of target particles divided by the beam section. This expression is somewhat misleading because the number of particles in the target seen by the beam depends on its section. We then express the luminosity in terms of the number of target particles per unit volume n_t and in terms of the length l of the target ($N_t = n_t\Sigma l$). Equation (1.43) becomes

$$\mathcal{L} = N_b n_t l = N_b\rho N_A 10^3 l \tag{1.44}$$

where ρ is the target density.

Example 1.9 An accelerator produces a beam of intensity $I = 10^{13}\,\mathrm{s^{-1}}$. The target is made up of liquid hydrogen ($\rho = 60\,\mathrm{kg\,m^{-3}}$) and $l = 10$ cm. Evaluate its luminosity.

$$\mathcal{L} = I\rho 10^3 l N_A = 10^{13} \times 60 \times 10^3 \times 0.1 \times 6 \times 10^{23} = 3.6 \times 10^{40}\,\mathrm{m^{-2}\,s^{-1}}.$$

We shall now recall a few concepts that should already be known to the reader. We start with the Fermi 'golden rule', which gives the interaction rate W per target particle

$$W = 2\pi |M_{fi}|^2 \rho(E) \tag{1.45}$$

where E is the *total* energy and $\rho(E)$ is the phase-space volume (or simply the phase space) available in the final state.

There are two possible expressions of phase space: the 'non-relativistic' expression used in atomic and nuclear physics, and the 'relativistic' one used in subnuclear physics. Obviously the rates W must be identical, implying that the matrix element M is different in the two cases. In the non-relativistic formalism neither the phase space nor the matrix element are Lorentz-invariant. Both factors are invariant in the relativistic formalism, a fact that makes things simpler.

We recall that in the non-relativistic formalism the probability that a particle i has the position $\mathbf{r}_i$ is given by the square modulus of its wave function, $|\psi(\mathbf{r}_i)|^2$. This is normalised by putting its integral over all volume equal to one.

The volume element dV is a scalar in three dimensions, but not in space-time. Under a Lorentz transformation $\mathbf{r} \to \mathbf{r}'$ the volume element changes as $dV \to dV' = \gamma\, dV$. Therefore, the probability density $|\psi(\mathbf{r}_i)|^2$ transforms as $|\psi(\mathbf{r}_i)|^2 \to |\psi'(\mathbf{r}_i)|^2 = |\psi(\mathbf{r}_i)|^2/\gamma$. To have a Lorentz-invariant probability density, we profit from the energy transformation $E \to E' = \gamma E$ and define the probability density as $|(2E)^{-1/2}\psi(\mathbf{r}_i)|^2$ (the factor 2 is due to a historical convention).

The number of phase-space states per unit volume is $d^3 p_i/h$ for each particle i in the final state. With n particles in the final state, the volume of the phase space is therefore

$$\rho_n(E) = (2\pi)^4 \int \prod_{i=1}^{n} \frac{d^3 p_i}{(h)^3 2E_i} \delta\left(\sum_{i=1}^{n} E_i - E\right) \delta^3\left(\sum_{i=1}^{n} \mathbf{p}_i - \mathbf{P}\right) \tag{1.46}$$

or, in NU (be careful! $\hbar = 1$ implies $h = 2\pi$)

$$\rho_n(E) = (2\pi)^4 \int \prod_{i=1}^{n} \frac{d^3 p_i}{2E_i (2\pi)^3} \delta\left(\sum_{i=1}^{n} E_i - E\right) \delta^3\left(\sum_{i=1}^{n} \mathbf{p}_i - \mathbf{P}\right) \tag{1.47}$$

where δ is the Dirac function. Now we consider the collision of two particles, say a and b, resulting in a final state with n particles. We shall give the expression for the cross section.

The cross section is normalised to one incident particle; therefore, we must divide by the incident flux. In the laboratory frame the target particles b are at rest, the beam particles a move with a speed of, say, $\boldsymbol{\beta}_a$. The flux is the number of particles inside a cylinder of unitary base and height $\boldsymbol{\beta}_a$.

Let us consider, more generally, a frame in which particles b also move, with velocity $\boldsymbol{\beta}_b$, that we shall assume parallel to $\boldsymbol{\beta}_a$. The flux of particles b is their number inside a cylinder of unitary base of height $\boldsymbol{\beta}_b$. The total flux is the number of particles in a cylinder of height $\boldsymbol{\beta}_a - \boldsymbol{\beta}_b$ (i.e. the difference between the speeds, which is not, as is often written, the relative speed). If E_a and E_b are the initial energies the normalisation factors of the initial particles are $1/(2E_a)$ and $1/(2E_b)$. It is easy to show, but we shall only give the result, that the cross section is

$$\sigma = \frac{1}{2E_a 2E_b |\boldsymbol{\beta}_a - \boldsymbol{\beta}_b|} \int |M_{fi}|^2 (2\pi)^4 \prod_{i=1}^{n} \frac{d^3 p_i}{(2\pi)^3 2E_i}$$
$$\times \delta\left(\sum_{i=1}^{n} E_i - E\right) \delta^3\left(\sum_{i=1}^{n} \mathbf{p}_i - \mathbf{P}\right). \tag{1.48}$$

The case of a decay is simpler, because in the initial state there is only one particle of energy E. The probability of transition per unit time to the final state f of n particles is

$$\Gamma_{if} = \frac{1}{2E} \int |M_{fi}|^2 (2\pi)^4 \prod_{i=1}^{n} \frac{d^3 p_i}{(2\pi)^3 2E_i} \delta\left(\sum_{i=1}^{n} E_i - E\right) \delta^3\left(\sum_{i=1}^{n} \mathbf{p}_i - \mathbf{P}\right). \tag{1.49}$$

With these expressions, we can calculate the measurable quantities, cross sections and decay rates, once the matrix elements are known. The Standard Model gives the rules to evaluate all the matrix elements in terms of a set of constants. Even if we do not have the theoretical instruments for such calculations, we shall be able to understand the physical essence of the principal predictions of the model and to study their experimental verification.

Now let us consider an important case, the **two-body phase space**. Let c and d be the two final-state particles of a collision or decay. We choose the centre of mass frame, in which calculations are easiest. Let E_c and E_d be the energies of the two particles, $E = E_c + E_d$ the total energy, and $\mathbf{p}_f = \mathbf{p}_c = -\mathbf{p}_d$ the momentum. We must evaluate the integral

$$\int |M_{fi}|^2 \frac{d^3 p_c}{(2\pi)^3 2E_c} \frac{d^3 p_d}{(2\pi)^3 2E_d} (2\pi)^4 \delta(E_c + E_d - E) \delta^3(\mathbf{p}_c + \mathbf{p}_d).$$

Having the energies and the absolute values of the momenta of the final particles fixed, the matrix element can depend only on the angles. Consider the phase-space integral

$$\rho_2 = \int \frac{d^3 p_c}{(2\pi)^3 2E_c} \frac{d^3 p_d}{(2\pi)^3 2E_d} (2\pi)^4 \delta(E_c + E_d - E) \delta^3(\mathbf{p}_c + \mathbf{p}_d).$$

Integrating over d^3p_d we obtain

$$\rho_2 = \frac{1}{(4\pi)^2} \int \frac{d^3p_c}{E_c E_d(p_c)} \delta(E_c + E_d(p_c) - E)$$

$$= \frac{1}{(4\pi)^2} \int \frac{p_f^2 \, dp_f \, d\Omega_f}{E_c E_d(p_f)} \delta(E_c + E_d(p_f) - E).$$

Using the remaining δ-function we obtain straightforwardly

$$\frac{1}{(4\pi)^2} \frac{p_f^2}{E_c E_d(p_f)} \frac{d\,p_f}{d(E_c + E_d(p_f))} d\Omega_f = \frac{1}{(4\pi)^2} \frac{p_f^2}{E_c E_d(p_f)} \frac{1}{\frac{d}{dp_f}(E_c + E_d(p_f))} d\Omega_f.$$

But $\dfrac{dE_c}{dp_f} = \dfrac{p_f}{E_c}$ and $\dfrac{dE_d}{dp_f} = \dfrac{p_f}{E_d}$, hence $\dfrac{1}{(4\pi)^2} \dfrac{p_f^2}{E_c E_d} \dfrac{1}{\frac{p_f}{E_c} + \frac{p_f}{E_d}} d\Omega_f = \dfrac{p_f}{E} \dfrac{d\Omega_f}{(4\pi)^2}$. Now let

us consider the decay of a particle of mass m. With $E = m$, (1.49) gives

$$\Gamma_{a,cd} = \frac{1}{2m} \frac{p_f}{E} \int |M_{a,cd}|^2 \frac{d\Omega_f}{(4\pi)^2}. \tag{1.50}$$

By integrating the above equation on the angles, we obtain

$$\Gamma_{a,cd} = \frac{p_f}{8\pi m^2} \overline{|M_{a,cd}|^2} \tag{1.51}$$

where the angular average of the absolute square of the matrix element appears.

Now let us consider the cross section of the process $a + b \to c + d$, in the centre of mass frame. Again let E_a and E_b be the initial energies, E_c and E_d the final ones. The total energy is $E = E_a + E_b = E_c + E_d$. Let $\mathbf{p}_i = \mathbf{p}_a = -\mathbf{p}_b$ be the initial momenta and $\mathbf{p}_f = \mathbf{p}_c = -\mathbf{p}_d$ the final ones.

Let us restrict ourselves to the case in which neither the beam nor the target is polarised and in which the final polarisations are not measured. Therefore, in the evaluation of the cross section we must sum over the final spin states and average over the initial ones. Using (1.48) we have

$$\frac{d\sigma}{d\Omega_f} = \frac{1}{2E_a 2E_b |\boldsymbol{\beta}_a - \boldsymbol{\beta}_b|} \overline{\sum_{\text{initial}} \sum_{\text{final}} |M_{fi}|^2} \frac{1}{(4\pi)^2} \frac{p_f}{E}. \tag{1.52}$$

We evaluate the difference between the speeds

$$|\boldsymbol{\beta}_a - \boldsymbol{\beta}_b| = \beta_a + \beta_b = \frac{p_i}{E_a} + \frac{p_i}{E_b} = \frac{p_i E}{E_a E_b}.$$

Hence

$$\frac{d\sigma}{d\Omega_f} = \frac{1}{(8\pi)^2} \frac{1}{E^2} \frac{p_f}{p_i} \overline{\sum_{\text{initial}} \sum_{\text{final}} |M_{fi}|^2}. \tag{1.53}$$

The average over the initial spin states is the sum over them divided by their number. If s_a and s_b are the spins of the colliding particles, then the spin multiplicities are $2s_a + 1$ and $2s_b + 1$. Hence

$$\frac{d\sigma}{d\Omega_f} = \frac{1}{(8\pi)^2} \frac{1}{E^2} \frac{p_f}{p_i} \frac{1}{(2s_a + 1)(2s_b + 1)} \sum_{\text{initial}} \sum_{\text{final}} |M_{fi}|^2. \qquad (1.54)$$

1.7 Hadrons, leptons and quarks

The particles can be classified, depending on their characteristics, into different groups. We shall give here the names of these groups and summarise their properties.

The particles of a given type, the electrons for example, are indistinguishable. Take for example a fast proton hitting a stationary one. After the collision, that we assume to be elastic, there are two protons moving in general in different directions with different energies. It is pointless to try to identify one of these as, say, the incident proton.

First of all, we can distinguish the particles of integer spin, in units $\hbar$ $(0, \hbar, 2\hbar, \ldots)$, that follow Bose statistics and are called **bosons** and the semi-integer spin particles $\left(\frac{1}{2}\hbar, \frac{3}{2}\hbar, \frac{5}{2}\hbar, \ldots\right)$ that follow Fermi–Dirac statistics and are called **fermions**. We recall that the wave function of a system of identical bosons is symmetric under the exchange of any pair of them, while the wave function of a system of identical fermions is antisymmetric.

Matter is made up of atoms. Atoms are made of **electrons** and **nuclei** bound by the electromagnetic force, whose quantum is the photon.

The **photons** (from the Greek word *phos* meaning *light*) are massless. Their charge is zero and therefore they do not interact among themselves. Their spin is equal to one; they are bosons.

The **electrons** have negative electric charge and spin 1/2; they are fermions. Their mass is small, $m_e = 0.511$ MeV, in comparison with that of the nuclei. As far as we know they do not have any structure, they are elementary.

Nuclei contain most of the mass of the atoms, hence of the matter. They are positively charged and made of protons and neutrons. **Protons** (from *proton* meaning *the first*, in Greek) and **neutrons** have similar masses, slightly less than a GeV. The charge of the proton is positive, opposite and exactly equal to the electron charge; neutrons are globally neutral, but contain charges, as shown, for example, by their non-zero magnetic moment. As anticipated, protons and neutrons are collectively called **nucleons**. Nucleons have spin 1/2; they are fermions. Protons are stable, within the limits of present measurements; the reason is that they have another conserved 'charge' beyond the electric charge, the 'baryonic number', which we shall discuss in Chapter 3.

In 1935, Yukawa formulated a theory of the strong interactions between nucleons (Yukawa 1935). Nucleons are bound in nuclei by the exchange of a zero spin particle, the quantum of the nuclear force. Given the finite range of this force, its mediator must be massive. Given the value of the range, about 10^{-15} m, its mass should be intermediate between the electron and the proton masses; therefore it was called the **meson** (*that which is in the middle*). More specifically, it is the π meson, also called the **pion**. We shall describe its properties in the next chapter. Pions come in three charge states: π^+, π^- and π^0. Unexpectedly, from 1946 onwards, other mesons were discovered in cosmic radiation, the K mesons, which come in two different charge doublets, K^+ and K^0, and their antiparticles, K^- and $\bar{K}^0$.

In the same period other particles were discovered that, like the nucleons, have half-integer spin and baryonic number. They are somewhat more massive than nucleons and are called **baryons** (*that which is heavy or massive*). Notice that nucleons are included in this category.

Baryons and mesons are not point-like; instead they have structure and are composite objects. The components of both of them are the **quarks**. In a first approximation, the baryons are made up of three quarks, the mesons of a quark and an antiquark. Quarks interact via one of the fundamental forces, the strong force, that is mediated by the **gluons** (from *glue*). As we shall see, there are eight different gluons; all are massless and have spin one. Baryons and mesons have a similar structure and are collectively called **hadrons** (*hard*, *strong* in Greek). All hadrons are unstable, with the exception of the lightest one, the proton.

Shooting a beam of electrons or photons at an atom we can free the electrons it contains, provided the beam energy is large enough. Analogously we can break a nucleus into its constituents by bombarding it, for example, with sufficiently energetic protons. The two situations are similar with quantitative, not qualitative, differences: in the first case a few eV are sufficient, in the second several MeV are needed. However, nobody has ever succeeded in breaking a hadron and extracting the quarks, whatever the energy and type of the bombarding particles. We have been forced to conclude that quarks do not exist in a free state; they exist only inside the hadrons. We shall see how the Standard Model explains this property, which is called 'quark confinement'.

The spin of the quarks is 1/2. There are three quarks with electric charge $+2/3$ (in units of the elementary charge), called up-type, and three with charge $-1/3$ called down-type. In order of increasing mass the up-type are: 'up' u, 'charm' c and 'top' t, the down-type are: 'down' d, 'strange' s and 'beauty' b. Nucleons, hence nuclei, are composed of up and down quarks, *uud* the proton, *udd* the neutron.

The electrons are also members of a class of particles of similar properties, the **leptons** (*light* in Greek, but there are also heavy leptons). Their spin is 1/2. There are three charged leptons, the electron e, the muon μ and the tau τ, and three neutral leptons, the **neutrinos**, one for each of the charged leptons. The electron is stable, the μ and the τ are unstable, and all the neutrinos are stable.

For every particle there is an antiparticle with the same mass and the same lifetime and all charges of opposite values: the positron for the electron, the antiproton, the antiquarks, etc.

One last consideration: astrophysical and cosmological observations have shown that 'ordinary' matter, baryons and leptons, makes up only a small fraction of the total mass of the Universe, no more than 20%. We do not know what the rest is made of. There is still a lot to understand beyond the Standard Model (see Chapter 10).

1.8 The fundamental interactions

Each of the interactions is characterised by one, or more, 'charge' that, like the electric charge, is the source and the receptor of the interaction. The Standard Model is the theory that describes all the fundamental interactions, except gravitation. For the latter, we do not yet have a microscopic theory, but only a macroscopic approximation, so-called general relativity. We anticipate here that the intensity of the interactions depends on the energy scale of the phenomena under study.

The source and the receptor of the gravitational interaction is the energy-momentum tensor; consequently this interaction is felt by all particles. However, gravity is extremely weak at all the energy scales experimentally accessible and we shall neglect its effects.

Let us find the orders of magnitude by the following dimensional argument. The fundamental constants, the Newton constant G_N of gravity, the speed of light c, the Lorentz transformations, and the Planck constant $\hbar$ of quantum mechanics, can be combined in an expression with the dimensions of mass which is called the Planck mass

$$M_P = \sqrt{\frac{\hbar c}{G_N}} = \sqrt{\frac{1.06 \times 10^{-34}\,\text{J s} \times 3 \times 10^8\,\text{m s}^{-1}}{6.67 \times 10^{-11}\,\text{m}^3\,\text{kg}^{-1}\,\text{s}^{-2}}} \tag{1.55}$$
$$= 2.18 \times 10^{-8}\,\text{kg} = 1.22 \times 10^{19}\,\text{GeV}.$$

It is enormous, not only in comparison to the energy scale of the Nature around us on Earth (eV) but also of nuclear (MeV) and subnuclear (GeV) physics. We shall never be able to build an accelerator to reach such an energy scale. We must search for quantum features of gravity in the violent phenomena naturally occurring in the Universe.

All the known particles have weak interactions, with the exception of photons and gluons. This interaction is responsible for beta decay and for many other types of decays. The weak interaction is mediated by three spin one mesons, W^+, W^- and Z^0; their masses are rather large, in comparison to, say, the proton mass (in round numbers $M_W \approx 80$ GeV, $M_Z \approx 90$ GeV). Their existence becomes evident at energies comparable to those masses.

All charged particles have electromagnetic interactions. This interaction is transmitted by the photon, which is massless. Quarks and gluons have strong interactions; the leptons do not. The corresponding charges are called 'colours'. The interaction amongst quarks in a hadron is confined inside the hadron. If two hadrons, two nucleons for example, come close enough (typically 1 fm) they interact via the 'tails' of the colour field that, shall we say, leaks out of the hadron. The phenomenon is analogous to the van der Waals force that is due to the electromagnetic field leaking out from a molecule. Therefore the nuclear (Yukawa) forces are not fundamental.

As we have said, the charged leptons more massive than the electron are unstable; the lifetime of the muon is about 2 μs, that of the τ, 0.3 ps. These are large values on the scale of elementary particles, characteristic of weak interactions.

All mesons are unstable: the lifetimes of $\pi^\pm$ and of $K^\pm$ are 26 ns and 12 ns respectively; they are weak decays. In the 1960s, other larger mass mesons were discovered; they have strong decays and extremely short lifetimes, of the order of 10^{-23}–10^{-24} s.

All baryons, except for the proton, are unstable. The neutron has a beta decay into a proton with a lifetime of 886 s. This is exceptionally long even for the weak interaction standard because of the very small mass difference between neutrons and protons. Some of the other baryons, the less massive ones, decay weakly with lifetimes of the order of 0.1 ns, others, the more massive ones, have strong decays with lifetimes of 10^{-23}–10^{-24} s.

Example 1.10 Consider an electron and a proton standing at a distance r. Evaluate the ratio between the electrostatic and the gravitational forces. Does it depend on r?

$$F_{\text{electrost.}}(ep) = \frac{1}{4\pi\varepsilon_0} \frac{q_e^2}{r^2} \qquad F_{\text{gravit.}}(ep) = G_N \frac{m_e m_p}{r^2}.$$

$$\frac{F_{\text{electrost.}}(ep)}{F_{\text{gravit.}}(ep)} = \frac{q_e^2}{4\pi\varepsilon_0 G_N m_e m_p}$$

$$= \frac{(1.6\times10^{-19})^2}{4\pi\times8.8\times10^{-12}\times6.67\times10^{-11}\times9.1\times10^{-31}\times1.7\times10^{-27}} \approx 10^{39}$$

independent of r.

1.9 The passage of radiation through matter

The Standard Model has been developed and tested by a number of experiments, some of which we shall describe. This discussion is not possible without some knowledge of the physics of the passage of radiation through matter, of the main particle detectors and the sources of high-energy particles.

When a high-energy charged particle or a photon passes through matter, it loses energy that excites and ionises the molecules of the material. It is through experimental observation of these alterations of the medium that elementary particles are detected. Experimental physicists have developed a wealth of detectors aimed at measuring different characteristics of the particles (energy, charge, speed, position, etc.). This wide and very interesting field is treated in specialised courses and books. Here we shall only summarise the main conclusions relevant for the experiments we shall discuss in the text and not including, in particular, the most recent developments.

Ionisation loss

The energy loss of a relativistic charged particle more massive than the electron passing through matter is due to its interaction with the atomic electrons. The process results in a trail of ion–electron pairs along the path of the particle. These free charges can be detected. Electrons also lose energy through bremsstrahlung in the Coulomb fields of the nuclei.

The expression of the average energy loss per unit length of charged particles other than electrons is known as the Bethe–Bloch equation (Bethe 1930). We give here an approximate expression, which is enough for our purposes. If z is the charge of the particle, ρ the density of the medium, Z its atomic number and A its atomic mass, the equation is

$$-\frac{dE}{dx} = K\frac{\rho Z}{A}\frac{z^2}{\beta^2}\left[\ln\left(\frac{2mc^2\gamma^2\beta^2}{I}-\beta^2\right)\right] \tag{1.56}$$

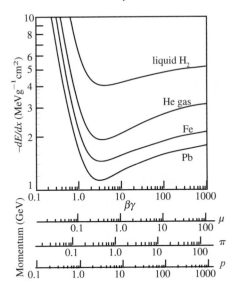

Fig. 1.6.   Specific average ionisation loss for relativistic particles of unit charge. (Simplified from Yao *et al.* 2006 by permission of Particle Data Group and the Institute of Physics)

where m is the electron mass (the hit particle), the constant K is given by

$$K = \frac{4\pi a^2 (\hbar c)^2 N_A (10^3\,\text{kg})}{mc^2} = 30.7\,\text{keV m}^2\,\text{kg}^{-1} \tag{1.57}$$

and I is an average ionisation potential. For $Z > 20$ it is approximately $I \approx 12\,Z\,\text{eV}$. The energy loss is a universal function of $\beta\gamma$ in a very rough approximation, but there are important differences in the different media, as shown in Fig. 1.6. The curves are drawn for particles of charge $z = 1$; for larger charges, multiply by z^2.

All the curves decrease rapidly at small momenta (roughly as $1/\beta^2$), reach a shallow minimum for $\beta\gamma = 3$–4 and then increase very slowly. The energy loss of a minimum ionising particle (mip) is $(0.1$–$0.2\,\text{MeV m}^2\,\text{kg}^{-1})\rho$.

The Bethe–Bloch formula is only valid in the energy interval corresponding to approximately $0.05 < \beta\gamma < 500$. At lower momenta, the particle speed is comparable to the speed of the atomic electrons. In these conditions a, possibly large, fraction of the energy loss is due to the excitation of atomic and molecular levels, rather than to ionisation. This fraction must be detected as light, coming from the de-excitation of those levels or, in a crystal, as phonons.

At energies larger than a few hundred GeV for pions or muons, much larger for protons, another type of energy loss becomes more important than ionisation, the bremsstrahlung losses in the nuclear fields. Consequently, dE/dx for muons and pions grows dramatically at energies larger than or around one TeV.

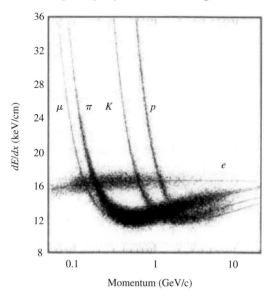

Fig. 1.7.   *dE/dx* measured in a TPC at SLAC. (Aihara *et al.* 1988)

Notice that the Bethe–Bloch formula gives the *average* energy loss, while the measured quantity is the energy loss for a given length. The latter is a random variable with a frequency function centred on the expectation-value given by the Bethe–Bloch equation. The variance, called the straggling, is quite large. Figure 1.7 shows a set of measurements of the ionisation losses as functions of the momentum for different particles. Notice, in particular, the dispersion around the average values.

Energy loss of the electrons

Figure 1.7 shows that electrons behave differently from other particles. As anticipated, electrons and positrons, due to their small mass, lose energy not only by ionisation but also by bremsstrahlung in the nuclear Coulomb field. This happens at several MeV.

As we have seen in Example 1.4, the process $e^- \rightarrow e^- + \gamma$ cannot take place in vacuum, but can happen near a nucleus. The reaction is

$$e^- + N \rightarrow e^- + N + \gamma \tag{1.58}$$

where N is the nucleus. The case of positrons is similar

$$e^+ + N \rightarrow e^+ + N + \gamma. \tag{1.59}$$

Classically, the power radiated by an accelerating charge is proportional to the square of its acceleration. In quantum mechanics, the situation is similar: the probability of radiating a photon is proportional to the acceleration squared.

Therefore, this phenomenon is much more important close to a nucleus than to an atomic electron. Furthermore, for a given external field, the probability is inversely proportional to the mass squared. We understand that for the particle immediately more massive than the electron, the muon that is 200 times heavier, the bremsstrahlung loss becomes important at energies larger by four orders of magnitude.

Comparing different materials, the radiation loss is more important if Z is larger. More specifically, the materials are characterised by their radiation length X_0. The radiation length is defined as the distance over which the electron energy decreases to $1/e$ of its initial value due to radiation, namely

$$-\frac{dE}{E} = \frac{dx}{X_0}. \tag{1.60}$$

The radiation length is roughly inversely proportional to Z and hence to the density. A few typical values are: air at n.t.p. $X_0 \approx 300 \, \text{m}$; water $X_0 \approx 0.36 \, \text{m}$; carbon $X_0 \approx 0.2 \, \text{m}$; iron $X_0 \approx 2 \, \text{cm}$; lead $X_0 \approx 5.6 \, \text{mm}$. We show in Fig. 1.8 the electron energy loss in lead; in other materials the behaviour is similar. At low energies the ionisation loss dominates, at high energies the radiation loss becomes more important. The crossover, when the two losses are equal, is called the critical energy. With a good approximation it is given by

$$E_c = 600 \, \text{MeV}/Z. \tag{1.61}$$

For example, the critical energy of lead, which has $Z = 82$, is $E_c = 7$ MeV.

Energy loss of the photons

At energies of the order of dozens of electronvolts, the photons lose energy mainly by the photoelectric effect on atomic electrons. Above a few keV, the Compton effect becomes important. When the production threshold of the electron–positron

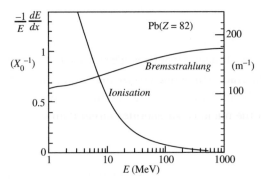

Fig. 1.8.   Relative energy loss of electrons in lead. (Adapted from Yao *et al.* 2006 by permission of Particle Data Group and the Institute of Physics)

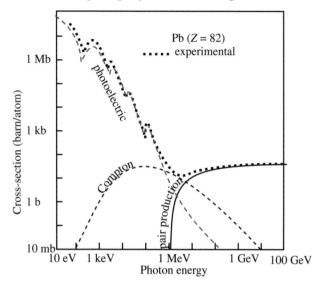

Fig. 1.9. Photon cross sections in Pb versus energy; total and calculated contributions of the three principal processes. (Adapted from Yao *et al.* 2006 by permission of Particle Data Group and the Institute of Physics)

pairs is crossed, at 1.022 MeV, this channel rapidly becomes dominant. The situation is shown in Fig. 1.9 in the case of lead.

In the pair production process

$$\gamma + N \rightarrow N + e^- + e^+ \tag{1.62}$$

a photon disappears, it is absorbed. The attenuation length of the material is defined as the length that attenuates the intensity of a photon beam to $1/e$ of its initial value. The attenuation length is closely related to the radiation length, being equal to $(9/7)X_0$. Therefore, X_0 determines the general characteristics of the propagation of electrons, positrons and photons.

Energy loss of the hadrons

High-energy hadrons passing through matter do not lose energy by ionisation only. Eventually they interact with a nucleus by the strong interaction. This leads to the disappearance of the incoming particle, the production of secondary hadrons and the destruction of the nucleus. At energies larger than several GeV, the total cross sections of different hadrons become equal within a factor of 2 or 3. For example, at 100 GeV the cross sections $\pi^+ p$, $\pi^- p$, $\pi^+ n$, $\pi^- n$ are all about 25 mb, those for pp and pn about 40 mb. The collision length λ_0 of a material is defined as the distance over which a neutron beam (particles that do not have electromagnetic interactions) is attenuated by $1/e$ in that material.

Typical values are: air at n.t.p. $\lambda_0 \approx 750\,\text{m}$; water $\lambda_0 \approx 0.85\,\text{m}$; carbon $\lambda_0 \approx 0.38\,\text{m}$; iron $\lambda_0 \approx 0.17\,\text{m}$; lead $\lambda_0 \approx 0.17\,\text{m}$. Comparing with the radiation length we see that collision lengths are larger and do not depend heavily on the material, provided this is solid or liquid. These observations are important in the construction of calorimeters (see Section 1.11).

1.10 Sources of high-energy particles

The instruments needed to study the elementary particles are sources and detectors. We shall give, in both cases, only the pieces of information that are necessary for the following discussions. In this section, we discuss the sources, in the next the detectors.

There is a natural source of high-energy particles, the cosmic rays; the artificial sources are the accelerators and the colliders.

Cosmic rays

In 1912, V. F. Hess, flying aerostatic balloons at high altitudes, discovered that charged particle radiation originated outside the atmosphere, in the cosmos (Hess 1912). Fermi formulated a theory of the acceleration mechanism in 1949 (Fermi 1949). Until the early 1950s, when the first high-energy accelerators were built, cosmic rays were the only source of particles with energy larger than a GeV. The study of cosmic radiation remains, even today, fundamental for both subnuclear physics and astrophysics.

We know rather well the energy spectrum of cosmic rays, which is shown in Fig. 1.10. It extends up to 100 EeV (10^{20} eV), 12 orders of magnitude on the energy scale and 32 orders of magnitude on the flux scale. To make a comparison, notice that the highest-energy accelerator, the LHC at CERN, has a centre of mass energy of 14 TeV, corresponding to 'only' 0.1 EeV. At these extreme energies the flux is very low, typically one particle per square kilometre per century. The Pierre Auger observatory in Argentina has an active surface area of 3000 km^2 and is starting to explore the energy range above EeV. In this region, one may well discover phenomena beyond the Standard Model.

The initial discoveries in particle physics, which we shall discuss in the next chapter, used the spectrum around a few GeV, where the flux is largest, tens of particles per square metre per second. In this region the primary composition, namely at the top of the atmosphere, consists of 85% protons, 12% alpha particles, 1% heavier nuclei and 2% electrons.

A proton or a nucleus penetrating the atmosphere eventually collides with a nucleus of the air. This strong interaction produces pions, less frequently K mesons and, even

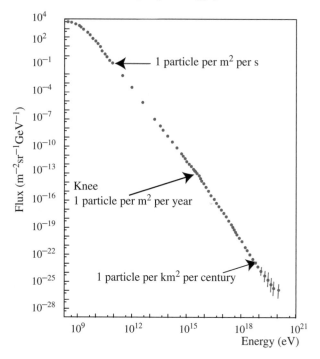

Fig. 1.10. The cosmic ray flux.

more rarely, other hadrons. The hadrons produced in the first collision generally have enough energy to produce other hadrons in a further collision, and so on. The average distance between collisions is the collision length ($\lambda_0 = 750$ m at n.t.p.). The primary particle gives rise to a 'hadronic shower': the number of particles in the shower initially grows, then, when the average energy becomes too small to produce new particles, decreases. This is because the particles of the shower are unstable. The charged pions, which have a lifetime of only 26 ns, decay through the reactions

$$\pi^+ \rightarrow \mu^+ + \nu_\mu \qquad \pi^- \rightarrow \mu^- + \bar{\nu}_\mu. \tag{1.63}$$

The muons, in turn, decay as

$$\mu^+ \rightarrow e^+ + \bar{\nu}_\mu + \nu_e \qquad \mu^- \rightarrow e^- + \nu_\mu + \bar{\nu}_e. \tag{1.64}$$

The muon lifetime is 2 µs, much larger than that of the pions. Therefore, the composition of the shower becomes richer and richer in muons while travelling through the atmosphere.

The hadronic collisions produce not only charged pions but also π^0. These latter decay quickly with the electromagnetic reaction

$$\pi^0 \rightarrow \gamma + \gamma. \tag{1.65}$$

The photons, in turn, give rise to an 'electromagnetic shower', which overlaps geometrically with the hadronic shower but has different characteristics. Actually, the photons interact with the nuclei producing a pair

$$\gamma + N \rightarrow e^+ + e^- + N. \tag{1.66}$$

The electron and the positron, in turn, can produce a photon by bremsstrahlung

$$e^\pm + N \rightarrow e^\pm + N + \gamma. \tag{1.67}$$

In addition, the new photon can produce a pair, and so on. The average distance between such events is the radiation length, which for air at n.t.p. is $X_0 = 300$ m. Figure 1.11 shows the situation schematically.

In the first part of the shower, the number of electrons, positrons and photons increases, while their average energy diminishes. When the average energy of the electrons decreases below the critical energy, the number of particles in the shower has reached its maximum and gradually decreases.

In 1932 B. Rossi discovered that cosmic radiation has two components: a 'soft' component that is absorbed by a material of modest thickness, for example a few centimetres of lead, and a 'hard' component that penetrates through a material of large thickness (Rossi 1933). From the above discussion we understand that the soft component is the electromagnetic one, the hard component is made up mostly of muons.

There is actually a third component, which is extremely difficult to detect: the neutrinos and antineutrinos ($\nu_e, \bar{\nu}_e, \nu_\mu$ and $\bar{\nu}_\mu$ to be precise) produced in the reactions (1.63) and (1.64). Neutrinos have only weak interactions and can cross the whole Earth without being absorbed. Consequently, observing them requires

Fig. 1.11.   Sketch of an electromagnetic shower.

detectors with sensitive masses of a thousand tons or more. These observations have led, in the past few years, to the discovery that neutrinos have non-zero masses.

Accelerators

Several types of accelerators have been developed. We shall discuss here only the synchrotron, the acceleration scheme that has made the most important contributions to subnuclear physics. Synchrotrons can be built to accelerate protons or electrons. Schematically, in a synchrotron, the particles travel in a pipe, in which high vacuum is established. The 'beam pipe' runs inside the gaps of dipole magnets forming a ring. The orbit of a particle of momentum p in a uniform magnetic field B is a circumference of radius R. These three quantities are related by an equation that we shall often use (see Problem 1.21)

$$p(\mathrm{GeV}) = 0.3B(\mathrm{T})R(\mathrm{m}). \tag{1.68}$$

Other fundamental components are the accelerating cavities. In them a radio-frequency electromagnetic field (RF) is tuned to give a push to the bunches of particles every time they go through. Actually, the beam does not continuously fill the circumference of the pipe, but is divided in bunches, in order to allow the synchronisation of their arrival with the phase of the RF.

In the structure we have briefly described, the particle orbit is unstable; such an accelerator cannot work. The stability can be guaranteed by the 'principle of phase stability', independently discovered by V. Veksler in 1944 in Russia (then the USSR) (Veksler 1944) and by E. McMillan in 1945 in the USA (McMillan 1945). In practice, stability is reached by alternating magnetic elements that focus and defocus in the orbit plane (Courant & Synder 1958). The following analogy can help. If you place a rigid stick vertically upwards on a horizontal support, it will fall; the equilibrium is unstable. However, if you place it on your hand and move your hand quickly to and fro, the stick will not fall.

The first proton synchrotron was the Cosmotron, operational at the Brookhaven National Laboratory in the USA in 1952, with 3 GeV energy. Two years later, the Bevatron was commissioned at Berkeley, also in the USA. The proton energy was 7 GeV, designed to be enough to produce antiprotons. In 1960 two 30 GeV proton synchrotrons became operational, the CPS (CERN Proton Synchrotron) at CERN, the European Laboratory at Geneva, and the AGS (Alternate Gradient Synchrotron) at Brookhaven.

The search for new physics has demanded that the energy frontier be moved towards higher and higher values. To build a higher-energy synchrotron one needs to increase the length of the ring or increase the magnetic field, or both.

The next generation of proton synchrotrons was ready at the end of the 1960s: the Super Proton Synchrotron (SPS) at CERN (450 GeV) and the Main Ring at Fermilab near Chicago (500 GeV). Their radius is about 1 km.

The synchrotrons of the next generation reached higher energies using field intensities of several tesla with superconducting magnets. These are the Tevatron at Fermilab, built in the same tunnel as the Main Ring with maximum energy of 1 TeV, and the proton ring of the HERA complex at DESY (Hamburg in Germany) with 0.8 TeV.

The high-energy experiments generally use the so-called secondary beams. The primary proton beam, once accelerated at the maximum energy, is extracted from the ring and driven onto a target. The strong interactions of the protons with the nuclei of the target produce all types of hadrons. Beyond the target, a number of devices are used to select one type of particle, possibly within a certain energy range. In such a way, one can build beams of pions, K mesons, neutrons, antiprotons, muons and neutrinos. A typical experiment steers the secondary beam of interest into a secondary target where the interactions to be studied are produced. The target is followed by a set of detectors to measure the characteristics of these interactions. These experiments are said to be on a 'fixed target' as opposed to those at the storage rings that we shall soon discuss. Figure 1.12 shows, as an example, the secondary beam configuration at Fermilab in the 1980s.

Storage rings

The ultimate technique to reach higher-energy scales is that of storage rings, or colliders as they are also called. Consider a fixed-target experiment with target particle of mass m_t and a beam of energy E_b and an experiment using two beams colliding from opposite directions in the centre of mass frame, each of energy E^*.

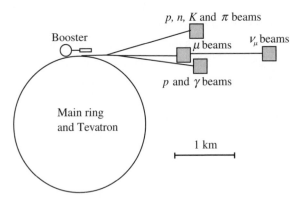

Fig. 1.12.   The Tevatron beams. The squares represent the experimental halls.

Equations (1.21) and (1.22) give the condition needed to have the same total centre of mass energy in the two cases

$$E^* = \sqrt{m_t E_b/2}. \tag{1.69}$$

We see that to increase the centre of mass energy at a fixed target by an order of magnitude we must increase the beam energy by two orders; with colliding beams, by only one.

A collider consists of two accelerator structures with vacuum pipes, magnets and RF cavities, in which two beams of particles travel in opposite directions. They may be both protons, or protons and antiprotons, or electrons and positrons, or electrons and protons, or also nuclei and nuclei. The two rings intercept each other at a few positions along the circumference. The phases of the bunches circulating in the two rings are adjusted to make them meet at the intersections. Then, if the number of particles in the bunches is sufficient, collisions happen at every crossing. Notice that the same particles cross repeatedly a very large number of times.

The first *pp* storage ring became operational at CERN in 1971: it was called ISR (Intersecting Storage Rings) and is shown in Fig. 1.13. The protons are first accelerated up to 3.5 GeV in the small synchrotron called the 'booster', transferred to the PS and accelerated up to 31 GeV. Finally they are transferred in bunches, alternately in the two storage rings. The filling process continues until the intensities reach the design values. The machine regime is then stable and the experiments can collect data for several hours.

The centre of mass energy is very important but it is useless if the interaction rate is too small. The important parameter is the luminosity of the collider.

We can think of the collision as taking place between two gas clouds, the bunches, that have densities much lower than that of condensed matter. To overcome this problem it is necessary:

1. to focus both beams in the intersection point to reduce their transverse dimensions as much as possible, in practice to a few μm or less;

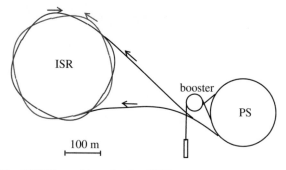

Fig. 1.13.   The CERN machines in the 1970s.

2. to reduce the random motion of the particles in the bunch. The fundamental technique, called 'stochastic cooling' was developed at CERN by S. van der Meer in 1968.

The luminosity is proportional to the product of the numbers of particles, n_1 and n_2, in the two beams. Notice that in a proton–antiproton collider the number of antiprotons is smaller than that of protons, due to the energetic cost of the antiprotons. The luminosity is also proportional to the number of crossings in a second f and inversely proportional to the section Σ at the intersection point

$$\mathcal{L} = f \frac{n_1 n_2}{\Sigma}. \tag{1.70}$$

In a particle–antiparticle collider (e^+e^- or $\bar{p}p$) the structure of the accumulator can be simplified. As particles and antiparticles have opposite charges and exactly the same mass, a single magnetic structure is sufficient to keep the two beams circulating in opposite directions. The first example of such a structure (ADA) was conceived and built by B. Touschek at Frascati in Italy as an electron–positron accumulator. Before discussing ADA, we shall complete our review of the hadronic machines.

In 1976, C. Rubbia, C. P. McIntire and D. Cline (Rubbia *et al.* 1976) proposed to transform the CERN SPS from a simple synchrotron to a proton–antiproton collider. The enterprise had limited costs, because the magnetic structure was left substantially as it was, while it was necessary to improve the vacuum substantially. It was also necessary to develop further the stochastic cooling techniques, already known from the ISR. Finally, the centre of mass energy ($\sqrt{s} = 540$ GeV) and the luminosity ($\mathcal{L} = 10^{28}$ cm^{-2} s^{-1}) necessary for the discovery of the bosons W and Z, the mediators of the weak interactions, were reached.

In 1987 at Fermilab, a proton–antiproton ring based on the same principles became operational. Its energy was larger, $\sqrt{s} = 2$ TeV and the luminosity $\mathcal{L} = 10^{31}$–10^{32} cm^{-2} s^{-1}.

In 2008, the next generation collider, LHC (Large Hadron Collider), should start operation at CERN. It has been built in the 27 km long tunnel that previously hosted LEP. The magnetic ring is made of superconducting magnets built with the most advanced technology to obtain the maximum possible magnetic field, 8 T. The centre of mass energy is 14 TeV, the design luminosity is $\mathcal{L} = 10^{33}$–10^{34} cm^{-2} s^{-1}.

Example 1.11 We saw in Example 1.9 that a secondary beam from an accelerator of typical intensity $I = 10^{13}$ s^{-1} impinging on a liquid hydrogen target with $l = 10$ cm gives a luminosity $\mathcal{L} = 3.6 \times 10^{36}$ cm^{-2} s^{-1}. We now see that this is much higher than that of the highest luminosity colliders. Calculate the luminosity for such a beam on a gas target, for example air in normal conditions ($\rho = 1$ kg m^{-3}). We obtain

$$\mathcal{L} = I \rho l N_A 10^3 = 10^{13} \times 10^3 \times 0.1 \times 6 \times 10^{23} = 6 \times 10^{38} \, \text{m}^{-2} \, \text{s}^{-1}.$$

This is similar to the LHC luminosity.

The proton–antiproton collisions are not simple processes because the two colliding particles are composite, not elementary, objects. The fundamental processes, the quark–quark or quark–antiquark collisions, which are the ones we are interested in, take place in a 'dirty' environment due to the rest of the proton and the antiproton. Furthermore, these processes happen only in a very small fraction of the collisions.

Electrons and positrons are, in contrast, elementary non-composite particles. When they collide they often annihilate; matter disappears in a state of pure energy. Moreover, this state has well-defined quantum numbers, those of the photon. B. Touschek, fascinated by these characteristics, was able to put into practice the dream of generating collisions between matter and antimatter beams. As a first test, in 1960 Touschek proposed building at Frascati (Touschek 1960) a small storage ring (250 MeV + 250 MeV), which was called ADA (Anello Di Accumulazione meaning Storage Ring in Italian). The next year ADA was working (Fig. 1.14).

The development of a facility suitable for experimentation was an international effort, mainly by the groups led by F. Amman in Frascati, G. I. Budker in Novosibirsk and B. Richter in Stanford. Then, everywhere in the world, a large number of e^+e^- rings of increasing energy and luminosity were built. Their contribution to particle physics was and still is enormous.

The maximum energy for an electron–positron collider, more than 200 GeV, was reached with LEP at CERN. Its length was 27 km. With LEP the practical energy limit of circular electron machines was reached. The issue is the power radiated by the electrons due to the centripetal acceleration, which grows dramatically with increasing energy. The next generation electron–positron collider will have a linear structure; the necessary novel techniques are currently under development.

HERA, operational at the DESY laboratory at Hamburg since 1991 (and up to 2007), is a third type of collider. It is made up of two rings, one for electrons, or positrons, that are accelerated up to 30 GeV, and one for protons that reach 920 GeV energy (820 GeV in the first years). The scattering of the point-like electrons on the protons informs us about the deep internal structure of the latter. The high centre of mass energy available in the head-on collisions makes HERA the 'microscope' with the highest existing resolving power.

Fig. 1.14. ADA at Frascati. (© INFN)

1.11 Particle detectors

The progress in our understanding of the fundamental laws of Nature is directly linked to our ability to develop instruments to detect particles and measure their characteristics, with ever increasing precision and sensitivity. We shall give here only a summary of the principal classes of detectors.

The quantities that we can measure directly are the electric charge, the magnetic moment (that we shall not discuss), the lifetime, the velocity, the momentum and the energy. The kinematic quantities are linked by the fundamental equations

$$\mathbf{p} = \mathbf{m}\gamma\beta \tag{1.71}$$

$$E = m\gamma \tag{1.72}$$

$$m^2 = E^2 - p^2. \tag{1.73}$$

We cannot measure the mass directly, to do so we measure *two* quantities: energy and momentum, momentum and velocity, etc.

Let us review the principal detectors.

Scintillation detectors

There are several types of scintillation counters, or, simply, 'scintillators'. We shall restrict ourselves to the plastic and organic liquid ones.

Scintillation counters are made up with transparent plastic plates with a thickness of a centimetre or so and of the required area (up to square metres). The material is doped with molecules that emit light at the passage of an ionising particle. The light is guided by a light guide glued, on the side of the plate, to the photocathode of a photomultiplier. One typically obtains 10 000 photons per MeV of energy deposit. Therefore the efficiency is close to 100%. The time resolution is very good and can reach 0.1 ns or even less.

Two counters at a certain distance on the path of a particle are used to measure its time of flight between them and, knowing the distance, its velocity.

Plastic counters are also used as the sensitive elements in the 'calorimeters', as we shall see.

A drawback of plastic (and crystal) scintillators is that their light attenuation length is not large. Consequently, when assembled in large volumes, the light collection efficiency is poor.

Broser and Kallmann discovered in 1947 (Broser & Kallmann 1947) that naphthalene emits fluorescence light under ionising radiation. In the next few years, different groups (Reynolds *et al.* 1950, Kallmann 1950, Ageno *et al.* 1950) discovered that binary and ternary mixtures of organic liquids and aromatic molecules had high scintillation yields, i.e. high numbers of photons per unit of energy loss (of the order of 10 000 photons/MeV), and long (up to tens of metres) attenuation lengths. These discoveries opened the possibility of building large scintillation detectors at affordable cost. The liquid scintillator technique has been, and is, of enormous importance, in particular for the study of neutrinos, including their discovery (Section 2.4).

Nuclear emulsions

Photographic emulsions are made of an emulsion sheet deposited on a transparent plastic or glass support. The emulsions contain grains of silver halides, the sensitive element. Once exposed to light the emulsions are developed, with a chemical process that reduces to metallic silver only those grains that have absorbed photons. It became known as early as 1910 that ionising radiation produces similar effects. Therefore, a photographic plate, once developed, shows as trails of silver grains the tracks of the charged particles that have gone through it.

In practice, normal photographic emulsions are not suitable for scientific experiments because of their small thickness and low efficiency. The development of emulsions as a scientific instrument, the 'nuclear emulsion', was mainly due to C. F. Powell and G. Occhialini at Bristol in co-operation with the Ilford Laboratories, immediately after World War II. In 1948 Kodak developed the first emulsion sensitive to minimum ionising particles; with these, Lattes, Muirhead, Occhialini and Powell discovered the pion (Chapter 2).

Nuclear emulsions have a practically infinite 'memory'; they integrate all the events during the time they are exposed. This is often a drawback. On the positive side, they have an extremely fine granularity, of the order of several micrometres. The coordinates of points along the track are measured with sub-micrometre precision.

Emulsions are a 'complete' instrument: the measurement of the 'grain density' (their number per unit length) gives the specific ionisation dE/dx, hence $\beta\gamma$; the 'range', i.e. the total track length to the stop point (if present), gives the initial energy; the multiple scattering gives the momentum.

On the other hand, the extraction of the information from the emulsion is a slow and time-consuming process. With the advent of accelerators, bubble chambers and, later, time projection chambers replaced the emulsions as visualising devices. But emulsions remain, even today, unsurpassed in spatial resolution and are still used when this is mandatory.

Cherenkov detectors

In 1934 P. A. Cherenkov (Cherenkov 1934) and S. I. Vavilov (Vavilov 1934) discovered that gamma rays from radium induce luminous emission in solutions. The light was due to the Compton electrons produced by the gamma rays, as discovered by Cherenkov who experimentally elucidated all the characteristics of the phenomenon. I. M. Frank and I. E. Tamm gave the theoretical explanation in 1937 (Frank & Tamm 1937).

If a charged particle moves in a transparent material with a speed v larger than the phase velocity of light, namely if $v > c/n$ where n is the refractive index, it generates a wave similar to the shock wave made by a supersonic jet in the atmosphere. Another, visible, analogy is the wave produced by a duck moving on the surface of a pond. The wave front is a triangle with the vertex at the duck, moving forward rigidly with it. The rays of Cherenkov light are directed normally to the V-shaped wave, as shown in Fig. 1.15(a).

The wave is the envelope of the elementary spherical waves emitted by the moving source at subsequent moments. In Fig. 1.15(b) we show the elementary

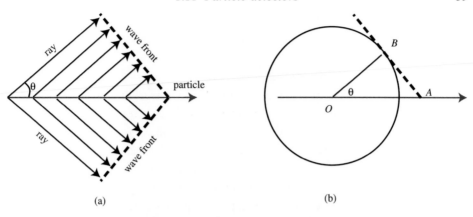

Fig. 1.15. The Cherenkov wave geometry.

wave emitted t seconds before. Its radius is then $OB = ct/n$; in the meantime the particle has moved by $OA = vt$. Hence

$$\theta = \cos^{-1}\left(\frac{1}{\beta n}\right) \tag{1.74}$$

where $\beta = v/c$.

The spectrum of the Cherenkov light is continuous with important fractions in the visible and in the ultraviolet.

Consider the surface limiting the material in which the particle travels. Its intersection with the light cone is a circle or, more generally, an ellipse, called the 'Cherenkov ring'. We can detect the ring by covering the surface with photomultipliers (PMs). If the particle travels, say, towards that surface, the photomultipliers see a ring gradually shrinking in time. From this information, we determine the trajectory of the particle. The space resolution is given by the integration time of the PMs, 30 cm for a typical value of 1 ns.

From the radius of the ring, we measure the angle at the vertex of the cone, hence the particle speed. The thickness of the ring, if greater than the experimental resolution, gives information on the nature of the particle. For example a muon travels straight, an electron scatters much more, giving a thicker ring.

Example 1.12   Super-Kamiokande is a large Cherenkov detector based on the technique described. It contains 50 000 t of pure water. Figure 1.16 shows a photo taken while it was being filled. The PMs, being inspected by the people on the boat in the picture, cover the entire surface. The diameter of each PM is half a metre. The detector, in a laboratory under the Japanese Alps, is dedicated to the search for astrophysical neutrinos and proton decay.

Fig. 1.16.   Inside Super-Kamiokande, being filled with water. People on the boat
are checking the photomultipliers. (Courtesy of Kamioka Observatory – Institute
of Cosmic Ray Research, University of Tokyo)

Figure 1.17 shows an example of an event consisting of a single charged track.
The dots correspond to the PMs that gave a signal; the colour, in the original,
codes the arrival time.

The Cherenkov counters are much simpler devices of much smaller dimensions. The light is collected by one PM, or by a few, possibly using mirrors. In its
simplest version the counter gives a 'yes' if the speed of the particle is $\beta > 1/n$, a
'no' in the opposite case. In more sophisticated versions one measures the angle
of the cone, hence the speed.

Example 1.13   Determine for a water-Cherenkov ($n = 1.33$): (1) the threshold
energy for electrons and muons; (2) the radiation angle for an electron of 300
MeV; (3) whether a K^+ meson with a momentum of 550 MeV gives light.

1. Threshold energy for an electron:
$$E = \gamma m = \frac{m}{\sqrt{1-(1/n)^2}} = \frac{0.511 \text{ MeV}}{\sqrt{1-(1/1.33)^2}} = 0.775 \text{ MeV}.$$
Threshold energy for a μ: $E = \frac{106 \text{ MeV}}{\sqrt{1-(1/1.33)^2}} = 213 \text{ MeV}.$

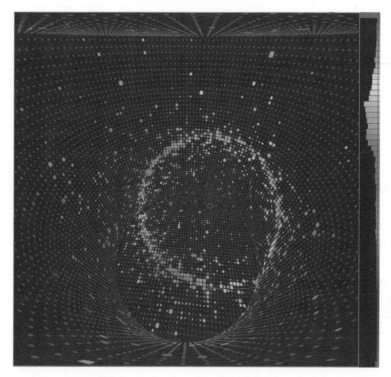

Fig. 1.17. A Cherenkov ring in Super-Kamiokande. (Courtesy of Super Kamiokande Collaboration)

2. The electron is above threshold. The angle is

$$\theta = \cos^{-1}\left(\frac{1}{\beta n}\right) = \cos^{-1}(1/1.33) = 41.2°.$$

3. Threshold energy for a K^+: $E = \frac{494\,\text{MeV}}{\sqrt{1-(1/1.33)^2}} = 749\,\text{MeV}$. The corresponding momentum is:

$$p = \sqrt{E^2 - m_K^2} = \sqrt{749^2 - 494^2} = 563\,\text{MeV}.$$ Therefore at 550 MeV a K^+ does not make light.

Cloud chambers

In 1895 C. T. R. Wilson, fascinated by atmospheric optical phenomena, such as the glories and the coronae he had admired from the observatory that existed on top of Ben Nevis in Scotland, started laboratory research on cloud formation. He built a container with a glass window, filled with air and saturated water vapour. The volume could be suddenly expanded, bringing the vapour to a supersaturated state. Very soon, Wilson understood that condensation nuclei other than dust particles were present in the air. Maybe, he thought, they are electrically charged atoms or ions. The hypothesis was confirmed by irradiating the volume with the X-rays that had recently been discovered. By the end of 1911, Wilson had developed his device to the point of observing the first tracks of alpha and beta particles (Wilson 1912). Actually, an ionising particle crossing the chamber leaves a trail of ions, which seeds many droplets when the chamber is expanded. By flashing light and taking a picture one can record the track. By 1923 the Wilson chamber had been perfected (Wilson 1933).

If the chamber is immersed in a magnetic field **B**, the tracks are curved. Measuring the curvature radius R, one determines the momentum p by Eq. (1.68).

The expansion of the Wilson chamber can be triggered. If we want, for example, to observe charged particles coming from above and crossing the chamber, we put one Geiger counter (see later) above and another below the chamber. We send the two electronic signals to a coincidence circuit, which commands the expansion. Blackett and Occhialini discovered the positron–electron pairs in cosmic radiation with this method in 1933. The coincidence circuit had been invented by B. Rossi in 1930 (Rossi 1930).

Bubble chambers

The bubble chamber was invented by D. Glaser in 1952 (Glaser 1952), but it became a scientific instrument only with L. Alvarez (see Nobel lecture Alvarez 1972) (see Example 1.14). The working principle is similar to that of the cloud chamber, with the difference that the fluid is a liquid which becomes superheated during expansion. Along the tracks, a trail of gas bubbles is generated.

Differently from the cloud chamber, the bubble chamber must be expanded *before* the arrival of the particle to be detected. Therefore, the bubble chambers cannot be used to detect random events such as cosmic rays, but are a perfect instrument at an accelerator facility, where the arrival time of the beam is known exactly in advance.

The bubble chamber acts at the same time both as target and as detector. From this point of view, the advantage over the cloud chamber is the higher density of liquids compared with gases, which makes the interaction probability larger. Different liquids can be used, depending on the type of experiment: hydrogen to have a target nucleus as simple as a proton, deuterium to study interactions on neutrons, liquids with high atomic numbers to study the small cross section interactions of neutrinos.

Historically, bubble chambers have been exposed to all available beams (protons, antiprotons, pions, K mesons, muons, photons and neutrinos). In a bubble chamber, all the charged tracks are visible. Gamma rays can also be detected if they 'materialise' into e^+e^- pairs. The 'heavy liquid' bubble chambers are filled with a high-Z liquid (for example a freon) to increase the probability of the process. All bubble chambers are in a magnetic field to provide the measurement of the momenta.

Bubble chambers made enormous contributions to particle physics: from the discovery of unstable hadrons, to the development of the quark model, to neutrino physics and the discovery of 'neutral' currents, to the study of the structure of nucleons.

Example 1.14 The Alvarez bubble chambers.

The development of bubble chamber technology and of the related analysis tools took place at Berkeley in the 1950s in the group led by L. Alvarez. The principal device was a large hydrogen bubble chamber 72″ long, 20″ wide and 15″ deep (1.8 m × 0.5 m × 0.4 m). The chamber could be filled with liquid hydrogen if the targets of the interaction were to be protons or with deuterium if they were to be neutrons. The uniform magnetic field had the intensity of 1.5 T.

In the example shown in Fig. 1.18, one sees, in a 10″ bubble chamber, seven beam tracks, which are approximately parallel and enter from the left (three more are due to an interaction before the chamber). The beam particles are π^- produced at the Bevatron.

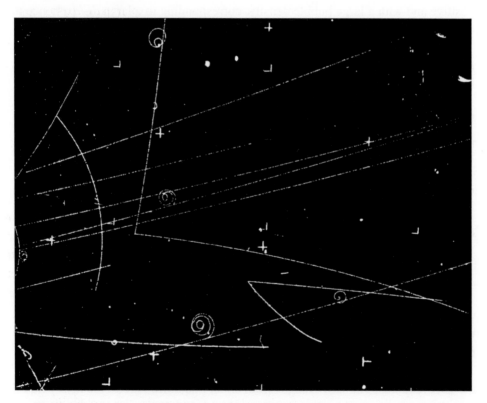

Fig. 1.18.   A picture of the Berkeley 10 inch bubble chamber. (From Alavarez 1972)

The small curls one sees coming out of the tracks are due to atomic electrons that during the ionisation process received an energy high enough to produce a visible track. Moving in the liquid they gradually lose energy and the radius of their orbit decreases accordingly. They are called 'δ-rays'.

The second beam track, counting from below, disappears soon after entering. A pion has interacted with a proton with all neutrals in the final state. A careful study shows that the primary interaction is

$$\pi^- + p \rightarrow K^0 + \Lambda^0 \tag{1.75}$$

followed by the two decays

$$K^0 \rightarrow \pi^+ + \pi^- \tag{1.76}$$

$$\Lambda^0 \rightarrow \pi^- + p. \tag{1.77}$$

We see in the picture two V-shaped events, called V^0s, the decays of two neutral particles into two charged particles. Both are clearly coming from the primary vertex. One of the tracks is a proton, as can be understood by the fact that it is positive and with a large bubble density, corresponding to a large dE/dx, hence to a low speed.

For every expansion, three pictures are taken with three cameras in different positions, obtaining a stereoscopic view of the events. The quantitative analysis implies the following steps:

- the measurement of the coordinates of the three vertices and of a number of points along each of the tracks in the three pictures;
- the spatial reconstruction of the tracks, obtaining their directions and curvatures, namely their momenta;
- the kinematic 'fit'. For each track, one calculates the energy, assuming in turn the different possible masses (proton or pion for example). The procedure then constrains the measured quantities, imposing energy and momentum conservation at each vertex. The problem is overdetermined. In this example, one finds that reactions (1.75), (1.76) and (1.77) 'fit' the data.

Notice that the known quantities are sufficient to allow the reconstruction of the event even in the presence of one (but not more) neutral unseen particles. If the reaction had been $\pi^- + p \rightarrow K^0 + \Lambda^0 + \pi^0$ we could have reconstructed it.

The resolution in the measurement of the coordinates is typically one-tenth of the bubble radius. The latter ranges from about one millimetre in the heavy liquid chambers, to a tenth of a millimetre in the hydrogen chambers, to about 10 μm in

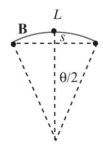

Fig. 1.19. Geometry of the track of a charged particle in a magnetic field.

the rapid cycling hydrogen chamber LEBC (Allison *et al.* 1974a) that was used to detect picosecond lifetime particles such as the charmed mesons.

Example 1.15 In general, the curvature radius R of a track in a magnetic field in a cloud chamber is computed by finding the circle that best fits a set of points measured along the track. Knowing the field B, Eq. (1.68) gives the momentum p. How can we proceed if we measure only three points, as in Fig. 1.19?

The measurements give directly the sagitta s. This can be expressed, with reference to the figure, as $s = R(1 - \cos \theta/2) \approx R\theta^2/8$. Furthermore, $\theta \approx L/R$ and we obtain

$$s \approx \frac{L^2}{8R} = 0.3 \frac{BL^2}{8p} \tag{1.78}$$

that gives us p.

Ionisation detectors

An ionisation detector contains two electrodes and a fluid, liquid or gas, in between. The ion pairs produced by the passage of a charged particle drift toward the electrodes in the electric field generated by the voltage applied to the electrodes. Electrons drift faster than ions and the intensity of their current is consequently larger.

For low electric field intensity, the electron current intensity is proportional to the primary ionisation. Its measurement at one of the electrodes determines dE/dx, which gives a measurement of the factor $\beta\gamma$, and hence the velocity of the particle. If we know the mass of the particle, we can calculate its momentum; if we do not, we can measure the momentum independently and determine the mass.

At higher field intensities, the process of secondary ionisation sets in, giving the possibility of amplifying the initial charge. At very high fields (say MV/m), the amplification process becomes catastrophic, producing a discharge in the detector.

The Geiger counter

The simplest ionisation counter is shown schematically in Fig. 1.20. It was invented by H. Geiger in 1908 at Manchester and later modified by W. Mueller (Geiger & Mueller 1928). The counter consists of a metal tube, usually earthed, bearing a central, insulated, metallic wire, with a diameter of the order of 100 μm. A high potential, of the order of 1000 V, is applied to the wire. The tube is filled with a gas mixture, typically argon and alcohol (to quench the discharge).

The electrons produced by the passage of a charged particle drift towards the wire where they enter a very intense field. They accelerate and produce secondary ionisation. An avalanche process starts that triggers the discharge of the capacitance. The process is independent of the charge deposited by the particle; consequently, the response is of the yes/no type. The time resolution is limited to about a microsecond by the variation from discharge to discharge of the temporal evolution of the avalanche.

Multi-wire chambers

Multi-wire proportional chambers (MWPC) were developed by G. Charpak and collaborators at CERN starting in 1967 (Charpak *et al.* 1968). Their scheme is shown in Fig. 1.21. The anode is a plane of metal wires (thickness from 20 μm to

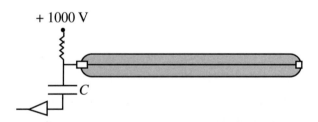

Fig. 1.20.   The Geiger counter.

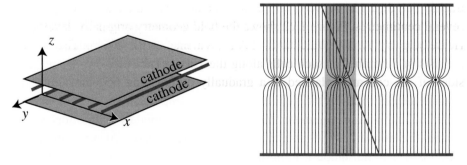

Fig. 1.21.   Geometry of the MWPC.

50 μm), drawn parallel and equispaced with a pitch of typically 2 mm. The anode plane is enclosed between two cathode planes, which are parallel and at the same distance of several millimetres, as shown in the figure.

The MWPC are employed in experiments on secondary beams at an accelerator, in which the particles to be detected leave the target within a limited solid angle around the forward direction. The chambers are positioned perpendicularly to the average direction. This technique allows large areas (several square metres) to be covered with detectors whose data can be transferred directly to a computer, differently from bubble chambers. The figure shows the inclined trajectory of a particle. The electric field shape divides the volume of the chamber into cells, one for each sensitive wire. The ionisation electrons produced in the track segment belonging to a given cell will drift towards the wire of that cell, following the field lines. In the neighbourhood of the anode wire, the charge is amplified, in the proportional regime. Typical amplification factors are of the order of 10^5.

Every wire is serviced by a charge amplifier for its read-out. Typically, thousands of electronic channels are necessary. The coordinate perpendicular to the wires, x in the figure, is determined by the position of the wire (or wires) that gives a signal above threshold. The coordinate z, normal to the plane, is known by construction. To measure the third coordinate y (at least) a second chamber is needed with wires in the x direction. The spatial resolution is the variance of a uniform distribution with the width of the spacing. For example, for 2 mm pitch, $\sigma = 2/\sqrt{12} = 0.6$ mm.

Drift chambers

Drift chambers are similar to MWPC, but provide two coordinates. One coordinate, as in the MWPC, is given by the position of the wire giving the signal; the second, perpendicular to the wire in the plane of the chamber, is obtained by measuring the time taken by the electron to reach it (drift time). The chambers are positioned perpendicularly to the average direction of the tracks. The distance between one of the cathodes and the anode is typically of several centimetres. Figure 1.22 shows the field geometry originally developed at Heidelberg by A. H. Walenta in 1971 (Walenta *et al.* 1971). The chamber consists of a number of such cells along the x-axis. The 'field wires' on the two sides of the cell are polarised at a gradually diminishing potential to obtain a uniform electric field.

In the uniform field, and with a correct choice of the gas mixture, one obtains a constant drift velocity. Given the typical value of the drift velocity of 50 mm/μs, measuring the drift time with a 4 ns precision, one obtains a spatial resolution in z of 200 μm.

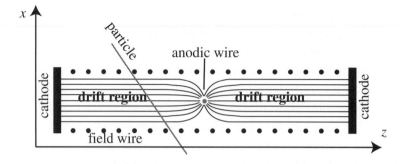

Fig. 1.22.   A drift chamber geometry.

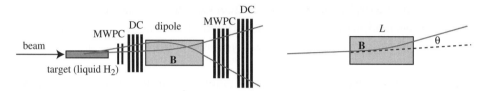

Fig. 1.23.   A simple spectrometer.

One can also measure the induced charge by integrating the current from the wire, obtaining a quantity proportional to the primary ionisation charge and so determining dE/dx.

Figure 1.23 shows an example of the use of MWPC and drift chambers (DC) in a fixed-target spectrometer, used to measure the momenta and the sign of the charges of the particles. A dipole magnet deflects each particle, by an angle inversely proportional to its momentum, toward one or the other side depending on the sign of its charge. The poles of the magnet are located above and below the plane of the drawing, at the position of the rectangle. The figure shows two tracks of opposite sign. One measures the track directions before and after the magnet as accurately as possible using multi-wire and drift chambers. The angle between the directions and the known value of the field gives the momenta.

The geometry is shown on the right of the figure. To simplify, we assume **B** to be uniform in the magnet, of length L, and zero outside it. We also consider only small deflection angles. With these approximations the angle is $\theta \approx L/R$ and, recalling (1.68)

$$\theta \approx 0.3 \frac{BL}{p}. \tag{1.79}$$

The quantity BL, more generally $\int B \, dl$, is called the 'bending power' with reference to the magnet, or 'rigidity' with reference to the particle. Consider for example a magnet of bending power $\int B\,dl = 1\,\mathrm{T\,m}$. A particle of momentum $p = 30$ GeV is bent by 10 mrad, corresponding to a lateral shift, for example at 5 m after the magnet, of 50 mm. This shift can be measured with good precision with a resolution of 100 μm.

The dependence on momentum of the deflection angle makes a dipole magnet a dispersive element similar to a prism in the case of light.

Time projection chambers (TPC) have sensitive volumes of cubic metres and give three-dimensional images of the ionising tracks. Their development was due to D. Nygren at Berkeley (Nygren 1981) and independently to W. W. Allison *et al.* at Oxford (Allison *et al.* 1974b), who built structures with long drift distances, of the order of a metre, in the 1970s. Two coordinates are measured in the same way as in a drift chamber. The third coordinate, the one along the wire, can be determined by measuring the charge at both ends. The ratio of the two charges gives the third coordinate with a resolution that is typically 10% of the wire length.

Cylindrical TPCs of different design are practically always used in collider experiments, in which the tracks leave the interaction point in all the directions. These 'central detectors' are immersed in a magnetic field to allow the momenta to be measured.

Silicon microstrip detectors

Microstrip detectors were developed in the 1970s. They are based on a silicon wafer, a hundred micrometres or so thick and with surfaces of several square centimetres. A ladder of many n-p diodes is built on the surface of the wafer in the shape of parallel strips with a pitch of tens of micrometres. The strips are the equivalent of the anode wires in an MWPC and are read-out by charge amplifiers. The device is reverse biased and is fully depleted. A charged particle produces electron–hole pairs that drift and are collected at the strips. The spatial resolution is very good, of the order of 10 μm.

The silicon detectors played an essential role in the study of charmed and beauty particles. These have lifetimes of the order of a picosecond and are produced with typical energies of a few GeV and decay within millimetres from the production point. To separate the production and decay vertices, devices are built made up of a number, typically four or five, of microstrip planes. The detectors are located just after the target in a fixed-target experiment, around the interaction point in a collider experiment. We shall see how important this 'vertex detector' is in the discussion of the discovery of the top quark in Section 4.10 and of the physics of the *B* mesons in Section 8.5.

Calorimeters

In subnuclear physics, the devices used to measure the energy of a particle or a group of particles are called calorimeters. The measurement is destructive, as all the energy must be released in the detector. One can distinguish two types of calorimeters: electromagnetic and hadronic.

Electromagnetic calorimeters

An electron, or a positron, travelling in a material produces an electromagnetic shower as we discussed in Section 1.10. We simply recall the two basic processes: bremsstrahlung

$$e^{\pm} + N \rightarrow e^{\pm} + N + \gamma \tag{1.80}$$

and pair production

$$\gamma + N \rightarrow e^{+} + e^{-} + N. \tag{1.81}$$

The average distance between such events is about the radiation length of the material.

In a calorimeter, one uses the fact that the total length of the charged tracks is proportional to their initial energy. This length is, in turn, proportional to the ionisation charge. This latter, or a quantity proportional to it, is measured.

In Fig. 1.24, an electromagnetic shower in a cloud chamber is shown. The longitudinal dimensions of the shower are limited by a series of lead plates, each 12.7 mm thick. The initial particle is a photon, as recognised from the absence of tracks in the first sector. The shower initiates in the first plate and completely develops in the chamber. The absorption is due practically only to the lead, for which $X_0 = 5.6$ mm, which is much shorter than that of the gas in the chamber. The total lead thickness is $8 \times 12.7 = 101.6$ mm, corresponding to 18 radiation lengths. In general, a calorimeter must be deep enough to completely absorb the shower: 15–25 radiation lengths, depending on the energy.

The calorimeter that we have described is of the 'sampling' type, because only a fraction of the deposited energy is detected. The larger part, which is deposited in the lead, is not measured. Calorimeters of this type are built by assembling sandwiches of lead plates (typically 1 mm thick) alternated with plastic scintillator plates (several mm thick). The scintillation light (proportional to the ionisation charge deposited in the detector) is collected and measured. The energy resolution is ultimately determined by the number N of the shower particles that are detected. The fluctuation is $\sqrt{N}$. Therefore, the resolution $\sigma(E)$ is proportional to $\sqrt{E}$. The relative resolution improves as the energy increases.

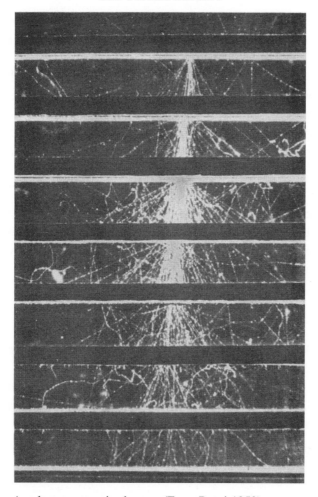

Fig. 1.24. An electromagnetic shower. (From Rossi 1952)

Typical values are

$$\frac{\sigma(E)}{E} = \frac{15 - 18\%}{\sqrt{E(\text{GeV})}}.$$ (1.82)

Hadronic calorimeters

Hadronic calorimeters are used to measure the energy of a hadron or a group of hadrons. As we shall see in Chapter 6, the quarks appear in a high-energy collision as a hadronic 'jet', namely as a group of hadrons travelling within a narrow solid angle. Hadronic calorimeters are the main instrument for measuring the jet energy, which is essentially the quark energy.

Hadronic calorimeters are in principle similar to electromagnetic ones. The main difference is that the average distance between interactions is the interaction λ_0.

A common type of hadronic calorimeter is made like a sandwich of metal plates (iron for example) and plastic scintillators. To absorb the shower completely 10–15 interaction lengths ($\lambda_0 = 17$ cm for iron) are needed. Typical values of the resolution are

$$\frac{\sigma(E)}{E} = \frac{40 - 60\%}{\sqrt{E(\text{GeV})}}. \tag{1.83}$$

The main reason for the rather poor resolution is that the hadronic shower always contains an electromagnetic component, due to the photons from the decay of the π^0s and to the difference in the response to the electromagnetic and hadronic components.

Problems

Introduction A common problem is the transformation of a kinematic quantity between the centre of mass (CM) and the laboratory (L) frames. There are two basic ways to proceed; either explicitly performing the Lorentz transformations or using invariant quantities, namely s, t or u. Depending on the case, one or the other, or a combination of the two, may be more convenient.

Let us find some useful expressions for a generic two-body scattering

$$a + b \rightarrow c + d.$$

We start with s expressed in the initial state and in the L frame

$$s = (E_a + m_b)^2 - p_a^2 = m_a^2 + m_b^2 + 2E_a m_b.$$

If s and the masses are known, the beam energy is

$$E_a = \frac{s - m_a^2 - m_b^2}{2m_b}. \tag{P1.1}$$

Now consider the quantities in the CM frame. From energy conservation we have

$$E_a^* = \sqrt{p_a^{*2} + m_a^2} = \sqrt{s} - E_b^*$$
$$p_a^{*2} + m_a^2 = s - 2E_b^*\sqrt{s} + E_b^{*2}$$
$$2E_b^*\sqrt{s} = s + \left(E_b^{*2} - p_a^{*2}\right) - m_a^2 = s + m_b^2 - m_a^2.$$

And we obtain

$$E_b^* = \frac{s + m_b^2 - m_a^2}{2\sqrt{s}}. \tag{P1.2}$$

By analogy, for the other particle we write

$$E_a^* = \frac{s + m_a^2 - m_b^2}{2\sqrt{s}}. \tag{P1.3}$$

From the energies, we immediately have the CM initial momentum

$$p_a^* = p_b^* = \sqrt{E_{a/b}^{*2} - m_{a/b}^2}. \tag{P1.4}$$

The same arguments in the final state give

$$E_c^* = \frac{s + m_c^2 - m_d^2}{2\sqrt{s}} \tag{P1.5}$$

$$E_d^* = \frac{s + m_d^2 - m_c^2}{2\sqrt{s}} \tag{P1.6}$$

$$p_c^* = p_d^* = \sqrt{E_{c/d}^{*2} - m_{c/d}^2}. \tag{P1.7}$$

Now consider t, and write explicitly (1.26)

$$\begin{aligned} t &= m_c^2 + m_a^2 + 2p_a p_c \cos\theta_{ac} - 2E_a E_c \\ &= m_d^2 + m_b^2 + 2p_b p_d \cos\theta_{bd} - 2E_b E_d. \end{aligned} \tag{P1.8}$$

In the CM frame we extract the expressions of the angles

$$\cos\theta_{ac}^* = \frac{t - m_a^2 - m_c^2 + 2E_a^* E_c^*}{2p_a^* p_c^*} \tag{P1.9}$$

$$\cos\theta_{bd}^* = \frac{t - m_b^2 - m_d^2 + 2E_b^* E_d^*}{2p_b^* p_d^*}. \tag{P1.10}$$

In the L frame, where $p_b = 0$, t has a very simple expression

$$t = m_b^2 + m_d^2 - 2m_b E_d \tag{P1.11}$$

that gives E_d, if t is known

$$E_d = \frac{m_b^2 + m_d^2 - t}{2m_b}. \tag{P1.12}$$

We can find E_c using energy conservation

$$E_c = m_b + E_a - E_d = \frac{s + t - m_a^2 - m_d^2}{2m_b}. \tag{P1.13}$$

Finally, let us also write u explicitly as

$$\begin{aligned} u &= m_d^2 + m_a^2 + 2p_a p_d \cos\theta_{ad} - 2E_a E_d \\ &= m_c^2 + m_b^2 + 2p_b p_c \cos\theta_{bc} - 2E_b E_c. \end{aligned} \tag{P1.14}$$

In the L frame the expression of u is also simple

$$u = m_b^2 + m_c^2 - 2m_b E_c \tag{P1.15}$$

which gives E_c if u is known.

From (P1.13) and (P1.15) Eq. (1.28) follows immediately.

1.1. Estimate the energy of a Boeing 747 (mass $M = 400$ t) at cruising speed (850 km/h) and compare it with the energy released in a mosquito–anti-mosquito annihilation.

1.2. Three protons have momenta equal in absolute value, $p = 3$ GeV, and directions at $120°$ from one another. What is the mass of the system?

1.3. Consider the weak interaction lifetimes of $\pi^{\pm}$: $\tau_\pi = 26$ ns, of $K^{\pm}$: $\tau_K = 12$ ns and of the Λ: $\tau_\Lambda = 0.26$ ns and compute their widths.

1.4. Consider the strong interaction total widths of the following mesons: ρ, $\Gamma_\rho = 149$ MeV; ω, $\Gamma_\omega = 8.5$ MeV; φ, $\Gamma_\varphi = 4.3$ MeV; K^*, $\Gamma_{K^*} = 51$ MeV; J/ψ, $\Gamma_{J/\psi} = 93$ keV; and of the baryon Δ, $\Gamma_\Delta = 118$ MeV and compute their lifetimes.

1.5. An accelerator produces an electron beam with energy $E = 20$ GeV. The electrons diffused at $\theta = 6°$ are detected. Neglecting their recoil motion, what is the minimum structure in the proton that can be resolved?

1.6. In the collision of two protons the final state contains a particle of mass m besides the protons.

a. Give an expression for the minimum (threshold) energy E_p for the process to happen and for the corresponding momentum p_p if the target proton is at rest.

b. Give the expression of the minimum energy E_p^* for the process to happen and of the corresponding momentum p_p^* if the two protons collide with equal and opposite velocities.

c. How large are the threshold energies in the cases (a) and (b) if the produced particle is a pion? How large is the kinetic energy in the first case?

1.7. Consider the process $\gamma + p \rightarrow p + \pi^0$ (π^0 photoproduction) with the proton at rest.

a. Find the minimum energy of the photon E_γ.

The Universe is filled by 'background electromagnetic radiation' at the temperature of $T = 3$ K. The corresponding Planck energy distribution peaks at 0.37 meV. Consider the highest energy photons with energy $E_{\gamma,3K} \approx 1$ meV.

b. Find the minimum energy E_p of the cosmic ray protons needed to induce π^0 photoproduction.

c. If the cross section, just above threshold, is $\sigma = 0.6$ mb and the background of high-energy photon density is $\rho \approx 10^6$ m^{-3}, find the attenuation length. Is it small or large on the cosmological scale?

1.8. The Universe contains two types of electromagnetic radiation: (a) the 'microwave background' at $T = 3$ K, corresponding to photon energies $E_{\gamma,3K} \approx 1$ meV, (b) the Extragalactic Background Light (EBL) due to the stars, with a spectrum which is mainly in the infrared. The Universe is opaque to photons whose energy is such that the cross section for pair production $\gamma + \gamma \rightarrow e^+ + e^-$ is large. This already happens just above threshold (see Fig. 1.9). Compute the two threshold energies, assuming in the second case the photon wavelength $\lambda = 1$ μm.

1.9. The Bevatron was designed to have sufficient energy to produce antiprotons. What is the minimum energy of the proton beam for such a process? Take into account that because of baryonic number conservation (see Section 3.7) the reaction is $p + p \rightarrow p + p + \bar{p} + p$.

1.10. In the LHC at CERN, two proton beams collide head on with energies $E_p = 7$ TeV. What energy would be needed to obtain the same CM energy with a proton beam on a fixed hydrogen target? How does it compare with cosmic ray energies?

1.11. Consider a particle of mass M decaying into two bodies of masses m_1 and m_2. Give the expressions for the energies and momenta of the decay products in the CM frame.

1.12. Evaluate the energies and momenta in the CM frame for the two final particles of the decays $\Lambda \rightarrow p\pi^-$, $\Xi^- \longrightarrow \Lambda\pi^-$.

1.13. Find the expressions for the energies and momenta of the final particles of the decay $M \rightarrow m_1 + m_2$ in the CM if m_2 mass is zero.

1.14. In a monochromatic π beam with momentum p_π, a fraction of the pions decay in flight as $\pi \to \mu \nu_\mu$. We observe that in some cases the muons move backwards. Find the maximum value of p_π for this to happen.

1.15. A Λ hyperon decays as $\Lambda \to p + \pi^-$; its momentum in the L frame is $p_\Lambda = 2$ GeV. Take the direction of the Λ in the L frame as the x-axis. In the CM frame the angle of the proton direction with x is $\theta_p^* = 30°$. Find

a. Energy and momentum of the Λ and the π in the CM frame
b. The Lorentz parameters for the L–CM transformation
c. Energy and momentum of the π, angle and momentum of the Λ in the L frame.

1.16. Consider the collision of a ball with an equal ball at rest. Compute the angle between the two final directions at non-relativistic speeds.

1.17. A proton with momentum $p_1 = 3$ GeV elastically diffuses on a proton at rest. The diffusion angle of one of the protons in the CM is $\theta_{ac}^* = 10°$. Find

a. The kinematic quantities in the L frame
b. The kinematic quantities in the CM frame
c. The angle between the final proton directions in the L frame; is it $90°$?

1.18. A 'charmed' meson D^0 decays $D^0 \to K^- \pi^+$ at a distance from the production point of $d = 3$ mm. Measuring the total energy of the decay products one finds $E = 30$ GeV. How long did the D live in proper time? How large is the π^+ momentum in the D rest-frame?

1.19. The primary beam of a synchrotron is extracted and used to produce a secondary monochromatic π^- beam. One observes that at the distance $l = 20$ m from the production target 10% of the pions have decayed. Find the momentum and energy of the pions.

1.20. A π^- beam is brought to rest in a liquid hydrogen target. Here π^0 are produced by the 'charge exchange' reaction $\pi^- + p \to \pi^0 + n$. Find: the energy of the π^0; the kinetic energy of the n; the velocity of the π^0; and the distance travelled by the π^0 in a lifetime.

1.21. A particle of mass m, charge $q = 1.6 \times 10^{-19}$ C and momentum p moves in a circular orbit at a constant speed (in absolute value) in the magnetic field **B** normal to the orbit. Find the relationship between m, p and B.

1.22. We wish to measure the total $\pi^+ p$ cross section at 20 GeV incident momentum. We build a liquid hydrogen target ($\rho = 60$ kg/m^3) with $l = 1$ m. We measure the flux before the target and that after the target with two scintillation counters. Measurements are made with the target empty and with the target full. By normalising the fluxes after the target to the same incident flux, we obtain in the two cases $N_0 = 7.5 \times 10^5$ and $N_H = 6.9 \times 10^5$ respectively. Find the cross section and its statistical error (ignoring the uncertainty of the normalisation).

1.23. In the Chamberlain *et al.* experiment that discovered the antiproton, the antiproton momentum was approximately 1.2 GeV. What is the minimum refractive index in order to have the antiprotons above threshold in a Cherenkov counter? How wide is the Cherenkov angle if $n = 1.5$?

1.24. Consider two particles with masses m_1 and m_2 and the same momentum p. Evaluate the difference Δt between the times taken to cross the distance L. Let us define the base with two scintillation counters and measure Δt with 300 ps resolution. How much must L be if we want to distinguish π from K at two standard deviations if their momentum is 4 GeV?

1.25. A Cherenkov counter containing nitrogen gas at pressure Π is located on a charged particle beam with momentum $p = 20$ GeV. The dependence of the refractive index on the pressure Π is given by the law $n - 1 = 3 \times 10^{-9}\Pi$ (Pa). The Cherenkov detector must see the π and not the K. In which range must the pressure be?

1.26. Superman is travelling on a Metropolis avenue at high speed. At a crossroads, seeing that the lights are green, he continues. However, he is stopped by the police, claiming he had crossed on red. Assuming both to be right, what was the speed of Superman?

Further reading

Alvarez, L. (1968); Nobel Lecture, *Recent Developments in Particle Physics* http://nobelprize.org/nobel_prizes/physics/laureates/1968/alvarez-lecture.pdf

Blackett, P. M. S. (1948); Nobel Lecture, *Cloud Chamber Researches in Nuclear Physics and Cosmic Radiation* http://nobelprize.org/nobel_prizes/physics/laureates/1948/blackett-lecture.pdf

Bonolis, L. (2005); *Bruno Touscheck vs. Machine Builders: AdA, the first matter-antimatter collider. La Rivista del Nuovo Cimento* **28** no. 11

Charpak, G. (1992); Nobel Lecture, *Electronic Imaging of Ionizing Radiation with Limited Avalanches in Gases* http://nobelprize.org/nobel_prizes/physics/laureates/1992/charpak-lecture.html

Glaser, D. A. (1960); Nobel Lecture, *Elementary Particles and Bubble Chamber* http://nobelprize.org/nobel_prizes/physics/laureates/1960/glaser-lecture.pdf

Hess, V. F. (1936); Nobel Lecture, *Unsolved Problems in Physics: Tasks for the Immediate Future in Cosmic Ray Studies* http://nobelprize.org/nobel_prizes/physics/laureates/1936/hess-lecture.html

Kleinknecht, K. (1998); *Detectors for Particle Radiation*. Cambridge University Press

Lederman, L. M. (1991); *The Tevatron. Sci. Am.* **264** no. 3, 48

Meyers, S. & Picasso, E. (1990); *The LEP collider. Sci. Am.* **263** no. 1, 54

Okun, L. B. (1989); *The concept of mass. Physics Today* June, 31

Rees, J. R. (1989); *The Stanford linear collider. Sci. Am.* **261** no. 4, 58

Rohlf, J. W. (1994); *Modern Physics from a to* Z^0. John Wiley & Sons. Chapter 16

van der Meer, S. (1984); Nobel Lecture, *Stochastic Cooling and the Accumulation of Antiprotons* http://nobelprize.org/nobel_prizes/physics/laureates/1984/meer-lecture.html

Wilson, C. R. T. (1927); Nobel Lecture, *On the Cloud Method of Making Visible Ions and the Tracks of Ionising Particles* http://nobelprize.org/nobel_prizes/physics/laureates/1927/wilson-lecture.html

2

Nucleons, leptons and bosons

2.1 The muon and the pion

Only a few elementary particles are stable: the electron, the proton, the neutrinos and the photon. Many more are unstable. The particles that decay by weak interactions live long enough to travel macroscopic distances between their production and decay points. Therefore, we can detect these particles by observing their tracks or measuring their time of flight. Distances range from a fraction of a millimetre to several metres. In this chapter, we shall study the simplest properties of these particles and discuss the corresponding experimental discoveries.

As already recalled, in 1935 H. Yukawa formulated a theory of the strong interactions between nucleons inside nuclei (Yukawa 1935). The mediator of the interaction is the π meson, or pion. It must have three charge states, positive, negative and neutral, because the nuclear force exists between protons, between neutrons and between protons and neutrons. As the nuclear force has a finite range, $\lambda \approx 1$ fm, Yukawa assumed a potential between nucleons of the form

$$\phi(r) \propto \frac{e^{-r/\lambda}}{r}. \tag{2.1}$$

From the uncertainty principle, the mass m of the mediator is inversely proportional to the range of the force. In NU, $m = 1/\lambda$. With $\lambda = 1$ fm, we obtain $m \approx 200$ MeV.

Two years later, Anderson and Neddermeyer (Anderson & Neddermeyer 1937) and Street and Stevenson (Street & Stevenson 1937), discovered that the particles of the penetrating component of cosmic rays have masses of just this order of magnitude. Apparently, the Yukawa particle had been discovered, but the conclusion was wrong.

In 1942 Rossi and Nereson (Rossi & Nereson 1942) measured the lifetime of penetrating particles to be $\tau = 2.15 \pm 0.10$ µs.

The crucial experiment showing that the penetrating particle is not the π meson was carried out in 1947 in Rome by M. Conversi, E. Pancini and O. Piccioni (Conversi *et al.* 1947). The experiment aimed at investigating whether the absorption of positive and negative particles in a material was the same or different. Actually, a negative particle can be captured by a nucleus and, if it is the quantum of nuclear forces, quickly interacts with it rather than decaying. In contrast, a positive particle is repelled by a nucleus and will decay as in vacuum. The two iron blocks, F_1 and F_2 in the upper part of Fig. 2.1, are magnetised in opposite directions normal to the drawing and are used to focus the particles of one sign or, inverting their positions, the other. The 'trigger logic' of the experiment is the following. The Geiger counters A and B, above and below the magnetised blocks, must discharge at the same instant ('fast' coincidence); one of the C counters under the absorber must fire not immediately but later, after a delay Δt in the range $1\,\mu s < \Delta t < 4.5\,\mu s$ ('delayed' coincidence). This logic guarantees the following: first that the energy of the particle is large enough to cross the blocks and small enough to stop in the absorber; second that, in this energy range and with the chosen geometry, only particles of one sign can hit both A and B; and finally that the particle decays in a time compatible with the lifetime value of Rossi and Nereson.

Figure 2.1(b) shows the trajectory of two particles of the 'right' sign in the right energy range, which discharges A and B but not C; Fig. 2.1(c) shows two particles of the 'wrong' sign. Neither of them gives a trigger signal because one discharges A and not B, the other discharges both but also C.

In a first experiment in 1945, the authors used an iron absorber. The result was that the positive particles decay as in vacuum, the negative particles do not decay, exactly as expected.

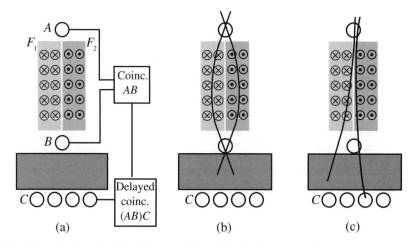

Fig. 2.1.   A sketch of the Conversi, Pancini, Piccioni experiment.

The authors repeated the experiment in 1946 with a carbon absorber, finding, to their surprise, that the particles of both signs decay (Conversi *et al.* 1947). A systematic search showed that in materials with low atomic numbers the penetrating particles are not absorbed by nuclei. However, calculation soon showed that the pions should interact so strongly as to be absorbed by any nucleus, even by small ones. In conclusion, the penetrating particles of the cosmic rays are not the Yukawa mesons.

In the same years, G. Occhialini and C. F. Powell, working at Bristol, exposed emulsion stacks at high altitudes in the mountains (up to 5500 m on the Andes). In 1947 they published, with Lattes and Muirhead, the observation of events in which a more massive particle decays into a less massive one (Lattes *et al.* 1947). The interpretation is that two particles are present in cosmic rays, the first is the π, the second, which was called μ or muon, is the penetrating particle. They observed that the muon range was equal in all the events (about 100 μm), showing that the pion decays into two bodies, the μ and a neutral undetected particle.

The final proof came in 1949, when the Bristol group, using the new Kodak emulsions sensitive to minimum ionising particles, detected events in which the complete chain of decays $\pi\mu e$ was visible. An example is shown in Fig. 2.2.

We know now that the charged pion decays are

$$\pi^+ \to \mu^+ + \nu_\mu \qquad \pi^- \to \mu^- + \bar{\nu}_\mu \tag{2.2}$$

and those of the muons are

$$\mu^+ \to e^+ + \nu_e + \bar{\nu}_\mu \qquad \mu^- \to e^- + \nu_\mu + \bar{\nu}_e. \tag{2.3}$$

In these expressions we have specified the types of neutrinos, something that was completely unknown at the time. We shall discuss neutrinos in Section 2.4.

Other experiments showed directly that pions interact strongly with nuclei, transforming a proton into a neutron and vice versa:

$$\pi^+ + {}^A_Z N \to {}^{A-1}_Z N + p \qquad \pi^- + {}^A_Z N \to {}^{A-1}_{Z-1} N + n. \tag{2.4}$$

Fig. 2.2.   A $\pi\mu e$ decay chain observed in emulsions. (From Brown *et al.* 1949)

In conclusion, the pions are the Yukawa particles. It took a quarter of a century to understand that the Yukawa force is not the fundamental strong nuclear interaction and that the pion is a composite particle. The fundamental interaction occurs between the quarks, mediated by the gluons, as we shall see in Chapter 6.

We shall dedicate Section 2.3 to the measurement of the pion quantum numbers. We summarise here that pions exist in three charge states: π^+, π^0 and π^-. The π^+ and the π^- are each the antiparticle of the other, while the π^0 is its own antiparticle. The π^0 decays practically always (99%) in the channel $\pi^0 \rightarrow \gamma\gamma$.

A mystery was left however: the μ. It was identical to the electron, but for its mass, 106 MeV, about 200 times as big. What is the reason for a heavier brother of the electron? 'Who ordered that?' asked Rabi. Even today, we have no answer.

2.2 Strange mesons and hyperons

Nature had other surprises in store.

In 1943 Leprince-Ringuet and l'Héritier (Leprince-Ringuet & l'Héritier 1944), working in a laboratory on the Alps with a 'triggered' cloud chamber in a magnetic field $B = 0.25$ T, discovered a particle with a mass of 506 ± 61 MeV.

Other surprises were to follow. Soon after the discovery of the pion, in several laboratories in the UK, France and the USA, cosmic ray events were found in which particles with masses similar to that of Leprince-Ringuet decayed, apparently, into pions. Some were neutral and decayed into two charged particles (plus possibly some neutral ones) and were called V^0 because of the shape of their tracks (see Fig. 2.3), others were charged, decaying into a charged daughter particle (plus neutrals) and were named θ, still others decayed into three charged particles, called τ.

It took a decade to establish that θ and τ are exactly the same particle, while the V^0s are its neutral counterparts. These particles are the K mesons, also called 'kaons'.

In 1947 Rochester and Butler published the observation of the associated production of a pair of such unstable particles (Rochester & Butler 1947). It was soon proved experimentally that those particles are always produced in pairs; the masses of the two partners turned out to be different, one about 500 MeV (a K meson), the other greater than that of the nucleon. The more massive ones were observed to decay into a nucleon and a pion. These particles belong to the class of the hyperons. The lightest are the Λ^0 and the Σs that have three charge states, Σ^+, Σ^0 and Σ^-. We discussed in Section 1.11 a clear example seen many years later in a bubble chamber. Figure 1.18 shows the associated production $\pi^- + p \rightarrow K^0 + \Lambda^0$, followed by the decays $K^0 \rightarrow \pi^+ + \pi^-$ and $\Lambda^0 \rightarrow p + \pi^-$.

The new particles had very strange behaviour. There were two puzzles (plus a third to be discussed later). Why were they always produced in pairs? Why were

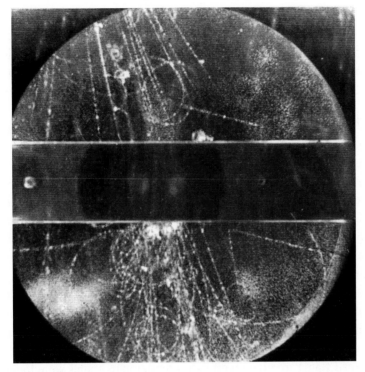

Fig. 2.3. A V^0, below the plate on the right, in a cloud chamber picture. (Rochester & Butler 1947)

they produced by 'fast' strong interaction processes, as demonstrated by the large cross section, while they decayed only 'slowly' with lifetimes typical of weak interactions? In other words, why do fully hadronic decays such as $\Lambda^0 \rightarrow p + \pi^-$ not proceed strongly? The new particles were called 'strange particles'.

The solution was given by Nishijima (Nakato & Nishijima 1953) and independently by Gell-Mann (Gell-Mann 1953). They introduced a new quantum number S, the 'strangeness', which is additive, like electric charge. Strangeness is conserved by strong and electromagnetic interactions but not by weak interactions. The 'old' hadrons, the nucleons and the pions, have $S = 0$, the hyperons have $S = -1$, the K mesons have $S = \pm 1$.

The production by strong interactions from an initial state with $S = 0$ can happen only if two particles of opposite strangeness are produced. The lowest mass strange particles, the K mesons, and the hyperons can decay for energetic reasons only into non-strange final states; therefore, they cannot decay strongly.

If the mass of a strange meson or of a hyperon is large enough, final states of the same strangeness are energetically accessible. This happens if the sum of the masses of the daughters is smaller than that of the mother particle. These particles

Table 2.1 *The K mesons*

	Q	S	m (MeV)	τ (ps)	Principal decays (BR in %)
K^+	+1	+1	494	12	$\mu^+\nu_\mu$(63), $\pi^+\pi^+\pi^-$(21), $\pi^+\pi^0$(5.6)
K^0	0	+1	(498)	n.a.	
K^-	−1	−1	494	12	$\mu^-\bar{\nu}_\mu$, $\pi^-\pi^-\pi^+$, $\pi^-\pi^0$
$\bar{K}^0$	0	−1	(498)	n.a.	

n.a. means not applicable

exist and decay by strong interactions with extremely short lifetimes, of the order of 10^{-24} s. In practice, they decay at the point where they are produced and do not leave an observable track. We shall see in Chapter 4 how to detect them.

We shall not describe the experimental work done with cosmic rays and later with beams from accelerators, rather we shall summarise the main conclusions on the metastable strange particles, which we define as those that are stable against strong interactions and decay weakly or electromagnetically.

The *K* mesons are the only metastable strange mesons. There are four of them. Table 2.1 gives their characteristics; in the last column the principal decay channels of the charged states are given with their approximate branching ratios (BR). The *K* mesons have spin zero.

There are two charged *K* mesons, the K^+ with $S = +1$ and its antiparticle, the K^- that has the same mass, the same lifetime and opposite charge and strangeness. The decay channels of one contain the antiparticles of the corresponding channels of the other.

We anticipate a fundamental law of physics, $\mathcal{CPT}$ invariance. $\mathcal{CPT}$ is the product of three operations, time reversal ($\mathcal{T}$), parity ($\mathcal{P}$), i.e. the inversion of the coordinate axes, and particle–antiparticle conjugation ($\mathcal{C}$). $\mathcal{CPT}$ invariance implies that a particle and its antiparticle have the same mass, lifetime and spin and all 'charges' of opposite value.

While the neutral pion is its own antiparticle, the neutral *K* meson is not, K^0 and $\bar{K}^0$ are distinguished because of their opposite strangeness. We anticipate that K^0 and $\bar{K}^0$ form an extremely interesting quantum two-state system that we shall study in Chapter 8. We mention here only that they are not the eigenstates of the mass and the lifetime. This is the reason for the 'n.a.' entries in Table 2.1.

Now let us consider the metastable hyperons. Three types of hyperons were discovered in cosmic rays, some with more than one charge status (six states in total). These are (see Table 2.2) the Λ^0, three Σs all with strangeness $S = -1$ and two Ξs with strangeness $S = -2$. All have spin $J = 1/2$. In the last column, the principal decays are shown. All but one are weak.

Table 2.2 *The metastable strange hyperons*

	Q	S	m (MeV)	τ (ps)	$c\tau$ (mm)	Principal decays (BR in %)
Λ	0	-1	1116	263	79	$p\pi^-(64)$, $n\pi^0(36)$
Σ^+	$+1$	-1	1189	80	24	$p\pi^0(51.6)$, $n\pi^+(48.3)$
Σ^0	0	-1	1193	7.4×10^{-8}	2.2×10^{-8}	$\Lambda\gamma(100)$
Σ^-	-1	-1	1197	148	44.4	$n\pi^-(99.8)$
Ξ^0	0	-2	1315	290	87	$\Lambda\pi^0(99.5)$
Ξ^-	-1	-2	1321	164	49	$\Lambda\pi^-(99.9)$

The neutral Σ^0 hyperon has a mass larger than the other neutral one, the Λ^0, and the same strangeness. Therefore, the Gell-Mann and Nishijima scheme foresaw the decay $\Sigma^0 \rightarrow \Lambda^0 + \gamma$. This prediction was experimentally confirmed.

Notice that all the weak lifetimes of the hyperons are of the order of one hundred picoseconds; the electromagnetic lifetime of the Σ^0 is nine orders of magnitude smaller.

As we have already said, hadrons are not elementary objects, they contain quarks. We shall discuss this issue in Chapter 4. We have anticipated that the 'old' hadrons contain two types of quarks, u and d. Their strangeness is zero. The strange hadrons contain one or more quarks s or antiquarks $\bar{s}$. The quark s has strangeness $S=-1$ (pay attention to the sign!), its antiquark $\bar{s}$ has strangeness $S=+1$. The $S=+1$ hadrons, such as K^+, K^0, Λ and the Σs, contain one $\bar{s}$, those with $S=-1$, such as K^-, $\bar{K}^0$, Λ and the Σs contain one s quark, the Ξs with $S=-2$ contain two s quarks, etc.

2.3 The quantum numbers of the charged pion

For every particle we must measure all the relevant characteristics: mass, lifetime, spin, charge, strangeness, branching ratios for its decays in different channels and, as we shall discuss in the next chapter, intrinsic parity and, if completely neutral, charge conjugation. This enormous work took several decades of the last century. We shall discuss here only some measurements of the quantum numbers of the charged pion.

The mass The first accelerator with sufficient energy to produce pions was the Berkeley cyclotron that could accelerate alpha particles up to a kinetic energy of $E_k = 380$ MeV.

To determine the mass, two kinematic quantities must be measured, for example the energy E and the momentum p. The mass is then given by

$$m^2 = E^2 - p^2.$$

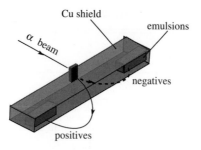

Fig. 2.4. A sketch of the Burfening *et al.* equipment for the pion mass measurement.

We show in Fig. 2.4 a sketch of the set-up of the pion mass measurement by Burfening and collaborators in 1951 (Burfening *et al.* 1951). Two emulsion stacks, duly screened from background radiation, are located in the cyclotron vacuum chamber, below the plane of the orbit of the accelerated alpha particles. When the alpha particles reach their final orbit they hit a small target and produce pions of both signs. The pions are deflected by the magnetic field of the cyclotron on one side or the other depending on their sign and penetrate the corresponding emulsion stack. After the exposure the emulsions are developed, the entrance point and direction of each pion track are measured. These, together with the known position of the target, give the pion momentum. The measurement of its range gives its energy.

The result of the measurement was

$$m_{\pi^+} = 141.5 \pm 0.6 \, \text{MeV} \qquad m_{\pi^-} = 140.8 \pm 0.7 \, \text{MeV}. \qquad (2.5)$$

The two values are equal within the errors. The present value is

$$m_{\pi^\pm} = 139.570 \, 18 \pm 0.000 \, 35 \, \text{MeV}. \qquad (2.6)$$

Lifetime To measure decay times of the order of several nanoseconds with good resolution we need electronic techniques and fast detectors. The first measurement with such techniques was due to O. Chamberlain and collaborators, as shown in Fig. 2.5 (Chamberlain *et al.* 1950).

The 340 MeV γ beam from the Berkeley synchrotron hit a paraffin (a proton-rich material) target and produced pions by the reaction

$$\gamma + p \to \pi^+ + n. \qquad (2.7)$$

Two scintillation counters were located, one after the other, on one side of the target. The logic of the experiment required that a meson crossed the first scintillator and stopped in the second. The positive particles were not absorbed by

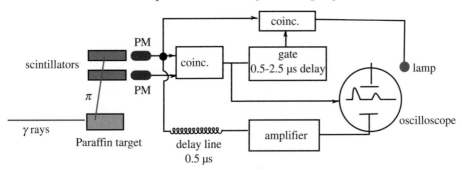

Fig. 2.5. A sketch of the detection scheme in the pion lifetime experiment of Chamberlain *et al.*

the nuclei and decayed at rest. The dominant decay channel is

$$\pi^+ \rightarrow \mu^+ + \nu_\mu. \tag{2.8}$$

The μ loses all its energy in ionisation, stops and after an average time of 2.2 μs decays

$$\mu^+ \rightarrow e^+ + \nu_e + \bar{\nu}_\mu. \tag{2.9}$$

To implement this logic, the electric pulses from the two photomultipliers that read the scintillators were sent to a coincidence circuit; this established that a particle had crossed the first counter and reached the second. A 'gate' circuit established the presence of a second pulse, from the second counter, with a delay of between 0.5 and 2.5 μs, meaning that a μ decayed. This confirmed that the primary particle was a π^+.

The signals from the second scintillator were sent, delayed by 0.5 μs, to an oscilloscope, whose sweep was triggered by the output of the fast coincidence. The gate signal, if present, lit a lamp located near the scope screen. Screen and lamp were photographed. The pictures show two pulses, one due to the arrival of the π and one due to its decay. They were well separated if their distance apart was > 22 ns.

In total 554 events were collected. As expected, the distribution of the times was exponential. The lifetime measurement gave $\tau = 26.5 \pm 1.2$ ns. The present value is $\tau = 26.033 \pm 0.005$ ns.

The spin A particle of spin s has $2s + 1$ degrees of freedom. As the probability of a reaction depends on the number of degrees of freedom, we can determine the spin by measuring such reaction probabilities. More specifically, we consider the ratio of the cross sections of the two processes, one the inverse of the other, at the same centre of mass energy

$$\pi^+ + d \rightarrow p + p \tag{2.10}$$

$$p + p \rightarrow \pi^+ + d. \tag{2.11}$$

We call them π^+ absorption and production respectively. Writing both reactions generically as $a + b \rightarrow c + d$, Eq. (1.54) gives the cross sections in the centre of mass system. As we are interested in the ratio of the cross sections at the same energy, we can neglect the common factors, including the energy E. We obtain

$$\frac{d\sigma}{d\Omega}(a + b \rightarrow c + d) \propto \frac{p_f}{p_i} \frac{1}{(2s_a + 1)(2s_b + 1)} \sum_{f,i} |M_{fi}|^2 \tag{2.12}$$

where the sum is over all the spin states, initial and final. The initial and final momenta are different in the two processes, but since the energy is the same, the initial momentum in one case is equal to the final one in the other. We can then write for the absorption $p_i = p_\pi$ and $p_f = p_p$, for the production $p_f = p_\pi$ and $p_i = p_p$, with the same values of p_π and p_p. We now write for the absorption process

$$\frac{d\sigma}{d\Omega}(\pi^+ d \rightarrow pp) \propto \frac{p_p}{p_\pi} \frac{1}{(2s_\pi + 1)(2s_d + 1)} \frac{1}{2} \sum_{f,i} |M_{fi}|^2. \tag{2.13}$$

Pay attention to the factor 1/2 that must be introduced to cancel the double counting implicit in the integration over the solid angle with two identical particles in the final state.

We now write for the production process

$$\frac{d\sigma}{d\Omega}(pp \rightarrow \pi^+ d) \propto \frac{p_\pi}{p_p} \frac{1}{(2s_p + 1)^2} \sum_{f,i} |M_{fi}|^2. \tag{2.14}$$

We give here, without proof, the 'detailed balance principle', which is a consequence of the time reversal invariance, which is satisfied by the strong interactions (see next chapter). The principle implies the equality

$$\sum_{f,i} |M_{fi}|^2 = \sum_{f,i} |M_{if}|^2.$$

Using this equation and knowing the spin of the proton, $s_p = 1/2$, and of the deuteron, $s_d = 1$, we obtain

$$\frac{\sigma(\pi^+ d \rightarrow pp)}{\sigma(pp \rightarrow \pi^+ d)} = \frac{(2s_p + 1)^2}{2(2s_\pi + 1)(2s_d + 1)} \frac{p_p^2}{p_\pi^2} = \frac{2}{3(2s_\pi + 1)} \frac{p_p^2}{p_\pi^2}. \tag{2.15}$$

The absorption cross section was measured by Durbin *et al.* (1951) and by Clark *et al.* (1951) at the laboratory kinetic energy $T_\pi = 24\,\text{MeV}$. The production cross section was measured by Cartwright *et al.* (1953) at the laboratory kinetic energy

$T_p = 341$ MeV. The CM energies are almost equal in both cases. From the measured values one obtains $2s_\pi + 1 = 0.97 \pm 0.31$, hence $s_\pi = 0$.

The neutral pion For the π^0, we shall only give the present values of the mass and the lifetime.

The mass of the neutral pion is smaller than that of the charged one by about 4.5 MeV

$$m_{\pi^0} = 134.9766 \pm 0.0006 \text{ MeV}. \tag{2.16}$$

The π^0 decays by electromagnetic interaction predominantly (99.8%) in the channel

$$\pi^0 \rightarrow \gamma\gamma. \tag{2.17}$$

Therefore, its lifetime is much shorter than that of the charged pions

$$\tau_{\pi^0} = (8.4 \pm 0.6) \times 10^{-17} \text{ s}. \tag{2.18}$$

2.4 Charged leptons and neutrinos

We know three charged leptons with identical characteristics. They differ in their masses and lifetimes, as shown in Table 2.3.

We give a few historical hints:

The electron was the first elementary particle to be discovered, by J. J. Thomson in 1897, in the Cavendish Laboratory at Cambridge. At that time, the cathode rays that had been discovered by Plücker in 1857 were thought to be waves, propagating in the ether. Thomson and his collaborators succeeded in deflecting the rays not only, as already known, by a magnetic field, but also by an electric field. By letting the rays pass through crossed electric and magnetic fields and adjusting the field intensities for null deflection, they measured the mass to charge ratio m/q_e and found it to have a universal value (Thomson 1897).

The muon, as we have seen, was discovered in cosmic rays by Anderson and Neddermeyer (1937), and independently by Street and Stevenson (1937); it was identified as a lepton by Conversi, Pancini and Piccioni in 1947 (Conversi *et al.* 1947).

The possibility of a third family of leptons, called the heavy lepton H_l and its neutrino ν_{H_l}, with a structure similar to the two known ones, was advanced by A. Zichichi, who developed in 1967 the search method that we shall now describe, built the experiment and searched for the H_l at ADONE (Bernardini *et al.* 1967). The H_l did indeed exist, but with a mass too large for ADONE. It was discovered at the SPEAR electron–positron collider in 1975 by M. Perl *et al.* (Perl *et al.* 1975). It

Table 2.3 *The charged leptons*

	m (MeV)	τ
e	0.511	$>4 \times 10^{26}$ yr
μ	105.6	2.2 μs
τ	1777	0.29 ps

was called τ, from the Greek word *triton*, meaning the third. The method was the following.

As we shall see in the next chapter, the conservation of the lepton flavours forbids the processes $e^+ e^- \rightarrow e^+ \mu^-$ and $e^+ e^- \rightarrow e^- \mu^+$. If a heavy lepton exists, the following reaction occurs

$$e^+ + e^- \rightarrow \tau^+ + \tau^- \tag{2.19a}$$

followed by the decays

$$\tau^+ \rightarrow e^+ + \nu_e + \bar{\nu}_\tau \qquad \tau^- \rightarrow \mu^- + \bar{\nu}_\mu + \nu_\tau \tag{2.19b}$$

and charge conjugated, resulting in the observation of $e^- \mu^+$ or $e^+ \mu^-$ pairs and apparent violation of the lepton flavours. The principal background is due to the pions that are produced much more frequently than the $e\mu$ pairs. Consequently, the experiment must provide the necessary discrimination power. Moreover, an important signature of the sought events is the presence of (four) neutrinos. Therefore, the two tracks and the direction of the beams do not belong to the same plane, due to the momenta of the unseen neutrinos. Such 'acoplanar' $e\mu$ pairs were finally found at SPEAR, when energy above threshold became available.

The neutrino was introduced as a 'desperate hypothesis', by W. Pauli in 1930, to explain the apparent violation of energy, momentum and angular momentum conservations in beta decays.

The first neutrino, the electron neutrino (ν_e) was discovered by F. Reines and collaborators in 1956 at the Savannah River reactor (Cowan *et al.* 1956). To be precise, they discovered the electron antineutrino, the one produced in fission reactions. We shall shortly describe this experiment.

The muon neutrino (ν_μ) was discovered, i.e. identified as a particle different from ν_e, by L. Lederman, M. Schwartz and J. Steinberger in 1962 at the proton accelerator AGS at Brookhaven (Danby *et al.* 1962). We shall briefly describe this experiment too.

The tau neutrino (ν_τ) was discovered by K. Niwa and collaborators with the emulsion technique at the Tevatron proton accelerator at Fermilab in 2000 (Kodama *et al.* 2001).

We shall now describe the discovery of the electron neutrino. The most intense sources of neutrinos on Earth are fission reactors. They produce electron antineutrinos with a continuum energy spectrum up to several MeV. The flux is proportional to the reactor power. The power of the Savannah River reactor in South Carolina (USA) was 0.7 GW. It was chosen by Reines because a massive building located underground, a dozen metres under the core, was available to the experiment. The $\bar{\nu}_e$ flux was about $\Phi = 10^{17}$ m^{-2} s^{-1}.

Electron antineutrinos can be detected by the inverse beta process but its cross section is extremely small,

$$\sigma(\bar{\nu}_e + p \rightarrow e^+ + n) \approx 10^{-47} (E_\nu/\text{MeV})^2 \text{m}^2. \qquad (2.20)$$

Notice that at low energy the cross section grows with the square of the energy.

An easily available material containing many protons is water. Let us evaluate the mass needed to have a counting rate of, say, $W = 10^{-3}$ Hz, or about one count every 20 s.

Let us evaluate in order of magnitude the quantity of water needed to have, for example, a rate of 10^{-3} Hz for reaction (2.20), on free protons. Taking a typical energy $E_\nu = 1$ MeV, the rate per target proton is $W_1 = \Phi\sigma = 10^{-30}$ s^{-1}. Consequently we need 10^{27} protons. Since a mole of H_2O contains $2N_A \approx 10^{24}$ protons, we need 1000 moles, hence 18 kg. In practice, much more is needed, taking all inefficiencies into account. Reines worked with 200 kg of water.

The main difficulty of the experiment is not the rate but the discrimination of the signal from the possibly much more frequent background sources that can simulate that signal. There are three principal causes: the neutrons that are to be found everywhere near a reactor, cosmic rays and the natural radioactivity of the material surrounding the detector and in the water itself.

Figure 2.6 is a sketch of the detector scheme used in 1955. It shows one of the two 100 litre water containers sandwiched between two liquid scintillator chambers, a technique that had been recently developed, as we saw in Section 1.11. An antineutrino from the reactor interacts with a proton, producing a neutron and a positron. The positron annihilates immediately with an electron, producing two gamma rays in opposite directions, both with 511 MeV energy. The Compton electrons produced by these gamma rays are detected in the liquid scintillators giving two simultaneous signals. This signature of the positron is not easily emulated by background effects.

A second powerful discrimination is given by the detection of the neutron. Water is a good moderator and the neutron slows down in several microseconds. Forty kilos of cadmium, which has a nucleus with a very high cross section for thermal neutron capture, is dissolved in the water. A Cd nucleus captures the neutron

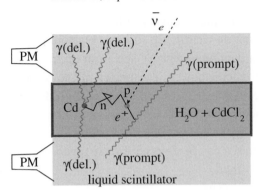

Fig. 2.6.   A sketch of the detection scheme of the Savannah River experiment.

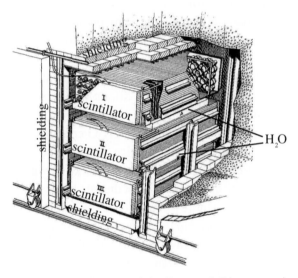

Fig. 2.7.   Sketch of the equipment of the Savannah River experiment. (Reines *et al.* 1996 © Nobel Foundation 1995)

resulting in an excited state that soon emits gamma rays which are detected by the scintillators as a delayed coincidence.

Figure 2.7 is a sketch of the equipment. The reduction of the cosmic ray background, due to the underground location, and the accurate design of the shielding structures were essential for the success of the experiment. Accurate control measurements showed that the observed event rate of $W = 3 \pm 0.2$ events/hour could not be due to background events. This was the experimental discovery of the neutrino, one quarter of a century after the Pauli hypothesis.

The second neutrino was discovered, as already recalled, at the AGS proton accelerator in 1962. The main problem was the extremely small neutrino cross

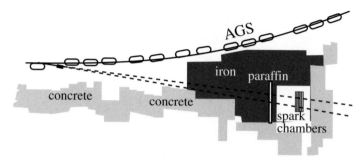

Fig. 2.8. Sketch of the Brookhaven neutrino experiment. (Danby *et al.* 1962 © Nobel Foundation 1988)

section. However, Pontecorvo (1959) and Schwartz (1960) independently calculated that the experiment was feasible.

Figure 2.8 is a sketch of the experiment. The intense proton beam is extracted from the accelerator pipe and sent against a beryllium target. Here a wealth of pions, of both signs, is produced. The pions decay as

$$\pi^+ \to \mu^+ + \nu \qquad \pi^- \to \mu^- + \bar{\nu}. \tag{2.21}$$

In these reactions, the neutrino and the antineutrino are produced in association with a muon. In the beta decays, neutrinos are produced in association with electrons. The aim of the experiment was to clarify whether these neutrinos are different or not. Therefore, we have not specified the type of the neutrinos in the above expressions.

To select only the neutrinos a 'filter' made of iron, 13.5 m long, is located after the target. It absorbs all particles, charged and neutral, apart from the neutrinos. The concrete blocks seen in the figure are needed to protect people from the intense radiation present near the target. To detect the neutrino interactions one needs a device working both as target and as tracking detector. Calculations show that its mass must be about 10 t, too much, at that time, for a bubble chamber. It was decided to use the spark chamber technique, invented by M. Conversi and A. Gozzini in 1955 (Conversi & Gozzini 1955) and developed by Fukui and Myamoto (Fukui & Myamoto 1959). A spark chamber element consists of a pair of parallel metal plates separated by a small gap (a few mm) filled with a suitable gas mixture. The chamber is made sensitive by suddenly applying a voltage to the plates after the passage of the particle(s), generating a high electric field (≈ 1 MV/m). The resulting discharge is located at the position of the ionisation trail and appears as a luminous spark that is photographed.

The neutrino detector consisted of a series of ten modules of nine spark chambers each. The aluminium plates had an area of 1.1×1.1 m^2 and a thickness of 2.5 cm, amounting to a total mass of 10 t.

After exposing the chambers to the neutrinos, photographs were scanned searching for muons from the reactions

$$v + n \rightarrow \mu^- + p \qquad \bar{v} + p \rightarrow \mu^+ + n \qquad (2.22)$$

and electrons from

$$v + n \rightarrow e^- + p \qquad \bar{v} + p \rightarrow e^+ + n. \qquad (2.23)$$

The two particles are easily distinguished because in the first case the photograph shows a long penetrating track, in the second, an electromagnetic shower. Many muon events were observed, but no electron event. The conclusion was that neutrinos produced in association with a muon produce, when they interact, muons, not electrons. It appears that two types of neutrinos exist, one associated with the electron, the other with the muon. The difference is called 'leptonic flavour'. The electron and the electron neutrino have positive electron flavour $\mathcal{L}_e = +1$, the positron and the electron antineutrino have negative electron flavour $\mathcal{L}_e = -1$; all of them have zero muonic flavour. The μ^- and the v_μ have positive muonic flavour $\mathcal{L}_\mu = +1$, the μ^+ and the $\bar{v}_\mu$ have negative muonic flavour $\mathcal{L}_\mu = -1$; all have zero electronic flavour. Electronic, muonic (and tauonic) flavours are also called electronic, muonic (and tauonic) numbers.

2.5 The Dirac equation

In this section we recall the basic properties of the Dirac equation.

In 1928 P. A. M. Dirac wrote the fundamental relativistic wave equation of the electron. The equation predicts all the electron properties known from atomic physics, in particular the value of the gyromagnetic ratio

$$g = 2. \qquad (2.24)$$

We recall that this dimensionless quantity is defined by the relationship between the spin s and the intrinsic magnetic moment μ_e from

$$\mu_e = g\mu_B s \qquad (2.25)$$

where μ_B is the Bohr magneton

$$\mu_B = \frac{q_e h}{2m_e} = 5.788 \times 10^{-11} \, \text{MeV T}^{-1}. \qquad (2.26)$$

The equation has apparently non-physical negative energy solutions. In December 1929, Dirac returned to the problem of trying to identify the 'holes' in the negative energy sea as positive particles, which he thought were the protons.

In November 1930, H. Weyl introduced the mathematical operator $\mathcal{C}$, the particle–antiparticle conjugation, finding that antiparticles and particles must have the same mass. This excluded the protons as antielectrons. In May 1931 Dirac (Dirac 1931) concluded that an as-yet undiscovered particle must exist, positive and with the same mass as the electron, the positron.

Two years later Anderson (Anderson 1933) discovered the positron.

The Dirac equation is

$$\left(i\gamma^{\mu}\partial_{\mu} - m\right)\psi(x) = 0 \tag{2.27}$$

where the sum on the repeated indices is understood.

In this equation, ψ is the Dirac bi-spinor

$$\psi(x) = \begin{pmatrix} \psi_1 \\ \psi_2 \\ \psi_3 \\ \psi_4 \end{pmatrix} = \begin{pmatrix} \varphi \\ \chi \end{pmatrix} \qquad \varphi = \begin{pmatrix} \varphi_1 \\ \varphi_2 \end{pmatrix} \qquad \chi = \begin{pmatrix} \chi_1 \\ \chi_2 \end{pmatrix}. \tag{2.28}$$

The two spinors φ and χ represent the particle and the antiparticle; the two components of each of them represent the two states of the third component of the spin $s_z = +1/2$ and $s_z = -1/2$. The four γ matrices are defined by the algebra they must satisfy and have different representations. We shall employ the Dirac representation, i.e.

$$\gamma^0 = \begin{pmatrix} 1 & 0 \\ 0 & -1 \end{pmatrix} \qquad \gamma^i = \begin{pmatrix} 0 & \sigma^i \\ -\sigma^i & 0 \end{pmatrix} \tag{2.29}$$

where the elements are 2×2 matrices and the σ are the Pauli matrices

$$\sigma^1 = \begin{pmatrix} 0 & 1 \\ 1 & 0 \end{pmatrix} \qquad \sigma^2 = \begin{pmatrix} 0 & -i \\ i & 0 \end{pmatrix} \qquad \sigma^3 = \begin{pmatrix} 1 & 0 \\ 0 & -1 \end{pmatrix}. \tag{2.30}$$

Now let us consider the solutions corresponding to free particles with mass m and definite four-momentum p_{μ}, namely the plane wave $\psi(x) = ue^{-ip^{\mu}x_{\mu}}$ where u is a bi-spinor

$$u = \begin{pmatrix} u_1 \\ u_2 \\ u_3 \\ u_4 \end{pmatrix}. \tag{2.31}$$

The equation becomes

$$\left(\gamma_{\mu}p^{\mu} - m\right)u = 0. \tag{2.32}$$

We now recall the definition of conjugate bi-spinor

$$\bar{u} = u^+ \gamma^0 = (u_1{}^* \quad u_2{}^* \quad -u_3{}^* \quad -u_4{}^*). \tag{2.33}$$

This satisfies the equation

$$\bar{u}\left(\gamma_\mu p^\mu - m\right) = 0. \tag{2.34}$$

A fifth important matrix is

$$\gamma^5 = \begin{pmatrix} 0 & 1 \\ 1 & 0 \end{pmatrix}. \tag{2.35}$$

With two bi-spinors, say a and b, and the five γ matrices, the following five covariant quantities, with the specified transformation properties, can be written

$$
\begin{array}{ll}
\bar{a}b & \text{scalar} \\
\bar{a}\gamma_5 b & \text{pseudoscalar} \\
\bar{a}\gamma_\mu b & \text{vector} \\
\bar{a}\gamma_\mu\gamma_5 b & \text{axial vector} \\
\frac{1}{2\sqrt{2}}\bar{a}\left(\gamma_a\gamma_\beta - \gamma_\beta\gamma_a\right)b & \text{tensor.}
\end{array}
\tag{2.36}
$$

These quantities are important because, in principle, each of them may appear in an interaction Lagrangian. Nature has chosen, however, to use only two of them, the vector and the axial vector, as we shall see.

   In the following, we shall assume, in accordance with the Standard Model, that the wave functions of all the spin 1/2 elementary particles obey the Dirac equation. However, we warn the reader that the extension of the Dirac theory to neutrinos is not supported by any experimental proof. Moreover, in 1937 E. Majorana (Majorana 1937) wrote a relativistic wave equation for neutral particles, different from the Dirac equation. The physical difference is that 'Dirac' neutrinos and antineutrinos are different particles, 'Majorana' neutrinos are two states of the same particle. We do not yet know which describes the nature of neutrinos.

2.6 The positron

In 1930 C. D. Anderson built a large cloud chamber, $17 \times 17 \times 3$ cm^3, and its magnet designed to provide a uniform field up to about 2 T. He exposed the chamber to cosmic rays. The chamber did not have a trigger and, consequently, only a small fraction of the pictures contained interesting events. Nevertheless, he observed tracks both negative and positive that turned out to be at the ionisation minimum from the number of droplets per unit length. Clearly, the negative tracks were electrons, but could the positive be protons, namely the only known positive particles?

Measuring the curvatures of the tracks, Anderson determined their momenta and, assuming they were protons, their energy. With this assumption, several tracks had a rather low kinetic energy, sometimes less than 500 MeV. If this were the case, the ionisation had to be much larger than the minimum. Those tracks could not be due to protons.

Cosmic rays come from above, but the particles that appeared to be positive if moving downwards, could have been negative going upwards, perhaps originating from an interaction in the material under the chamber. This was a rather extreme hypothesis because of the relatively large number of such tracks. The issue had to be settled by determining the direction of motion without ambiguity. To accomplish this, a plate of lead, 6 mm thick, was inserted across a horizontal diameter of the chamber. The direction of motion of the particles could then be ascertained due to the lower energy, and consequently larger curvature, after they had traversed the plate and suffered energy loss.

Figure 2.9 shows a single minimum ionising track with a direction which is clearly upward (!). Knowing the direction of the field (1.5 T in intensity), Anderson concluded that the track was positive. Measuring the curvatures at the two sides of the plate he obtained the momenta $p_1 = 63$ MeV and $p_2 = 23$ MeV. The expected energy loss could be easily calculated from the corresponding energy before the plate. Assuming the proton mass, the kinetic energy after the plate would be $E_{K2} = 200$ keV. This corresponds to a range in the gas of the chamber of 5 mm, to

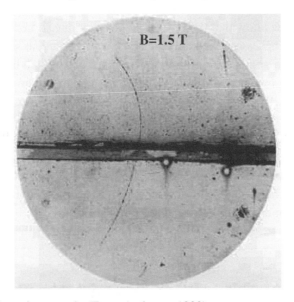

Fig. 2.9. A positron track. (From Anderson 1933)

be compared to the observed range of 50 mm. The difference is too large to be due to a fluctuation. On the contrary, assuming the electron mass, the expected range was compatible with 50 mm.

From the measurement of several events of the same type, Anderson concluded that the mass of the positive particles was equal to the electron mass to within 20% and published the discovery of the positron in September 1932.

At the same time, Blackett and Occhialini were also working with a Wilson chamber in a magnetic field. Their device had the added advantage of being triggered by a coincidence of Geiger counters at the passage of a cosmic ray (Rossi 1930) and of being equipped with two cameras to allow the spatial reconstruction of the tracks. They observed several pairs of tracks of opposite signs at the ionisation minimum originating from the same point. Measuring the curvature and the droplet density they measured the masses, which were equal to that of the electron. In conclusion, Blackett and Occhialini not only confirmed, in the spring of 1933, the discovery of the positron, but also discovered the production of e^+e^- pairs (Blackett & Occhialini 1933).

2.7 The antiproton

A quarter of a century after the discovery of the positron a fundamental question was still open: does the antiparticle of the proton exist? From the theoretical point of view, the Dirac equation did not give a unique answer, because, in retrospect, the proton, unlike the electron, is not a simple particle; its magnetic moment, in particular, is not as foreseen by the Dirac equation. The partner of the proton, the neutron, has a magnetic moment even if neutral. Antiprotons were searched for in cosmic rays, but not found. We now know that they exist, but are very rare. It became clear that the instrument really necessary was an accelerator with sufficient energy to produce antiprotons. Such a proton synchrotron was designed and built at Berkeley under the leadership of E. Lawrence and E. McMillan, with a maximum proton energy of 7 GeV. In the USA, the GeV was then called BeV (from billion, meaning one thousand million) and the accelerator was called Bevatron. After it became operational in 1954, the experiments at the Bevatron took the lead in subnuclear physics for several years.

As we shall see in the next chapter, the baryon number, defined as the difference between the number of nucleons and the number of antinucleons, is conserved in all interactions. Therefore, a reaction must produce a proton–antiproton pair and cannot produce an antiproton alone. The simplest reaction is

$$p + p \rightarrow p + p + \bar{p} + p. \tag{2.37}$$

The threshold energy (see Problem 1.9) is

$$E_p(\text{thr.}) = 7m_p = 6.6\,\text{GeV}. \tag{2.38}$$

The next instrument was the detector, which was built in 1955 by O. Chamberlain, E. Segrè, C. Wiegand and T. Ypsillantis (Chamberlain *et al.* 1955). The 7.2 GeV proton beam extracted from the Bevatron collided with an external target, producing a number of secondary particles. The main difficulty of the experiment was to detect the very few antiprotons that may be present amongst these secondaries. From calculations only one antiproton to every 100 000 pions was expected.

To distinguish protons from pions, one can take advantage of the large difference between their masses. As usual, this requires that two quantities be measured or defined. The choice was to build a spectrometer to define the momentum p accurately and to measure the speed. Then the mass is given by

$$m = \frac{p}{v}\sqrt{1 - v^2/c^2}. \tag{2.39}$$

We shall exploit the analogy between a spectrometer for particles and a spectrometer for light.

The spectrometer had two stages. Figure 2.10 is a sketch of the first. The particles produced in the target, both positive and negative, have a broad momentum spectrum. The first stage is designed to select negative particles with a momentum defined within a narrow band. The trajectory of one of these particles is drawn in the figure. The magnet is a dipole, which deflects the particles at an angle that, for the given magnetic field, is inversely proportional to the particle momentum (see Eq. (1.80)). Just as a prism disperses white light into its colours, the dipole disperses a non-monoenergetic beam into its components. A slit in a thick absorber transmits only the particles with a certain momentum, within a narrow range. Figure 2.11(a) shows the analogy with light. The sign of the accepted particle is decided by the polarity of the magnet.

However, as pointed out by O. Piccioni, this scheme does not work; every spectrometer, for particles as for light, must contain focussing elements. The reason becomes clear if we compare Fig. 2.11 (a) and (b), in which only two colours are shown for simplicity. If we use only a prism we do select a colour, but we transmit an

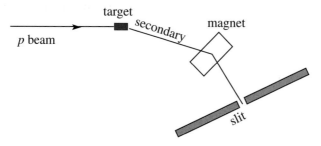

Fig. 2.10.   Sketch of the first stage, without focussing.

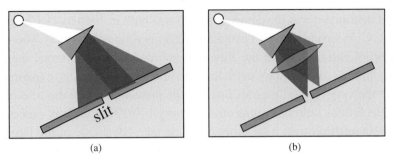

Fig. 2.11.   Principle of a focussing spectrometer.

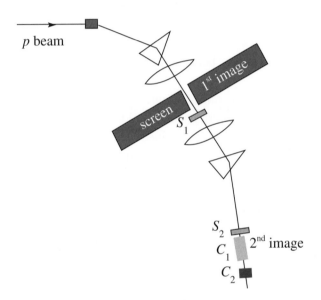

Fig. 2.12.   A sketch of the antiproton experiment.

extremely low intensity. As is well known in optics, to have appreciable intensity we must use a lens to produce an image of the source in the slit.

Figure 2.12 is a sketch of the final configuration, including the second stage that we shall now discuss. Summarising, the first stage produces a secondary source of well-defined momentum negative particles. The chosen central value of the momentum is $p = 1.19\,\mathrm{GeV}$. The corresponding speeds of pions and antiprotons are

$$\beta_\pi = \frac{p}{E_\pi} = \frac{1.19}{\sqrt{1.19^2 + 0.14^2}} = 0.99 \tag{2.40}$$

$$\beta_p = \frac{p}{E_p} = \frac{1.19}{\sqrt{1.19^2 + 0.938^2}} = 0.78. \tag{2.41}$$

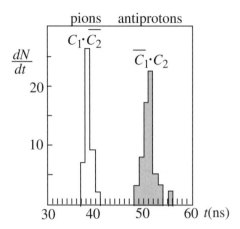

Fig. 2.13. Time of flight distribution between S_1 and S_2. (Adapted from Chamberlain *et al.* 1955)

The time of flight is measured between two scintillator counters S_1 and S_2 on a 12 m long base. The flight times expected from the above evaluated speeds are $t_\pi = 40$ ns and $t_p = 51$ ns. The difference $\Delta t = 11$ ns is easily measurable. The resolution is ± 1 ns.

A possible source of error is due to random coincidences. Sometimes two pulses separated by 11 ns might result from the passage of a pion in S_1 and a different one in S_2. Two Cherenkov counters are used to cure the problem. C_2 is used in the threshold mode, with threshold set at $\beta_p = 0.99$, to see the pions but not the antiprotons. C_1 has a lower threshold and both particles produce light, but at different angles. A spherical mirror focusses the antiproton light onto the photomultiplier and not that of the pions; in such a way, C_2 sees only the antiprotons. In conclusion, the pions are identified by the coincidence $C_1 \bar{C}_2$, the antiprotons by $C_2 \bar{C}_1$.

Figure 2.13 shows the time of flight distribution for the two categories. The presence of antiprotons (about 50) is clearly proved.

We know now that an antiparticle exists for every particle, both for fermions and for bosons.

Problems

2.1. Compute energies and momenta in the CM system of the decay products of $\pi \to \mu + \nu$.

2.2. Consider the decay $K \to \mu + \nu$. Find

    a. the energy and momentum of the μ and the ν in the reference frame of the K at rest;

    b. the maximum μ momentum in a frame in which the K momentum is 5 GeV.

2.3. A π^0 decays emitting one photon in the forward direction of energy E_1 = 150 MeV. What is the direction of the second photon? What is its energy E_2? What is the speed of the π^0?

2.4. Two muons are produced by a cosmic ray collision at an altitude of 30 km. Their two energies are $E_1 = 5$ GeV and $E_2 = 5$ TeV. What are the distances at which each of the muons sees the surface of the Earth in its rest reference frame? What are the distances travelled in the Earth reference frame in a lifetime?

2.5. A π^+ is produced at an altitude of 30 km by a cosmic ray collision with energy $E_\pi = 5$ GeV. What is the distance at which the pion sees the surface of the Earth in its rest reference frame? What is the distance travelled in the Earth reference frame in a lifetime?

2.6. A photon converts into an $e^+ e^-$ pair in a cloud chamber with magnetic field $B = 0.2$ T. In this case two tracks are observed with the same radius $\rho = 20$ cm. The initial angle between the tracks is zero. Find the energy of the photon.

2.7. Consider the following particles and their lifetimes:

$$\rho^0: 5 \times 10^{-24} \text{ s}, K^+: 1.2 \times 10^{-8} \text{ s}, \eta^0: 5 \times 10^{-19} \text{ s}, \mu^-: 2 \times 10^{-6} \text{ s}, \pi^0: 8 \times 10^{-17} \text{ s}.$$

Guess which interaction leads to the following decays: $\rho^0 \to \pi^+ + \pi^-$; $K^+ \to \pi^0 + \pi^+$; $\eta^0 \to \pi^+ + \pi^- + \pi^0$; $\mu^- \to e^- + \bar{\nu}_e + \nu_\mu$; $\pi^0 \to \gamma + \gamma$.

2.8. Consider the decay $\pi^0 \to \gamma\gamma$ in the CM. Assume a Cartesian coordinate system x^*, y^*, z^*, and the polar coordinates ρ^*, θ^*, ϕ^*. In this reference frame, the decay is isotropic. Give the expression for the probability per unit solid angle, $P(\cos\theta^*, \phi^*) = dN/d\Omega^*$, of observing a photon in the direction θ^*, ϕ^*. Then consider the L reference frame, in which the π^0 travels in the direction $z = z^*$ with momentum p and write the probability per unit solid angle, $P(\cos\theta, \phi)$, of observing a photon in the direction θ, ϕ.

2.9. Chamberlain *et al.* employed scintillators to measure the pion lifetime. Why did they not use Geiger counters?

2.10. Compute the ratio between the magnetic moments of the electron and the μ and between the electron and the τ.

2.11. We calculated the energy threshold for the reaction $p + p \to p + p + \bar{p} + p$ on free protons as targets in Problem 1.9. Repeat the calculation for protons that are bound in a nucleus and have a Fermi momentum of $p_f = 150$ MeV. For the incident proton use the approximation $p_p \approx E_p$.

2.12. We wish to produce a monochromatic beam with momentum $p = 20$ GeV and a momentum spread $\Delta p/p = 1\%$. The beam is 2 mm wide and we have a magnet with a bending power of $BL = 4$ T m and a slit $d = 2$ mm wide. Calculate the distance l between magnet and slit.

2.13. A hydrogen bubble chamber was exposed to a 3 GeV momentum π^- beam. We observe an interaction with secondaries that are all neutral and two V^0s pointing to the primary vertex. Measuring the two tracks of one of them, we find for the positive: $p^- = 121\,\text{MeV}$, $\theta^- = -18.2°$, and $\phi^- = 15°$ and for the negative: $p^+ = 1900\,\text{MeV}$, $\theta^+ = 20.2°$ and $\phi^+ = -15°$. θ and φ are the polar angles in a reference frame with polar axis z in the beam direction. What is the nature of the particle? Assume that the measurement errors give a $\pm 4\%$ resolution on the reconstructed mass of the V^0.

Further reading

Anderson, C. D. (1936); Nobel Lecture, *The Production and Properties of Positrons* http://nobelprize.org/nobel_prizes/physics/laureates/1936/anderson-lecture.pdf

Chamberlain, O. (1959); Nobel Lecture, *The Early Antiproton Work* http://nobelprize. org/nobel_prizes/physics/laureates/1959/chamberlain-lecture.pdf

Lederman, L. (1963); *The two-neutrino experiment. Sci. Am.* **208** no. 3, 60

Lederman, L. (1988); Nobel Lecture, *Observations in Particle Physics from Two Neutrinos to the Standard Model* http://nobelprize.org/nobel_prizes/physics/ laureates/1988/lederman-lecture.pdf

Lemmerich, J. (1998); *The history of the discovery of the electron.* Proceedings of the XVIII International Symposium on 'Lepton Photon Interactions 1997'. World Scientific

Perkins, D. H. (1998); *The discovery of the pion at Bristol in 1947.* Proceedings of 'Physics in Collision 17' 1997. World Scientific

Piccioni, O. (1989); *On antiproton discovery* in L. M. Brown *et al.* editors *Pions and quarks*. Cambridge University Press, p. 285

Powell, C. B. (1950); Nobel Lecture, *The Cosmic Radiation* http://nobelprize.org/ nobel_prizes/physics/laureates/1950/powell-lecture.pdf

Reines, F. (1995); Nobel Lecture, *The Neutrino: From Poltergeist to Particle* http:// nobelprize.org/nobel_prizes/physics/laureates/1995/reines-lecture.html

Rossi, B. (1952); *High-Energy Particles.* Prentice-Hall

Schwartz, M. (1988); Nobel Lecture, *The First High Energy Neutrino Experiment* http:// nobelprize.org/nobel_prizes/physics/laureates/1988/schwartz-lecture.pdf

3

Symmetries

3.1 Symmetries

The rules that limit the possibility of an initial state transforming into another state in a quantum process (collision or decay) are called **conservation laws** and are expressed in terms of the **quantum numbers** of those states. We shall not deal with the invariance under continuum transformations in space-time and the corresponding conservation of energy-momentum and of angular momentum, which are known to the reader. We shall consider the following types of quantum numbers.

Discrete additive If a quantum number is additive, the total quantum number of a system is the sum of the quantum numbers of its components. The 'charges' of all fundamental interactions fall into this category, the electric charge, the colour charges and the weak charges. They are conserved absolutely, as far as we know. The conservation of each of them corresponds to the invariance of the Lagrangian of that interaction under the transformations of a unitary group. The group is called the 'gauge group' and the invariance of the Lagrangian is called 'gauge invariance'. The gauge group of the electromagnetic interaction is $U(1)$, that of the strong interaction is $SU(3)$ and that of the electroweak interaction is $SU(2) \otimes U(1)$. Other quantum numbers in this category are the quark flavours, the baryon number, the lepton flavours and the lepton numbers. They do not correspond to a gauge symmetry and are not necessarily conserved (actually, quark and lepton flavours are not).

Internal symmetries The transformations are continuous and take place in a 'unitary space' defined by a symmetry group. These symmetries allow us to classify a number of particles in 'multiplets', the members of which have similar behaviour. An example of this is the charge independence of nuclear forces. The corresponding symmetry is the invariance under the transformations of the group $SU(2)$ and isotopic spin conservation.

Discrete multiplicative These transformations cannot be constructed starting from infinitesimal transformations. The most important are: parity $\mathcal{P}$, i.e. the inversion of the coordinate axes, particle–antiparticle conjugation $\mathcal{C}$, and time reversal $\mathcal{T}$. The eigenvalues of $\mathcal{P}$ and $\mathcal{C}$ are amongst the quantum numbers of the particles. Notice that applying these transformations twice brings the system back to its original state, in other words $\mathcal{P}^2 = 1$ and $\mathcal{C}^2 = 1$. The possible eigenvalues are then $P = \pm 1$ and $C = \pm 1$.

Several symmetries are 'broken', i.e. are not respected by all the interactions. Therefore, only those interactions that do not break them conserve the corresponding quantum numbers. Only experiments can decide whether a certain quantum number is conserved or not in a given interaction.

3.2 Parity

The parity operation $\mathcal{P}$ is the inversion of the three spatial coordinate axes. Note that, while in two dimensions the inversion of the axes is equivalent to a rotation, this is not true in three dimensions. The inversion of three axes is equivalent to the inversion of one, followed by a 180° rotation. An object and its mirror image are connected by a parity operation.

The following scheme will be useful. The $\mathcal{P}$ operation

inverts the coordinates	$\mathbf{r} \Rightarrow -\mathbf{r}$
does not change time	$t \Rightarrow t$
as a consequence	
it inverts momenta	$\mathbf{p} \Rightarrow -\mathbf{p}$
and does not change angular momenta	$\mathbf{r} \times \mathbf{p} \Rightarrow \mathbf{r} \times \mathbf{p}$
including spins	$\mathbf{s} \Rightarrow \mathbf{s}.$

More generally, scalar quantities remain unchanged, pseudoscalar ones change their sign, vectors change sign, and axial vectors do not.

We can talk of the parity of a state only if it is an eigenstate of $\mathcal{P}$. Vacuum is such a state and its parity is set positive by definition.

A single particle can be, but is not necessarily, in an eigenstate of $\mathcal{P}$ only if it is at rest. The eigenvalue P of $\mathcal{P}$ in this frame is called **intrinsic parity** (or simply parity), which can be positive ($P = +1$) or negative ($P = -1$).

The parity of bosons can always be defined without ambiguity. We shall see in Section 3.5 how it is measured in the case of the pion. Fermions have half-integer spins and angular momentum conservation requires them to be produced in pairs. Therefore only relative parities can be defined. Conventionally proton parity is assumed positive and the parities of the other fermions are given relative to the proton. Quantum field theory requires fermions and their antifermions to have opposite parities, and requires bosons and their antibosons to have the same parity.

Therefore, the parity of the antiproton is negative. The same is true for the positron.

Strange hyperons are produced in pairs together with another strange particle. This prevents the measurement of both parities. One might expect to be able to choose one hyperon and to refer its parity to that of the proton using a decay, say for example $\Lambda \to p\pi^-$. This does not work because the decays are weak processes and weak interactions, as we shall see, violate parity conservation. We then take by convention $P(\Lambda) = +1$.

Strange hyperons differ from non-strange ones because of the presence of a strange quark. More hadrons were discovered containing other quark types. The general rule at the quark level is that, by definition, all quarks have positive parity, antiquarks have negative parity.

The parity of the photon The photon is the quantum equivalent of the classical vector potential **A**. Therefore, its spin and parity, with a notation that we shall always employ, are $J^P = 1^-$. The same conclusion can be reached remembering that the transitions between atomic levels with a single photon emission are of the electric dipole type. For them the rule $\Delta l = \pm 1$ applies. Therefore, from a property of spherical harmonics that we shall soon recall, the two levels have opposite parities.

The parity of a two-particle system A system of two particles of intrinsic parities, say, P_1 and P_2, can be a parity eigenstate only in the centre of mass system. In this frame, let us call **p** the momentum and θ, ϕ the angles for one particle and $-$**p** the momentum of the other. We shall write these states as $|p, \theta, \phi\rangle$ or as $|\mathbf{p}, -\mathbf{p}\rangle$. Call $|p, l, m\rangle$ the state with orbital angular momentum l and third component m. The relationship between the two bases is

$$|p,\, l,\, m\rangle = \sum_{\theta,\phi} |p,\, \theta,\, \phi\rangle \langle p,\, \theta,\, \phi \mid p,\, l,\, m\rangle = \sum_{\theta,\,\phi} Y_l^{*m}(\theta,\, \phi)|\mathbf{p}, -\mathbf{p}\rangle. \quad (3.1)$$

The inversion of the axes in polar coordinates is $r \Rightarrow r$, $\theta \Rightarrow \pi - \theta$ and $\phi \Rightarrow \pi + \phi$. Spherical harmonics transform as

$$Y_l^{*m}(\theta, \phi) \Rightarrow Y_l^{*m}(\pi - \theta, \pi + \phi) = (-1)^l Y_l^{*m}(\theta, \phi). \quad (3.2)$$

Consequently

$$
\begin{aligned}
\mathcal{P}\,|p,\, l,\, m\rangle &= P_1 P_2 \sum_{\theta,\,\phi} Y_l^{*m}(\pi - \theta,\ \phi + \pi)|-\mathbf{p},\ \mathbf{p}\rangle \\
&= P_1 P_2 (-1)^l \sum_{\theta,\phi} Y_l^{*m}(\theta, \phi)|\mathbf{p}, -\mathbf{p}\rangle \\
&= P_1 P_2 (-1)^l\, |p, l, m\rangle. \quad (3.3)
\end{aligned}
$$

In conclusion, the parity of the system of two particles with orbital angular momentum l is

$$P = P_1 P_2 (-1)^l. \tag{3.4}$$

Let us see some important cases.

Parity of two mesons with the same intrinsic parity (for example, two π). Calling them m_1 and m_2, Eq. (3.4) simply gives

$$P(m_1, m_2) = (-1)^l. \tag{3.5}$$

For particles without spin such as pions, the orbital angular momentum is equal to the total momentum, $J = l$.

The possible values of parity and angular momentum are $J^P = 0^+, 1^-, 2^+, \ldots,$ provided the two pions are different.

If the two pions are equal, their status must be symmetrical, as requested by Bose statistics. Therefore, l and hence J must be even. The possible values are $J^P = 0^+, 2^+, \ldots$

Fermion–antifermion pair (for example, proton–antiproton). The two intrinsic parities are opposite in this case. Therefore, if again l is the orbital angular momentum, we have

$$P(f\bar{f}) = (-1)^{l+1}. \tag{3.6}$$

Example 3.1 Find the possible values of J^P for a spin 1/2 particle and its antiparticle if they are in an S wave state, or in a P wave state.

The total spin can be 0 (singlet) or 1 (triplet). In an S wave the orbital momentum is $l = 0$ and the total angular momentum can be $J = 0$ (in spectroscopic notation $^1S_0$) or $J = 1$ ($^3S_1$). Parity is negative in both cases. In conclusion $^1S_0$ has $J^P = 0^-$, $^3S_1$ has $J^P = 1^-$. The P wave has $l = 1$ hence positive parity. The possible states are: $^1P_1$ with $J^P = 1^+$, $^3P_0$ ($J^P = 0^+$), $^3P_1$ ($J^P = 1^+$) and $^3P_2$ ($J^P = 2^+$).

Parity conservation is not a universal law of physics. Strong and electromagnetic interactions conserve parity, weak interactions do not. We shall study parity violation in Chapter 7. The most sensitive tests for parity conservation in strong interactions are based on the search for reactions that can only proceed through parity violation.

Experimentally, we can detect parity violation effects if the matrix element is the sum of a scalar and a pseudoscalar term. Actually, if only one of them is present, the transition probability that is proportional to its absolute square is in any case a scalar, meaning it is invariant under the parity operation. However, if both terms are present, the transition probability is the sum of the two absolute squares, which are invariant under parity, and of their double-product, which changes sign. Let us then assume a matrix element of the type

$$M = M_S + M_{PS}. \tag{3.7}$$

A process that violates parity is the decay of an axial vector state into two scalars $1^+ \to 0^+ + 0^+$. An example is the $J^P = 1^+$ Ne excited state $^{20}\text{Ne}^*(Q = 13.2 \text{ MeV})$. If it decays into ^{16}O ($J^P = 0^+$) and an alpha particle ($J^P = 0^+$), parity is violated. To search for this decay we look for the corresponding resonance in the process

$$p + {}^{19}\text{F} \to [{}^{20}\text{Ne}^*] \to {}^{16}\text{O} + a.$$

The resonance was not found (Tonner 1957), a fact that sets the limit, for strong interactions

$$|M_S/M_{PS}|^2 \leq 10^{-8}. \tag{3.8}$$

3.3 Particle–antiparticle conjugation

The particle–antiparticle conjugation operator $\mathcal{C}$ acting on one particle state changes the particle into its antiparticle, leaving space coordinates, time and spin unchanged. Therefore, the sign of all the additive quantum numbers, electric charge, baryon number and lepton flavour is changed. It is useful to think that if a particle and its antiparticle annihilate then the final state is the vacuum, in which all 'charges' are zero. We shall also call this operator 'charge conjugation', as is often done for brevity, even if the term is somewhat imprecise.

Let us consider a state with momentum **p**, spin **s** and 'charges' $\{Q\}$. Then

$$\mathcal{C}|\mathbf{p}, \mathbf{s}, \{Q\}\rangle = C|\mathbf{p}, \mathbf{s}, \{-Q\}\rangle. \tag{3.9}$$

As we have seen, the possible eigenvalues are $C = \pm 1$.

Only 'completely' neutral particles, namely particles for which $\{Q\} = \{-Q\} = \{0\}$, are eigenstates of $\mathcal{C}$. In this case, the particle coincides with its antiparticle. We already know two cases, the photon and the π^0; we shall meet two more, the η and η' mesons. The eigenvalue C for such particles is called their intrinsic charge conjugation, or simply charge conjugation.

The charge conjugation of the photon Let us consider again the correspondence between the photon and the macroscopic vector potential **A**. If all the particle sources of the field are changed into their antiparticles, all the electric charges change sign and therefore **A** changes its sign. Consequently, the charge conjugation of the photon is negative

$$\mathcal{C}|\gamma\rangle = -|\gamma\rangle. \tag{3.10}$$

A state of n photons is an eigenstate of C. Since C is a multiplicative operator

$$C|n\gamma\rangle = (-1)^n|n\gamma\rangle. \tag{3.11}$$

The charge conjugation of the π^0 The π^0 decays into two photons by electromagnetic interaction, which conserves C, hence

$$C|\pi^0\rangle = +|\pi^0\rangle. \tag{3.12}$$

Charged pions are not C eigenstates, rather we have

$$C|\pi^+\rangle = +|\pi^-\rangle \qquad C|\pi^-\rangle = +|\pi^+\rangle. \tag{3.13}$$

The charge conjugation of the η meson The η too decays into two photons and consequently

$$C|\eta^0\rangle = +|\eta^0\rangle. \tag{3.14}$$

The tests of C conservation are based on searches for C-violating processes. Two examples for the electromagnetic interaction are the experimental limits for the π^0 from McDonough *et al.* (1988) and for the η from Nefkens *et al.* (2005)

$$\Gamma(\pi^0 \to 3\gamma)/\Gamma_{\text{tot}} \le 3.1 \times 10^{-8} \qquad \Gamma(\eta \to 3\gamma)/\Gamma_{\text{tot}} \le 4 \times 10^{-5}. \tag{3.15}$$

We shall see in Chapter 7 that weak interactions violate C conservation.

Particle–antiparticle pair A system of a particle and its antiparticle is an eigenstate of the particle–antiparticle conjugation in its centre of mass frame. Let us examine the various cases, calling l the orbital angular momentum.

Meson and antimeson (m^+, m^-) with zero spin (example, π^+ and π^-). The net effect of C is the exchange of the two mesons; as such it is identical to that of P. Hence

$$C|m^+, m^-\rangle = (-1)^l|m^+, m^-\rangle. \tag{3.16}$$

Meson and antimeson (M^+, M^-) with non-zero spin $s \ne 0$. The effect of C is again the exchange of the mesons, but now it is not the same as that of P, because C exchanges not only the positions but also the spins. Let us see what happens.

The wave function can be symmetric or antisymmetric under the exchange of the spins. Let us consider the example of two spin 1 particles. The total spin can have the values $s = 0$, 1 or 2. It is easy to check that the states of total spin $s = 0$ and $s = 2$ are symmetric, while the state with $s = 1$ is antisymmetric. Therefore, the spin exchange gives a factor $(-1)^s$. This conclusion is general, as one can

Table 3.1 J^{PC} for the spin 1/2 particle–antiparticle systems

	$^1S_0$	$^3S_1$	$^1P_1$	$^3P_0$	$^3P_1$	$^3P_2$
J^{PC}	0^{-+}	1^{--}	1^{+-}	0^{++}	1^{++}	2^{++}

show. In conclusion, we have

$$C|M^+, M^-\rangle = (-1)^{l+s}|M^+, M^-\rangle. \tag{3.17}$$

Fermion and antifermion $(f\bar{f})$. Let us start again with an example, namely two spin 1/2 particles. The total spin can be $s = 0$ or 1. This time, the state with total spin $s = 1$ is symmetric, the state with $s = 0$ is antisymmetric. Therefore, the factor due to the exchange of the spin is $(-1)^{s+1}$. This result too is general.

Fermions and antifermions have opposite intrinsic charge conjugations, hence a factor -1. In conclusion

$$C|f\bar{f}\rangle = (-1)^{l+s}|f\bar{f}\rangle. \tag{3.18}$$

The final result is identical to that of the mesons.

We call the reader's attention to the fact that the sum $l + s$ in the above expressions is the sum of two numbers, not the composition of the corresponding angular momenta, i.e. it is not in general the total angular momentum of the system.

Example 3.2 Find the eigenvalues of C for the system of a spin 1/2 particle and its antiparticle when they are in an S wave and when they are in a P wave.

The singlets have $S = 0$, hence $^1S_0$ has $C = +$, $^1P_1$ has $C = -$; the triplets have $S = 1$, hence $^3S_1$ has $C = -$, $^3P_0$, $^3P_1$ and $^3P_2$ have all $C = +$.

From the results obtained in Examples 3.1 and 3.2 we list in Table 3.1 the J^{PC} values for a fermion–antifermion pair. Notice that not all values are possible. For example, the states with $J^{PC} = 0^{+-}, 0^{--}, 1^{-+}$ cannot be composed of a fermion and its antifermion with spin 1/2.

3.4 Time reversal and CPT

The time reversal operator T inverts time leaving the coordinates unchanged. We shall not discuss it in any detail. We shall only mention that extremely general principles imply the invariance of the theories under the combined operations P, C and T. The result is independent of the order and is called CPT.

A consequence of CPT is that the mass and lifetime of a particle and its antiparticle must be identical, as already mentioned. The most sensitive tests of CPT invariance are based on the search of possible differences.

For example, a limit on CPT violation was set by searching for a possible difference between proton and antiproton masses. 'Antiprotonic $\bar{p}$ ^{4}He$^+$ atoms', namely atoms made up of a ^{4}He nucleus and a $\bar{p}$, were produced at CERN by the ASACUSA experiment. By studying the spectroscopy of the system, the following limit was established (Hori *et al.* 2003)

$$\left|m_p - m_{\bar{p}}\right|/m_p \leq 10^{-8}. \tag{3.19}$$

3.5 The parity of the pion

The parity of the π^- is determined by observing its capture at rest by deuterium nuclei, a process that is allowed only if the pion parity is negative, as we shall prove. The process is

$$\pi^- + d \rightarrow n + n. \tag{3.20}$$

In practice, one brings a π^- beam of low energy into a liquid deuterium target. The energy is so low that large fractions of the pions come to rest in the liquid after having suffered ionisation energy loss.

Once a π^- is at rest the following processes take place. Since they are negative, the pions are captured, within a time lag of a few picoseconds, in an atomic orbit, replacing an electron. The system is called a 'mesic atom'. The initial orbit has high values of both the quantum numbers n and l, but again very quickly (~ 1 ps), the pion reaches a principal quantum number n of about 7. At these values of n the wave function of those pions that are in an S orbit largely overlaps with the nucleus. In other words, the probability of the π^- being inside the nucleus is large, and they are absorbed.

The pions that initially are not in an S wave reach it anyway by the following process. The mesic atom is actually much smaller than a common atom, because $m_\pi \gg m_e$. Being so small, it eventually penetrates another molecule and becomes exposed to the high electric field present near a nucleus. The consequent Stark effect mixes the levels, repopulating the S waves. Then, almost immediately, the pion is absorbed. The conclusion is that the capture takes place from states with $l = 0$.

This theory was developed by T. B. Day, G. A. Snow and J. Sucher in 1960 (Day *et al.* 1960) and experimentally verified by the measurement of the X-rays emitted from the above-described atomic transitions.

Therefore, the initial angular momentum of the reaction (3.20) is $J = 1$, because the spins of the deuterium and the pion are 1 and 0 respectively and the orbital momentum is $l = 0$. The deuterium nucleus contains two nucleons, of positive intrinsic parity, in an S wave; hence its parity is positive. In conclusion, its initial parity is equal to that of the pion.

The final state consists of two identical fermions and must be antisymmetric in their exchange. If the two neutrons are in a spin singlet state, which is anti-symmetric in the spin exchange, the orbital momentum must be even, vice versa if the neutrons are in a triplet. Writing them explicitly, we have the possibilities $^1S_0, ^3P_{0,1,2}, ^1D_2, \ldots$ The angular momentum must be equal to the initial momentum, i.e. $J = 1$. There is only one choice, namely $^3P_1$. Its parity is negative. Therefore, if the reaction takes place the parity of π^- is negative.

Panowsky and collaborators (Panowsky *et al.* 1951) showed that the reaction (3.20) proceeds and that its cross section is not suppressed.

We shall not further discuss the experimental evidence, but only say that all pions are pseudoscalar particles.

3.6 Pion decay

Charged pions decay predominantly ($>99\%$) in the channel

$$\pi^+ \to \mu^+ + \nu_\mu \qquad \pi^- \to \mu^- + \bar{\nu}_\mu. \tag{3.21}$$

The second most probable channel is similar, with an electron in place of the muon

$$\pi^+ \to e^+ + \nu_e \qquad \pi^- \to e^- + \bar{\nu}_e. \tag{3.22}$$

Since the muon mass is only a little smaller than that of the pion, the first channel is energetically disfavoured relative to the second; however, its decay width is the larger one

$$\frac{\Gamma(\pi \to e\nu)}{\Gamma(\pi \to \mu\nu)} = 1.2 \times 10^{-4}. \tag{3.23}$$

We have seen in Section 1.6 that the phase space volume for a two-body system is proportional to the centre of mass momentum. The ratio of the phase space volumes for the two decays is then p_e^* / p_μ^*.

Calling the charged lepton generically l, energy conservation is written as $\sqrt{p_l^{*2} + m_l^2} + p_l^* = m_\pi$, which gives $p_l^* = \frac{m_\pi^2 - m_l^2}{2m_\pi}$. The ratio of the momenta is then

$$\frac{p_e^*}{p_\mu^*} = \frac{m_\pi^2 - m_e^2}{m_\pi^2 - m_\mu^2} = \frac{140^2 - 0.5^2}{140^2 - 106^2} = 2.3. \tag{3.24}$$

As anticipated, phase space favours the decay into an electron. Given the experimental value (3.23), the ratio of the two matrix elements must be very small. This observation gives us very important information on the space-time structure of weak interactions.

We do not have the theoretical instruments for a rigorous discussion, but we can find the most general matrix element using simple Lorentz invariance arguments. Leaving the possibility of parity violation open, the matrix element may be a scalar, a pseudoscalar or the sum of the two. We must build such quantities with the covariant quantities at our disposal.

Again let l be the charged lepton and ν_l its neutrino. The matrix element must contain their wave functions combined in a covariant quantity. The possible combinations are

$$
\begin{array}{ll}
\bar{l}\nu_l & \text{scalar (S)} \\
\bar{l}\gamma_5\nu_l & \text{pseudoscalar (PS)} \\
\bar{l}\gamma_\mu\nu_l & \text{vector (V)} \\
\bar{l}\gamma_\mu\gamma_5\nu_l & \text{axialvector (A)} \\
\frac{1}{2\sqrt{2}}\bar{l}\left(\gamma_a\gamma_\beta - \gamma_\beta\gamma_a\right)\nu_l & \text{tensor (T).}
\end{array} \tag{3.25}
$$

This part of the matrix element is the most important, because it represents the weak interaction Hamiltonian. It is called the 'weak current'. Only experiments can determine which of these terms are present in weak interactions. It took a long series of experiments to establish that only the 'vector current' V and the 'axial current' A are present in Nature. We shall examine some of these experiments in Chapter 7, limiting our discussion here to what we can learn from the pion decay.

Another factor of the matrix element is the wave function of the pion in its initial state, ϕ_π, which is a pseudoscalar.

The kinematic variables of the decay may also appear in the matrix element. Actually, only one of these quantities exists, the four-momentum of the pion, p^μ. Finally, a scalar constant can be present, called the pion decay constant, which we indicate by f_π.

We must now construct with the above listed elements the possible matrix elements, namely scalar or pseudoscalar quantities. There are two scalar quantities (the dots stand for uninteresting factors)

$$
M = \ldots f_\pi\phi_\pi\bar{l}\gamma_5\nu_l \qquad M = \ldots f_\pi\phi_\pi\bar{l}p^\mu\gamma_\mu\gamma_5\nu_l \tag{3.26}
$$

the pseudoscalar and axial vector current term respectively. There are also two pseudoscalar terms

$$
M = \ldots f_\pi\phi_\pi\bar{l}\nu_l \qquad M = \ldots f_\pi\phi_\pi\bar{l}p^\mu\gamma_\mu\nu_l \tag{3.27}
$$

the scalar and the vector current terms. We have used four of the covariant quantities; there is no possibility of using the fifth one, the tensor.

Let us start with the vector current term

$$M = \ldots f_\pi \phi_\pi \bar{l} p^\mu \gamma_\mu \nu_l. \tag{3.28}$$

The pion four-momentum is equal to the sum of those of the charged lepton and the neutrino $p^\mu = p^\mu_{\nu_l} + p^\mu_l$, hence

$$M = \ldots f_\pi \phi_\pi \bar{l} \left(p^\mu_{\nu_l} + p^\mu_l \right) \gamma_\mu \nu_l = \ldots f_\pi \phi_\pi \bar{l} \gamma_\mu p^\mu_{\nu_l} \nu_l + \ldots f_\pi \phi_\pi \bar{l} \gamma_\mu p^\mu_l \nu_l.$$

The wave functions of the final-state leptons, which are free particles, are solutions of the Dirac equation

$$\left(\gamma_\mu p^\mu_{\nu_l} - m_{\nu_l} \right) \nu_l = 0 \quad \Rightarrow \quad \gamma_\mu p^\mu_{\nu_l} \nu_l = 0$$

$$\bar{l} \left(\gamma_\mu p^\mu_l - m_l \right) = 0 \quad \Rightarrow \quad \bar{l} \gamma_\mu p^\mu_l = \bar{l} m_l.$$

In conclusion, we obtain

$$M = \ldots m_l f_\pi \phi_\pi \bar{l} \nu_l. \tag{3.29}$$

We see from the Dirac equation that the matrix element is proportional to the mass of the final lepton. Therefore, the ratio of the decay probabilities in the two channels is proportional to the ratio of their masses squared

$$m_e^2 / m_\mu^2 = 0.22 \times 10^{-4}. \tag{3.30}$$

This factor has the correct order of magnitude to explain the smallness of $\Gamma(\pi \to e\nu)/\Gamma(\pi \to \mu\nu)$. We shall complete the discussion at the end of this section.

Let us now examine the axial vector current term, namely

$$M = \ldots f_\pi \phi_\pi \bar{l} p^\mu \gamma_\mu \gamma_5 \nu_l. \tag{3.31}$$

Repeating the arguments of the vector case we obtain

$$M = \ldots m_l f_\pi \phi_\pi \bar{l} \gamma_5 \nu_l \tag{3.32}$$

and we again obtain the result (3.30).

Considering now the scalar and the pseudoscalar current terms, we see immediately that they do not contain the factor m_l^2. Therefore, they cannot explain the smallness of (3.23).

In conclusion, the observed small value of the ratio between the probabilities of a charged pion decaying into an electron or into a muon proves that, at least in this case, the weak interaction currents are of type *V*, or of type *A,* or of a mixture of the two. Notice that if both are present, parity is violated. We shall see in Chapter 7 that weak interactions do violate parity and that they do so maximally. The space-time structure of the so-called 'charged' currents, those that we are considering, is *V–A*. The matrix element of the leptonic decays of the pion is

$$M = \dots m_l f_\pi \phi_\pi \bar{l}(1 - \gamma_5) p^\mu \gamma_\mu \nu_l. \tag{3.33}$$

To obtain the decay probabilities we must integrate the absolute square of this quantity, for the electron and the muon, over phase space. We cannot do the calculation here and we give the result directly

$$\frac{\Gamma(\pi \to e\nu)}{\Gamma(\pi \to \mu\nu)} = \frac{p_e^* p_e^* m_e^2}{p_\mu^* p_\mu^* m_\mu^2} = \frac{m_e^2}{m_\mu^2}\left(\frac{m_\pi^2 - m_e^2}{m_\pi^2 - m_\mu^2}\right)^2 = 0.22 \times 10^{-5} \times 2.3^2$$
$$\approx 1.2 \times 10^{-4}. \tag{3.34}$$

We conclude that the *V–A* structure is in agreement with the experiment, but we cannot consider this as a definitive proof. The *V–A* hypothesis is proven by the results of many experiments.

Question 3.1 Knowing the experimental ratio for the K^+ meson

$$\Gamma(K \to e\nu)/\Gamma(K \to \mu\nu) = 1.6 \times 10^{-5} / 0.63 = 2.5 \times 10^{-5} \tag{3.35}$$

prove that the *V–A* hypothesis gives the correct prediction.

3.7 Quark flavours and baryonic number

The baryon number of a state is defined as the number of baryons minus the number of antibaryons

$$\mathcal{B} = N(\text{baryons}) - N(\text{antibaryons}). \tag{3.36}$$

Within the limits of experiments, all known interactions conserve the baryon number. The best limits come from the search for proton decay. In practice, one seeks a specific hypothetical decay channel and finds a limit for that channel. We shall consider the most plausible decay, namely

$$p \to e^+ + \pi^0. \tag{3.37}$$

Notice that this decay also violates the lepton number but conserves the difference $\mathcal{B}-\mathcal{L}$.

The present limit is huge, almost 10^{34} years, 10^{24} times the age of the Universe. To reach such levels of sensitivity one needs to control nearly 10^{34} protons for several years, ready to detect the decay of a single one, if it should happen.

The main problem when searching for rare phenomena, as in this case, is the identification and the drastic reduction, hopefully the elimination, of the 'backgrounds', namely of those natural phenomena that can simulate the events being sought (the 'signal'). The principal background sources are cosmic rays and nuclear radioactivity. In the case of proton decay, the energy of the decay products is of the order of a GeV. Therefore, nuclear radioactivity is irrelevant, because its energy spectra end at $10-15$ MeV. The shielding from cosmic rays is obtained by working in deep underground laboratories.

The sensitivity of an experiment grows with its 'exposure', the product of the sensitive mass and of the time for which data are taken.

The most sensitive detector is currently Super-Kamiokande which, as we have seen in Example 1.13, uses the Cherenkov water technique. It is located in the Kamioka Observatory at 1000 m below the Japanese Alps. The total water mass is 50 000 t. Its central part, in which all backgrounds are reduced, is defined as the 'fiducial mass' and amounts to 22 500 t. Let us calculate how many protons it contains. In H_2O the protons are 10/18 of all the nucleons, and we obtain

$$N_p = M \times 10^3 \times N_A(10/18) = 2.25 \times 10^7 \times 10^3 \times 6 \times 10^{23}(10/18)$$
$$= 7.5 \times 10^{33}.$$

After several years the exposure reached was $M\Delta t = 138\,000$ t yr, corresponding to $N_p\Delta t = 45 \times 10^{33}$ protons per year.

The irreducible background is due to neutrinos produced by cosmic rays in the atmosphere that penetrate underground. Their interactions must be identified and distinguished from the possible proton decay events. If an event is a proton decay (3.37) the electron gives a Cherenkov ring. The photons from the π^0 decay produce lower energy electrons that are also detected as rings. The geometrical aspect of an event, the number of rings, their type, etc., is called the event 'topology'. The first step in the analysis is the selection of the events with a topology compatible with proton decay. This sample contains, of course, background events.

Super-Kamiokande measures the velocity of a charged particle from the position of its centre and from the radius of its Cherenkov ring. Its energy can be inferred from the total number of photons. If the process is the one given in (3.37), then the particles that should be the daughters of the π^0 must have the right invariant mass, and the total energy of the event must be equal to the proton mass. No event was found satisfying these conditions.

We must still consider another experimental parameter: the detection efficiency. Actually, not every proton decay can be detected. The main reason is that the majority of the protons are inside an oxygen nucleus. Therefore, the π^0 from the decay of one of them can interact with another nucleon. If this interaction is accompanied by charge exchange, a process that happens quite often, in the final state we have a π^+ or a π^- and the π^0 is lost. Taking this and other less important effects into account the calculated efficiency is about 40%. The partial decay lifetime in this channel is at 90% confidence level

$$\tau \geq B\left(p \to e^+ \pi^0\right) \times 8.4 \times 10^{33} \text{ yr} \tag{3.38}$$

where B is the unknown branching ratio (Raaf 2006). Similar limits have been obtained for other decay channels, including $\mu^+ \pi^0$ and $K^+ \nu$.

Let us now consider the quarks. Since baryons contain three quarks, the baryon number of the quarks is $\mathcal{B} = 1/3$.

A correlated concept is the 'flavour': the quantum number that characterises the type of quark. We define the 'down quark number' N_d as the number of down quarks minus the number of antidown quarks, and similarly for the other flavours. Notice that the strangeness S of a system and the 'strange quark number' are exactly the same quantity. Three other quarks exist, each with a different flavour, called charm C, beauty B and top T. For historical reasons the flavours of the constituents of normal matter, the up and down quarks, do not have a name

$$\begin{aligned}
N_d &= N(d) - N(\bar{d}) & N_u &= N(u) - N(\bar{u}) \\
S = N_s &= N(s) - N(\bar{s}) & C = N_c &= N(c) - N(\bar{c}) \\
B = N_b &= N(b) - N(\bar{b}) & T = N_t &= N(t) - N(\bar{t}).
\end{aligned} \tag{3.39}$$

Strong and electromagnetic interactions conserve all the flavour numbers while weak interactions violate them.

3.8 Leptonic flavours and lepton number

The (total) lepton number is defined as the number of leptons minus the number of antileptons.

$$\mathcal{L} = N(\text{leptons}) - N(\text{antileptons}). \tag{3.40}$$

Let us also define the partial lepton numbers or, rather, the lepton flavour numbers: the electronic number (or flavour), the muonic number (or flavour) and the tauonic number (or flavour)

$$\mathcal{L}_e = N(e^- + \nu_e) - N(e^+ + \bar{\nu}_e) \tag{3.41}$$

$$\mathcal{L}_\mu = N(\mu^- + \nu_\mu) - N(\mu^+ + \bar{\nu}_\mu) \tag{3.42}$$

$$\mathcal{L}_\tau = N(\tau^- + \nu_\tau) - N(\tau^+ + \bar{\nu}_\tau). \tag{3.43}$$

Obviously, the total lepton number is the sum of these three

$$\mathcal{L} = \mathcal{L}_e + \mathcal{L}_\mu + \mathcal{L}_\tau. \tag{3.44}$$

All known interactions conserve the total lepton number.

The lepton flavours are conserved in all the observed collision and decay processes. The most sensitive tests are based, as usual, on the search for forbidden decays. The best limits are

$$\Gamma(\mu^\pm \to e^\pm + \gamma)/\Gamma_{\text{tot}} \leq 1.2 \times 10^{-11}$$
$$\Gamma(\mu^\pm \to e^\pm + e^+ + e^-)/\Gamma_{\text{tot}} \leq 1 \times 10^{-12} \tag{3.45}$$

which are very small indeed. However, experiments are being done to improve them, in search of possible contributions beyond the Standard Model.

The Standard Model does not allow any violation of the lepton flavour number. On the contrary, it has been experimentally observed that neutrinos produced with a certain flavour may later be observed to have a different flavour. This has been observed in two phenomena:

- The ν_μ flux produced by cosmic radiation in the atmosphere reduces to 50% over distances of several thousand kilometres, namely crossing part of the Earth. This cannot be due to absorption because cross sections are too small. Rather, the fraction that has disappeared is transformed into another neutrino flavour, presumably ν_τ.
- The thermonuclear reactions in the centre of the Sun produce ν_e; only one-half of these (or even less, depending on their energy) leave the surface as such. The electron neutrinos, coherently interacting with the electrons of the dense solar matter, transform, partially, in a quantum superposition of ν_μ and ν_τ.

These are the only phenomena so far observed in contradiction with the Standard Model. We shall come back to this in Chapter 10.

3.9 Isospin

A well-known symmetry property of nuclear forces is their charge independence: two nuclear states with the same spin and the same parity differing by the

Table 3.2 *The lowest isospins and the dimensions of the corresponding representations*

Dimension	**1**	**2**	**3**	**4**	**5**	...
I	0	1/2	1	3/2	2	...

exchange of a proton with a neutron have approximately the same energy. This property can be described in a formal and effective way as proposed by W. Heisenberg in 1932 (Heisenberg 1932). Heisenberg introduced the concept of isotopic spin or, for brevity, isospin. The proton and neutron should be considered two states of the same particle, the nucleon, which has isospin $I = 1/2$. The states that correspond to the two values of the third component are the proton with $I_z = +1/2$ and the neutron with $I_z = -1/2$.

The situation is formally equal to that of the angular momentum. The transformations in 'isotopic space' are analogous to the rotations in normal space. The charge independence of nuclear forces corresponds to their invariance under rotations in isotopic space.

The different values of the angular momentum (J) correspond to different representations of the group of the rotations in normal space. The dimensionality $2J + 1$ of the representation is the number of states with different values of the third component of their angular momentum. In the case of the isospin I, the dimensionality $2I + 1$ is the number of different particles, or nuclear levels, that can be thought of as different charge states of the same particle, or nuclear state. They differ by the third component I_z. The group is called an isotopic multiplet. Clearly, all the members of a multiplet must have the same mass, spin and parity. Table 3.2 shows the simplest representations.

There are several isospin multiplets in nuclear physics. We consider the example of the energy levels of the triplet of nuclei: ^{12}B (made of $5p + 7n$), ^{12}C ($6p + 6n$) and ^{12}N ($7p + 5n$). The ground states of ^{12}B and ^{12}N and one excited level of ^{12}C have $J^P = 1^+$. We lodge them in an $I = 1$ multiplet with $I_z = -1$, 0 and +1 respectively. All of them decay to the ^{12}C ground state: ^{12}B by β^- decay with 13.37 MeV, the excited ^{12}C level by γ decay with 15.11 MeV, and ^{12}N by β^+ decay with 16.43 MeV. If the isotopic symmetry were exact, namely if isospin were perfectly conserved, the energies would be identical. The symmetry is 'broken' because small differences, of the order of a MeV, are present. This is due to two reasons.

Firstly, the symmetry is broken by the electromagnetic interaction, which does not conserve isospin, even if it does conserve its third component. Secondly, the masses of the proton and of the neutron are not identical, but $m_n - m_p \approx 1.3$ MeV.

At the quark level, the mass of the d quark is a few MeV larger than that of the u, contributing to make the neutron, which is ddu, heavier than the proton, uud.

In subnuclear physics, it is convenient to describe the isospin invariance with the group $SU(2)$, instead of that of the three-dimensional rotations. The two are equivalent, but $SU(2)$ will make the extension to $SU(3)$ easier, as we shall discuss in the next chapter.

Just like nuclear levels, the hadrons are grouped in $SU(2)$ (or isospin) multiplets. This is not possible for non-strong-interacting particles, such as the photon and the leptons. Another useful quantum number defined for the hadrons is the flavour hypercharge (or simply hypercharge), which is defined as the sum of the baryon number and strangeness

$$Y = B + S. \tag{3.46}$$

Since the baryon number is conserved by all interactions, hypercharge is conserved in the same cases as strangeness. For mesons, the hypercharge is simply their strangeness. Here we are limiting our discussion to the hadrons made of the quarks u, d, and s only. The particles in the same multiplet are distinguished by the third component of the isospin, which is defined by the Gell-Mann and Nishijima relationship

$$I_z = Q - Y/2 = Q - (B + S)/2. \tag{3.47}$$

Let us see how the hadrons that we have already met are classified in isospin multiplets.

All the baryons we discussed have $J^P = 1/2^+$. They are grouped in the isospin multiplets shown in Fig. 3.1. The approximate values of the mass in MeV are reported next to each particle. The masses within each multiplet are almost but not exactly equal. The small differences are due to the same reasons as for the nucleons. All the members of a multiplet have the same hypercharge, which is reported in the figure next to every multiplet. We shall see more baryons in the next chapter.

For every baryon, there is an antibaryon with identical mass. The multiplets are the same, with opposite charge, strangeness, hypercharge and I_z.

Question 3.2   Draw the figure corresponding to Fig. 3.1 for its antibaryons.

All the mesons we have met have $J^P = 0^-$ and are grouped in the multiplets shown in Fig. 3.2. The π^- and the π^+ are each the antiparticle of the other and are members of same multiplet. The π^0 in the same multiplet is its own antiparticle. The situation is different for the K mesons, which form two doublets

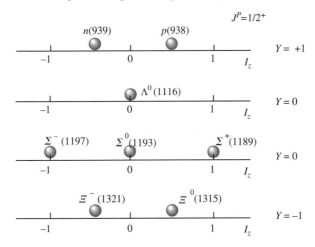

Fig. 3.1. $J^P = 1/2^+$ baryon isospin multiplets.

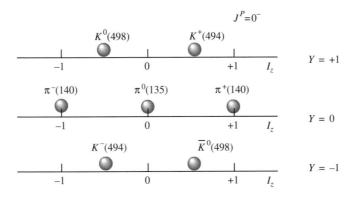

Fig. 3.2.   The pseudoscalar meson isospin multiplets.

containing the particles and their antiparticles respectively. We shall see more mesons in the next chapter.

3.10 The sum of two isospins: the product of two representations

The isospin concept is not only useful for classifying the hadrons, but also in constraining their dynamics in scattering and decay processes. If these proceed through strong interactions, both the total isospin and its third component are conserved; if they proceed through electromagnetic interactions only the third component is conserved; while if they proceed through weak interactions neither is conserved.

Isospin conservation implies definite relationships between the cross sections or the decay probabilities of different strong processes. Consider for example a reaction with two hadrons in the final state, and two in the initial one. The two initial hadrons belong to two isospin multiplets, and similarly the final ones. Changing the particles in each of these multiplets we have different reactions, with cross sections related by isospin conservation. We shall see some examples soon.

In the first step of the isospin analysis one writes both initial and final states as a superposition of states of total isospin. The reaction can proceed strongly only if there is at least one common value of the total isospin. In this case, we define a transition amplitude for each isospin value present in both initial and final states. The transition probability of each process of the set is a linear combination of the isospin amplitudes. We shall now see how.

The rules for isospin composition are the same as for angular momentum. After having recalled them, we shall introduce an alternative notation, which will be useful when dealing with the $SU(3)$ extension of the $SU(2)$ symmetry.

To be specific, let us consider a system of two particles, one of isospin 1 (for example a pion) and one of isospin 1/2 (for example a nucleon). The total isospin can be 1/2 or 3/2. We write this statement as $\mathbf{1} \otimes \mathbf{1/2} = \mathbf{1/2} \oplus \mathbf{3/2}$. This means that the product of the representation of $SU(2)$ corresponding to isospin 1 and the representation corresponding to isospin 1/2 is the sum of the representations corresponding to isospins $\mathbf{1/2}$ and $\mathbf{3/2}$.

The alternative is to label the representation with the number of its states $(2I + 1)$, instead of with its isospin (I). The above written relationship becomes $\mathbf{3} \otimes \mathbf{2} = \mathbf{2} \oplus \mathbf{4}$. Notice that we shall use a different font for this notation.

Let us start with a few important examples.

Example 3.3   Verify the conserved quantities in the reaction $\pi^- + p \rightarrow \pi^0 + n$. Is the process allowed?

The isospin decomposition of the initial state is $\mathbf{1} \otimes \mathbf{1/2} = \mathbf{1/2} \oplus \mathbf{3/2}$; that of the final state is, again, $\mathbf{1} \otimes \mathbf{1/2} = \mathbf{1/2} \oplus \mathbf{3/2}$. There are two common values of the total isospin, 1/2 and 3/2, hence the isospin can be conserved. For the third component, we initially have $I_z = -1 + 1/2 = -1/2$, and finally $I_z = 0 - 1/2 = -1/2$. The third component is conserved. The interaction can proceed strongly.

Example 3.4   Does the reaction $d + d \rightarrow {}^4\text{He} + \pi^0$ conserve isospin? In the initial state, the total isospin is given by $\mathbf{0} \otimes \mathbf{0} = \mathbf{0}$. In the final state, it is given by $\mathbf{0} \otimes \mathbf{1} = \mathbf{1}$. The reaction violates isospin conservation. Experimentally this reaction is not observed, with a limit on its cross section $< 10^{-2}$ of the value computed in the assumption of isospin violation by the strong interaction.

Example 3.5   Compute the isospin balance for $\Sigma^0 \rightarrow \Lambda + \gamma$. The γ does not have isospin. Thus the total isospin changes from the initial to the final state from $\mathbf{1} \rightarrow \mathbf{0}$, while the third component is 0 in both the initial and the final states. The interaction is electromagnetic.

Example 3.6   Compute the isospin balance for $\Lambda \rightarrow p + \pi^-$. The initial isospin is 0. In the final state, the isospin is $\mathbf{1/2} \otimes \mathbf{1} = \mathbf{1/2} \oplus \mathbf{3/2}$. The isospin cannot be conserved. The third component is initially $I_z = 0$ and finally $I_z = 1/2 - 1 = -1/2$; it is not conserved. Even if there are only hadrons in the process, it is a weak decay as shown by the violation of I and of I_z.

We consider now an example of isospin relationships among cross sections. Consider the four reactions: (1) $\pi^+ + p \rightarrow \pi^+ + p$, (2) $\pi^- + p \rightarrow \pi^0 + n$, (3) $\pi^- + p \rightarrow \pi^- + p$, (4) $\pi^- + n \rightarrow \pi^- + n$.

The isospin composition in the initial state is $\mathbf{1} \otimes \mathbf{1/2} = \mathbf{1/2} \oplus \mathbf{3/2}$. In this case, but not always, the composition in the final states is the same. In conclusion, the transition probabilities of the four processes are linear combinations of two isospin amplitudes $A_{1/2}$ and $A_{3/2}$. These are complex functions of the kinematic variables.

We shall now see the general rules for finding these linear combinations.

We have two bases. In the first base, the isospins and their third components of each particle are defined. We call these states $|I_1, I_{z1}; I_2, I_{z2}\rangle$. In the second base, the total isospin (I) and its third component (I_z) are defined, I_1 and I_2 also. We call these states $|I, I_z; I_1, I_2\rangle$. The relationship between the two bases is

$$|I, I_z; I_1, I_2\rangle = \sum_{I_{z1}, I_{z2}} |I_1, I_{z1}; I_2, I_{z2}\rangle\langle I_1, I_{z1}; I_2, I_{z2}|I, I_z; I_1, I_2\rangle. \tag{3.48}$$

The quantities $\langle I_1, I_{z1}; I_2, I_{z2}|I, I_z; I_1, I_2\rangle$ are the Clebsch–Gordan coefficients and can be found in Appendix 4.

Example 3.7   Find the expressions for the cross sections of the four reactions discussed above in terms of the two isospin amplitudes. We write for simplicity the kets in the second members as $|I, I_z\rangle$. Using the Clebsch–Gordan tables we find

$$|\pi^+ p\rangle = \left|\frac{3}{2}, +\frac{3}{2}\right\rangle \qquad |\pi^- p\rangle = \sqrt{\frac{1}{3}}\left|\frac{3}{2}, -\frac{1}{2}\right\rangle - \sqrt{\frac{2}{3}}\left|\frac{1}{2}, -\frac{1}{2}\right\rangle$$

$$|\pi^0 n\rangle = \sqrt{\frac{2}{3}}\left|\frac{3}{2}, -\frac{1}{2}\right\rangle + \sqrt{\frac{1}{3}}\left|\frac{1}{2}, -\frac{1}{2}\right\rangle \qquad |\pi^- n\rangle = \left|\frac{3}{2}, -\frac{3}{2}\right\rangle. \tag{3.49}$$

With a proportionality constant K equal for all we obtain

$$\sigma(\pi^+ p \to \pi^+ p) = K|A_{3/2}|^2 \tag{3.50}$$

$$\sigma(\pi^- p \to \pi^0 n) = K\left|\frac{\sqrt{2}}{3}A_{3/2} - \frac{\sqrt{2}}{3}A_{1/2}\right|^2 \tag{3.51}$$

$$\sigma(\pi^- p \to \pi^- p) = K\left|\frac{1}{3}A_{3/2} + \frac{2}{3}A_{1/2}\right|^2 \tag{3.52}$$

$$\sigma(\pi^- n \to \pi^- n) = K|A_{3/2}|^2. \tag{3.53}$$

In particular we arrive at the prediction $\sigma(\pi^+ p \to \pi^+ p) = \sigma(\pi^- n \to \pi^- n)$ for the same energy. It is experimentally well verified. From these cross sections we know $K|A_{3/2}|$. The other cross sections, and those of other processes like $\pi^+ + n \to \pi^+ + n$ and $\pi^+ + n \to \pi^0 + p$, depend on two unknowns, $|A_{1/2}|$ and $\arg(A_{3/2}^* A_{1/2})$.

At low energies all these cross sections show a large resonance, which was discovered by Fermi (Anderson *et al.* 1952), called $\Delta(1236)$, which has a maximum at $\sqrt{s} = 1236$ MeV (see Section 4.2). We know that its isospin is $I = 3/2$ by observing that the cross section is dominated by $|A_{3/2}|$. Actually, in this case we obtain from the above expressions

$$\sigma(\pi^+ p \to \pi^+ p):\sigma(\pi^- p \to \pi^- p):\sigma(\pi^- p \to \pi^0 n) = 9:1:2 \tag{3.54}$$

and the experimental values of the cross sections in mb are 195:22:45.

3.11 *G*-parity

G-parity is not a fundamental quantum number, however it is convenient when dealing with non-strange states with zero baryonic number. These states, typically the pion systems, are eigenstates of G.

The π^0 is an eigenstate of the charge conjugation C. The charged pions (see Eq. (3.13)) transform as

$$C|\pi^+\rangle = +|\pi^-\rangle \qquad C|\pi^-\rangle = +|\pi^+\rangle. \tag{3.55}$$

G is defined as C followed by a 180° rotation around the y-axis in isotopic space, namely

$$G \equiv \exp(-i\pi I_y)C. \tag{3.56}$$

The three π states are the components of a vector in isotopic space (iso-vector). The relationships between Cartesian components $|\pi_x\rangle$, $|\pi_y\rangle$ and $|\pi_z\rangle$ and charge states are

$$|\pi^+\rangle = \frac{1}{\sqrt{2}}\left(|\pi_x\rangle + i|\pi_y\rangle\right) \qquad |\pi^0\rangle = |\pi_z\rangle \qquad |\pi^-\rangle = \frac{1}{\sqrt{2}}\left(|\pi_x\rangle - i|\pi_y\rangle\right). \tag{3.57}$$

Let us apply C and then the rotation to these expressions:

$$
\begin{aligned}
&|\pi^+\rangle = \frac{1}{\sqrt{2}}\left(|\pi_x\rangle + i|\pi_y\rangle\right) \quad |\pi^-\rangle = \frac{1}{\sqrt{2}}\left(|\pi_x\rangle - i|\pi_y\rangle\right) \quad \frac{1}{\sqrt{2}}\left(-|\pi_x\rangle - i|\pi_y\rangle\right) = -|\pi^+\rangle \\
&|\pi^0\rangle = |\pi_z\rangle \qquad\qquad C \Rightarrow |\pi^0\rangle = |\pi_z\rangle \qquad\qquad e^{i\pi I_y} \Rightarrow -|\pi_z\rangle \qquad\qquad = -|\pi^0\rangle \\
&|\pi^-\rangle = \frac{1}{\sqrt{2}}\left(|\pi_x\rangle - i|\pi_y\rangle\right) \quad |\pi^+\rangle = \frac{1}{\sqrt{2}}\left(|\pi_x\rangle + i|\pi_y\rangle\right) \quad \frac{1}{\sqrt{2}}\left(-|\pi_x\rangle + i|\pi_y\rangle\right) = -|\pi^-\rangle.
\end{aligned}
$$

We see that all the charge states are eigenstates with negative eigenvalue

$$G|\pi\rangle = -|\pi\rangle. \tag{3.58}$$

For a system of n_π pions we have

$$G = (-1)^{n_\pi}. \tag{3.59}$$

It is easy to prove that all non-strange non-baryonic states are eigenstates of G. If their isospin is $I = 1$ the situation is identical to that of the pions. The neutral state has $I_z = 0$ and $G = -C$. If $I = 0$, obviously $G = C$.

Only the strong interaction conserves the G-parity because the electromagnetic and the weak interactions violate the isospin (and the latter also C).

Problems

3.1. For each interaction type, strong (S), electromagnetic (EM) and weak (W), insert a Y or N in the cell of every quantum number, depending on whether it is conserved or not (I isospin, I_z its third component, S strangeness, B baryon number, $\mathcal{L}$ lepton number, $\mathcal{T}$ time reversal, C particle–antiparticle conjugation, $\mathcal{P}$ parity, J angular momentum, J_z its third component).

	I	I_z	S	B	$\mathcal{L}$	$\mathcal{T}$	C	$\mathcal{P}$	J	J_z
S										
EM										
W										

3.2. Consider a π^- beam impinging on a liquid hydrogen target. Find the threshold energy for K^- production.

3.3. The existence of the antihyperons was proven by the discovery of an antilambda by M. Baldo-Ceolin and D. J. Prowse in 1958. A beam of negative pions with energy $E_\pi = 4.6$ GeV hit an emulsion stack. What is the final state containing a $\bar{\Lambda}$ that can be produced in a $\pi^- p$ collision at minimum energy? Find the threshold energy if the target protons are free and, approximately, if they are bound inside nuclei with a Fermi momentum $p_f = 150$ MeV. Assuming that the pion beam was produced at a distance $l = 8$ m upstream of the emulsion and that the number of produced pions was $N_0 = 10^6/\text{cm}^2$, how many pions/cm^2 reached the emulsion?

3.4. For each of the following reactions (a) establish whether it is allowed or not, (b) if it is not, give the reasons (there may be more than one), (c) establish the types of interaction that allow it: (1) $\pi^- p \rightarrow \pi^0 + n$; (2) $\pi^+ \rightarrow \mu^+ + \nu_\mu$; (3) $\pi^+ \rightarrow \mu^+ + \bar{\nu}_\mu$; (4) $\pi^0 \rightarrow 2\gamma$; (5) $\pi^0 \rightarrow 3\gamma$; (6) $e^+ + e^- \rightarrow \gamma$; (7) $p + \bar{p} \rightarrow \Lambda + \Lambda$; (8) $p + p \rightarrow \Sigma^+ + \pi^+$; (9) $n \rightarrow p + e^-$; (10) $n \rightarrow p + \pi^-$.

3.5. For each of the following reactions establish whether it is allowed or not; if it is not, give the reasons: (1) $\mu^+ \rightarrow e^+ + \gamma$; (2) $e^- \rightarrow \nu_e + \gamma$; (3) $p + p \rightarrow \Sigma^+ + K^+$; (4) $p + p \rightarrow p + \Sigma^+ + K^-$; (5) $p \rightarrow e^+ + \nu_e$; (6) $p + p \rightarrow \Lambda + \Sigma^+$; (7) $p + n \rightarrow \Lambda + \Sigma^+$; (8) $p + n \rightarrow \Xi^0 + p$; (9) $p \rightarrow n + e^+ + \nu_e$; (10) $n \rightarrow p + e^- + \nu_e$.

3.6. Give the reasons forbidding each of the following decays: (a) $n \rightarrow p + e^-$; (b) $n \rightarrow \pi^+ + e^-$; (c) $n \rightarrow p + \pi^-$; (d) $n \rightarrow p + \gamma$.

3.7. Which of the following processes is allowed and which forbidden by strangeness conservation? (a) $\pi^- + p \rightarrow K^- + p$; (b) $\pi^- + p \rightarrow K^+ + \Sigma^-$; (c) $K^- + p \rightarrow K^+ + \Xi^0 + \pi^-$; (d) $K^+ + p \rightarrow K^- + \Xi^0 + \pi^-$.

3.8. For each of the following reactions establish whether it is allowed or not; if it is not, give the reasons: (a) $p \rightarrow n + e^+$; (b) $\mu^+ \rightarrow \nu_\mu + e^+$; (c) $e^+ + e^- \rightarrow \nu_\mu + \bar{\nu}_\mu$; (d) $\nu_\mu + p \rightarrow \mu^+ + n$; (e) $\nu_\mu + n \rightarrow \mu^- + p$; (f) $\nu_\mu + n \rightarrow e^- + p$; (g) $e^+ + n \rightarrow p + \nu_e$; (h) $e^- + p \rightarrow n + \nu_e$.

3.9. Evaluate the ratios between the cross sections of the following reactions at the same energy assuming (unrealistically) that they proceed only through the $I = 3/2$ channel: $\pi^- p \rightarrow K^0 \Sigma^0$; $\pi^- p \rightarrow K^+ \Sigma^-$; $\pi^+ p \rightarrow K^+ \Sigma^+$.

3.10. Evaluate the ratios between the cross sections at the same energy of (1) $\pi^- p \rightarrow K^0 \Sigma^0$, (2) $\pi^- p \rightarrow K^+ \Sigma^-$, (3) $\pi^+ p \rightarrow K^+ \Sigma^+$, taking into account the contributions of both isospin amplitudes $A_{1/2}$ and $A_{3/2}$.

3.11. Evaluate the ratio between the cross sections of the reactions $\pi^- p \rightarrow \Lambda K^0$ and $\pi^+ n \rightarrow \Lambda K^+$, at the same energy.

3.12. Evaluate the ratio of the cross sections of the processes $p + d \rightarrow {}^3\text{He} + \pi^0$ and $p + d \rightarrow {}^3\text{H} + \pi^+$ at the same value of the CM energy $\sqrt{s}$ (${}^3\text{He}$ and ${}^3\text{H}$ are an isospin doublet).

3.13. Evaluate the ratio of cross sections $\sigma(pp \rightarrow d\pi^+)/\sigma(pn \rightarrow d\pi^0)$ at the same energy.

3.14. Evaluate the ratio of cross sections $\sigma(K^- + {}^4\text{He} \rightarrow \Sigma^0 + {}^3\text{H})/\sigma(K^- + {}^4\text{He} \rightarrow \Sigma^- + {}^3\text{He})$ at the same energy.

3.15. Express the ratios between the cross sections of (1) $K^-p \rightarrow \pi^+\Sigma^-$, (2) $K^-p \rightarrow \pi^0\Sigma^0$, (3) $K^-p \rightarrow \pi^-\Sigma^+$ in terms of the isospin amplitudes A_0, A_1 and A_2.

3.16. Express the ratio of cross sections of the elastic $\pi^-p \rightarrow \pi^-p$ and the charge exchange $\pi^-p \rightarrow \pi^0 n$ scatterings in terms of the isospin amplitudes $A_{1/2}$ and $A_{3/2}$.

3.17. A π^- is captured by a deuteron $d\,(J^P = 1^-)$ and produces the reaction $\pi^- + d \rightarrow n + n$. (a) If the capture is from an S wave, what is the total spin of the two neutrons and what is their orbital momentum? (b) Show that, if the capture is from a P state, the neutrons are in a singlet.

3.18. The positronium is an atomic system made by an e^- and an e^+ bound by the electromagnetic force.

   a. Determine the relationship that this condition imposes between the orbital momentum l, the total spin s and the charge conjugation C.

   b. Determine the relationship between l, s and n which allows the reaction $e^- e^+ \rightarrow n\gamma$ to occur without violating C.

   c. What is the minimum number of photons in which the ortho-positronium (${}^3S_1$) and the para-positronium (${}^1S_0$) can annihilate respectively?

3.19. Establish from which initial states of the $\bar{p}p$ system amongst ${}^1S_0$, ${}^3S_1$, ${}^1P_1$, ${}^3P_0$, ${}^3P_1$, ${}^3P_2$, ${}^1D_2$, ${}^3D_1$, ${}^3D_2$ and ${}^3D_3$ the reaction $\bar{p}p \rightarrow n\pi^0$ can proceed with parity conservation: (1) for any n; (2) for $n = 2$.

3.20. Consider the strong processes $\bar{K}K \rightarrow \pi^+\pi^-$ (where $\bar{K}K$ means both K^+K^- and $\bar{K}^0K^0$). (1) What are the possible angular momentum values if the initial total isospin is $I = 0$? (2) What are they if $I = 1$?

3.21. Consider the following $\bar{p}p$ initial states: ${}^1S_0$, ${}^3S_1$, ${}^1P_1$, ${}^3P_0$, ${}^3P_1$, ${}^3P_2$, ${}^1D_2$, ${}^3D_1$, ${}^3D_2$, ${}^3D_3$. Establish from which of these the reaction $\bar{p}p \rightarrow \pi^+\pi^-$ can proceed if the two πs are: (1) in an S wave; (2) in a P wave; (3) in a D wave.

3.22. The quark contents of the following charmed particles are: the hyperon Λ_c is udc, the D^+ meson is $c\bar{d}$ and the D^- meson is $\bar{c}d$. Which of the following reactions are allowed? (a) $\pi^+ p \to D^+ p$, (b) $\pi^+ p \to D^- \Lambda_c \pi^+ \pi^+$, (c) $\pi^+ p \to D^+ \Lambda_c$, (d) $\pi^+ p \to D^- \Lambda_c$.

3.23. The quark contents of the following particles are: the beauty hyperon $\Lambda_b = udb$, the charmed meson $D^0 = c\bar{u}$, the beauty mesons $B^+ = u\bar{b}$, $B^- = \bar{u}b$ and $B^0 = d\bar{b}$. Which of the following reactions are allowed? (a) $\pi^- p \to D^0 \Lambda_b$, (b) $\pi^- p \to B^0 \Lambda_b$, (c) $\pi^- p \to B^+ \Lambda_b \pi^-$, (d) $\pi^- p \to B^- \Lambda_b \pi^+$, and (e) $\pi^- p \to B^- B^+$.

3.24. An η meson decays into 2γ while moving in the x direction with energy $E_\eta = 5$ GeV.

1. If the two γs are emitted in the $+x$ and $-x$ directions, what are their energies?
2. If the two γs are emitted at equal and opposite angles $\pm\theta$ with x, what is the angle between the two?

3.25. The state $\Delta(1232)$ has isospin $I = 3/2$. (1) What is the ratio between the decay rates $\Delta^0 \to p\pi^-$ and $\Delta^0 \to n\pi^0$? (2) What would it have been if $I = 1/2$?

Further reading

Wigner, E. P. (1963); Nobel Lecture, *Events, Laws of Nature, and Invariance Principles* http://nobelprize.org/nobel_prizes/physics/laureates/1963/wigner-lecture.pdf

Wigner, E. P. (1964); *The Role of Invariance Principles in Natural Philosophy.* Proceedings of the International School of Physics 'Enrico Fermi' **29** p. 40. Academic Press

Wigner, E. P. (1965); *Violation of symmetry in physics. Sci. Am.* **213** no. 6, 28

4

Hadrons

4.1 Resonances

The particles we have discussed up to now are metastable, in other words they decay by weak or electromagnetic interactions. The distances between their production and decay points or between the corresponding times are long enough to be separately observable.

If the mass of a hadron is large enough, final states that can be reached by strong interaction, i.e. without violating any selection rule, become accessible to its decay. Therefore, they have extremely short lifetimes, of the order of 10^{-24} seconds, and decay, from the point of view of an observer, exactly where they were born. To fix the orders of magnitude, consider such a particle produced in the laboratory reference frame with a Lorentz factor as large as $\gamma = 300$. In a lifetime, it will travel one femtometre.

These extremely unstable hadrons can be observed as 'resonances' in two basic ways: in the process of 'formation' as a local maximum in the energy dependence of a cross section or in the 'production' process as a maximum in the (invariant) mass distribution of a few particles in the final states of a reaction.

Resonant phenomena are ubiquitous in physics, both at the macroscopic and microscopic levels. Even in very different physical situations, ranging from mechanics to electrodynamics, from acoustics to optics, from atomic to nuclear physics, etc., the fundamental characteristics of this phenomenon are the same. Actually, the classical and the quantum formalisms are also very similar. Resonances have an extremely important role in hadron spectroscopy.

To recollect the fundamental concepts, let us start from a naïve model of an atom. We imagine the atom to be made up of a massive central charge surrounded by a cloud of equal and opposite charges. Initially we assume that the system can be described by classical physics.

The system has a triple infinite set of normal modes, each with a proper frequency and a proper width. Let us concentrate on one of these and call ω_0 the proper angular frequency and Γ the proper width. At $t = 0$ we set the system in this mode and we then leave it to evolve freely in time. Let Ψ be a coordinate measuring distance of the system from equilibrium, for example the angle with the vertical in the case of the pendulum. The time dependence of Ψ is an exponentially damped sinusoidal function

$$\Psi(t) = \Psi_0 \exp\left(-\frac{\Gamma t}{2}\right) \cos \omega_0 t = \Psi_0 \exp\left(-\frac{t}{2\tau}\right) \cos \omega_0 t, \quad \tau \equiv 1/\Gamma. \quad (4.1)$$

Notice that the time constant is denoted by 2τ. This choice gives τ the meaning of the time constant of the intensity, which is the square of the amplitude. We have defined τ as the reciprocal of the width.

Let us now consider the forced oscillations of the system. We act on the oscillator with a periodic force, we slowly vary its angular frequency ω and, waiting for the system to reach the stationary regime at every change, we measure the oscillation amplitude. We obtain the well-known resonance curve. In the neighbourhood of the maximum, the response function is, with a generally good approximation

$$R(\omega) = \frac{\Gamma^2 \omega^2}{\left(\omega_0^2 - \omega^2\right)^2 + \omega^2 \Gamma^2}. \quad (4.2)$$

Comparing the two expressions, we conclude that: *the width of the resonance curve of the forced oscillator is equal to the reciprocal of the lifetime of its free oscillations.*

We also recall that the square of the Fourier transform of the decaying oscillations (4.1) is proportional to the response function (4.2).

If the width is small, $\Gamma \ll \omega_0$, (4.2) can be approximated in the neighbourhood of the peak with a simpler expression, called the Breit–Wigner function. Close to the resonance, the fastest varying factor is the denominator

$$\left(\omega_0^2 - \omega^2\right)^2 + \omega^2 \Gamma^2 = (\omega_0 - \omega)^2 (\omega_0 + \omega)^2 + \omega^2 \Gamma^2.$$

Here we see that the variation is due mainly to $\omega_0 - \omega$. We replace ω by ω_0 everywhere but in this term, obtaining

$$(\omega_0 - \omega)^2 (\omega_0 + \omega)^2 + \omega^2 \Gamma^2 = 4\omega_0^2 \left[(\omega_0 - \omega)^2 + (\Gamma/2)^2\right].$$

Substituting this into (4.2) we finally obtain the Breit–Wigner shape function

$$R(\omega) \approx L(\omega) = \frac{(\Gamma/2)^2}{(\omega_0 - \omega)^2 + (\Gamma/2)^2}. \quad (4.3)$$

Actually, the atom is a quantum, not a classical, object, but its correct quantum description brings us to the same conclusion with the only difference, as far as we are concerned here, that the atom has a series of excited levels instead of the classical normal modes. Each of them has a proper energy and a proper lifetime.

Another aspect that is similar in classical and quantum resonance phenomena is the behaviour of the phase. Namely, the energy dependence of the phase of the scattering amplitude in quantum mechanics is similar to the frequency dependence of the phase of the oscillation amplitude relative to that of the external force in classical mechanics. Both are small and about π respectively at frequencies far below and far above resonance. A rapid crossing of $\pi/2$ marks the resonance.

Coming back to our analogy and to approach the subnuclear physics processes, we consider the fundamental level of the atom, which we call A, and its excited levels, which we call A_i^*, as different particles. Actually, this is a totally correct point of view. Each of these has an energy, a lifetime – infinite for A, finite for the A_i^* – and an angular momentum.

Let us suppose that we know of the existence of A and we decide to set up an experiment to search for possible unstable particles A_i^*. We design a **resonance formation** experiment.

We prepare a transparent container that we fill with the A atoms and build a device capable of producing a collimated and monochromatic light beam, which has a frequency that we can vary. We send the beam through the container and measure the intensity of the diffused light (at a certain angle). If we do this varying the frequency, we find a resonance peak for each of the A_i^*. The line shapes are described by (4.3) and we can determine the proper frequencies and the widths of the A_i.

Note that two processes take place at the microscopic level: the formation of the A_i^* particle and its decay

$$\gamma + A \rightarrow A_i^* \rightarrow \gamma + A. \tag{4.4}$$

The time between the two processes may be too short to be measurable, but we can infer the lifetime of A_i^* from the width of the resonance.

The particle A_i^* has a well-defined angular momentum and parity, J^P. These can be determined, as can be easily understood, by measuring, in resonance, the diffused intensity as a function of the angle.

Similarly in subnuclear physics, we search for very unstable particles by measuring, as a function of the energy, the cross section of processes of the following type

$$a + b \rightarrow c + d + \cdots + f. \tag{4.5}$$

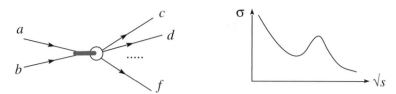

Fig. 4.1.   Schematic of a resonance formation study.

Clearly, in this process we can find particles with quantum numbers that are compatible with those of the initial and final states. A schematic is shown in Fig. 4.1.

Example 4.1   The πp resonance (1236) has the width $\Gamma \approx 120\,\text{MeV}$. What is its lifetime?

$$\tau = \frac{1}{\Gamma} = \frac{1}{120\,\text{MeV}} = \frac{1}{120\,(\text{MeV}) \times 1.52 \times 10^{21}\,(\text{s}^{-1}\,\text{MeV}^{-1})} = 5.4 \times 10^{-24}\,\text{s}.$$

We now give an expression for the cross section as a function of the centre of mass energy E, in the neighbourhood of a resonance. We shall often use it in the following. Let M_R be the mass of the resonance, Γ its width and J its spin. The expression is particularly simple in 'ultrarelativistic' conditions, namely if the energies are so large compared to the masses as to be practically equal to the corresponding momenta. We call Γ_i and Γ_f the partial widths of the initial and final states respectively. Let s_a and s_b be the spins of the particles a and b in the initial state. We state without proof that the following expression is valid in the Breit–Wigner approximation, namely if $\Gamma \ll M_R$

$$\sigma(E) = \frac{(2J+1)}{(2s_a+1)(2s_b+1)} \frac{4\pi}{E^2} \frac{\Gamma_i \Gamma_f}{(E-M_R)^2 + (\Gamma/2)^2}. \tag{4.6}$$

This expression is very useful for rough calculations but not when precision is needed.

The second class of experiments on unstable hadrons is based on **resonance production**.

Assume that we are searching for particles decaying into the stable or metastable particles c and d. Call M_R the mass and Γ the width of such a particle R. We select a process with a final state that contains those two particles (c and d) and, at least, another one, for example

$$a + b \rightarrow c + d + e. \tag{4.7}$$

If the unstable particle R decaying $R \rightarrow c + d$ exists, the above reaction proceeds, at least in a fraction of cases, through an intermediate state containing R, namely

$$a + b \rightarrow R + e \rightarrow c + d + e. \tag{4.8}$$

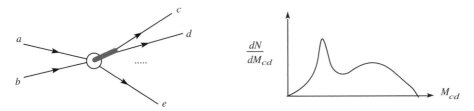

Fig. 4.2.   Schematic of a resonance production study.

In these cases, which are examples of the 'resonant process', the mass of the system $c + d$, call it M_{cd}, is expected to be equal to M_R or, better, to have a Breit–Wigner distribution peaked at M_R with width Γ. If, on the other hand, the reaction goes directly to the final state (non-resonant process), M_{cd} can have any value, within the constraints of energy and momentum conservation. Its distribution is smooth, without peaks, given substantially by the phase space factor.

We then measure for each event (4.7) the energies and the momenta of the final particles and compute

$$M_{cd} = \sqrt{(E_c + E_d)^2 - (\mathbf{p}_c + \mathbf{p}_d)^2}. \tag{4.9}$$

The resonance appears, as sketched in Fig. 4.2, as a peak on a smooth background in the M_{cd} distribution.

Obviously, by the same method one can search for resonances decaying in more than two particles, computing the mass of such systems. Notice however that the simple observation of a peak is not enough to establish a resonance. Much more detailed study is necessary.

4.2 The 3/2⁺ baryons

Up to now we have encountered eight baryons, all with spin 1/2 and positive parity. They are metastable because their masses are not large enough to allow strong decays. Many other baryons exist, both strange and non-strange, with both positive and negative parity and different spin values. These have larger masses and decay strongly. We shall consider only the ground-level baryons that have spin-parity 1/2⁺ or 3/2⁺.

The search for strongly decaying baryons follows the principles described in the previous section. Considering first the formation experiments, let us see what are the possible targets and beams. To form a baryon, we need a baryon target and a meson beam.

The target must be an elementary particle. In practice, we can have free protons using hydrogen, in the liquid phase for adequate luminosity. We cannot have free neutrons and we must use the simplest nucleus containing neutrons, i.e. deuterium.

Mesons can be used in a beam only if their lifetime is long enough. There are not many of them. With charged pions, both positive and negative, the formation of $S = 0$ baryons has been systematically studied, with K^- beams the formation of the $S = -1$ baryons, and finally with K^+ beams that of the $S = +1$ baryons.

This systematic investigation carried out by several experiments in all the accelerator laboratories in the 1960s led to the following well-established conclusions. Several dozen resonances exist in the pion–nucleon and K^-–nucleon systems. Their isospins have been measured by comparing the cross sections in the different charge states, as in Example 3.7. Their spins and parities have been determined from the angular differential cross sections.

On the other hand, no resonance exists in the K^+–nucleon system, namely no positively strange ($S = +1$) baryon ($\mathcal{B} = 1$) exists in Nature. This fact is explained by the quark structure of the hadrons.

Let us consider the isospin representations now. The pions are an isospin triplet, namely they are in the representation **3**, and the nucleons are a doublet, in the representation **2**. The composition rule gives $\mathbf{3} \otimes \mathbf{2} = \mathbf{2} \oplus \mathbf{4}$. Therefore, the pion–nucleon resonances can be doublets or quartets or, in other words, can have isospin 1/2 or 3/2. By convention the $I = 1/2$ states are called N(xxxx), the $I = 3/2$ ones are called Δ(xxxx) where xxxx is the mass in MeV.

Figure 4.3 shows the total $\pi^+ p$ cross section and the elastic cross section, namely that of

$$\pi^+ + p \rightarrow \pi^+ + p. \tag{4.10}$$

We see that at low energy the two cross sections are equal. This is because no other channel is open. At higher energies, gradually more channels open up: $\pi^+ + p \rightarrow \pi^+ + \pi^0 + p$, $\pi^+ + p \rightarrow \pi^+ + \pi^+ + \pi^- + p$, $\pi^+ + p \rightarrow \pi^+ + K^+ + K^- + p$, etc. and the elastic cross section is only a fraction of the total.

The first resonance was a completely unforeseen discovery of Fermi and collaborators (Anderson *et al.* 1952) working with positive and negative pion beams from the Chicago cyclotron. It is the huge peak in the elastic region, the $\Delta(1236)$ that we have already met. As we have already discussed its isospin is $I = 3/2$. Its four charge states are Δ^-, Δ^0, Δ^+, and Δ^{++}. The analysis of the angular differential cross sections established that the orbital momentum of the pion–nucleon system is $l = 1$ and that the total angular momentum is 3/2. The parity is $(-1)^l$ times the product of the pion and nucleon intrinsic parity, which is -1. In conclusion, $J^P = 3/2^+$.

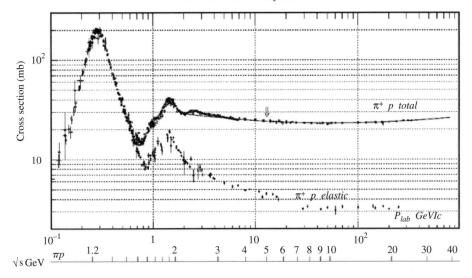

Fig. 4.3. π^+p total and elastic cross sections. (From Yao *et al.* 2006 by permission of Particle Data Group and the Institute of Physics)

Figure 4.3 shows many other peaks corresponding to other resonances. However, the vast majority of the resonances cannot be seen 'by eye', rather they are found by studying the energy dependence of the scattering amplitudes of defined angular momentum and isospin. Resonances are marked by a rapid transition of the phase through $\pi/2$, possibly superposed on a non-resonant contribution.

As a final observation that will be useful in the following, note that as the energy increases the resonances disappear and the cross section reaches a value of $\sigma_{\pi^+p} \approx$ 25 mb and is slowly increasing.

We now consider the hyperons with $S = -1$. They can be formed from, or decay into a $\bar{K}N$ system (where N stands for nucleon). The K meson and the nucleon are isospin doublets. Following the combination rule $\mathbf{2} \otimes \mathbf{2} = \mathbf{1} \oplus \mathbf{3}$ they can form hyperons with isospin $I = 0$ or $I = 1$. The former are called Λ(xxxx), the second Σ(xxxx), where xxxx is the mass in MeV.

An $S = -1$ hyperon with mass smaller than the sum of the K meson and proton masses ($494 + 938 = 1432$ MeV) cannot be observed as a resonance in the $K^- + p \rightarrow K^- + p$ cross section. Actually, the lowest mass $S = -1$ baryonic system is $\Lambda\pi^{\pm}$ ($m_\Lambda + m_\pi = 1115 + 140 = 1255$ MeV). To search in the mass range between 1255 and 1432 MeV we must use the production method, searching for $\Lambda\pi^{\pm}$ resonances in the final state of

$$K^- + p \rightarrow \Lambda + \pi^+ + \pi^-. \qquad (4.11)$$

This was done by Alvarez *et al.* (1963) using a K^- beam with 1.5 GeV momentum produced by the Bevatron. The detector was the $72''$ bubble chamber discussed in Section 1.11.

Notice that there are two $\Lambda\pi$ charge states in (4.11). Therefore, there are two possible intermediate states, which we call here Σ^*, leading to the same final state

$$
\begin{aligned}
K^+ + p &\rightarrow \Sigma^{*+} + \pi^- \rightarrow (\Lambda + \pi^+) + \pi^- \\
K^- + p &\rightarrow \Sigma^{*-} + \pi^+ \rightarrow (\Lambda + \pi^-) + \pi^+.
\end{aligned}
\tag{4.12}
$$

We must then consider both masses $\Lambda\pi^+$ and $\Lambda\pi^-$. In the plot in Fig. 4.4 every point is an event. The scales of the axes are the squares of the masses, not the masses, because, as we shall see in Section 4.3, the phase space volume is constant in those variables. Therefore, any non-uniformity corresponds to a dynamic feature. The contour of the plot is given by the energy and momentum conservation.

Looking at Fig. 4.4 we clearly see two perpendicular high-density bands. Each of the projections on the axes would show a peak corresponding to the band perpendicular to that axis, similar to that sketched in Fig. 4.2. The bands appear at the same value of the mass, $M = 1385$ MeV, and have the same width, $\Gamma = 35$ MeV. Finally, being $\Lambda\pi^+$ and $\Lambda\pi^-$ pure $I = 1$ states, the isospin of the observed hyperon is one.

The hyperon is called $\Sigma(1385)$. The analysis of the angular distributions of its daughters shows that $J^P = 3/2^+$.

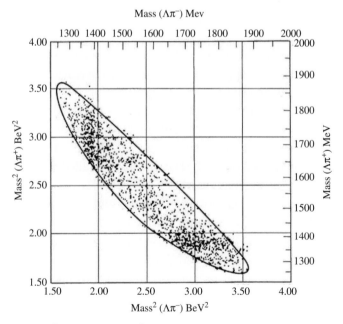

Fig. 4.4.   (Mass)^2 $\Lambda\pi^+$ vs. (Mass)^2 $\Lambda\pi^-$ for reaction (4.12). (Alvarez *et al.* 1963)

Hyperons with isospin $I = 0$ and strangeness $S = -1$ (such as the Λ) and $J^P = 3/2^+$ have been sought, but do not exist. Again, this fact is well explained by the quark model.

The search for $S = -2$ hyperons can be done only in production, looking for possible resonances in the $\Xi\pi$ systems in different charge states. They are generically called Ξ(xxxx). The lowest mass state, called Ξ(1530), which here we call Ξ^*, was found using a K^- beam with 1.8 GeV momentum from the Bevatron and the 72″ hydrogen bubble chamber (Pjerrou *et al.* 1962).

The Ξ^* was observed in the two charge states Ξ^{*0} and Ξ^{*-}. We focus on the former, which was observed in the two reactions

$$K^- + p \to \Xi^{*0} + K^0 \to (\Xi^- + \pi^+) + K^0 \tag{4.13}$$

$$K^- + p \to \Xi^{*0} + K^0 \to (\Xi^0 + \pi^0) + K^0. \tag{4.14}$$

Figure 4.5 shows the $\Xi^-\pi^+$ mass distribution. In this case, the reaction is completely dominated by the resonance. The dotted line is the estimate of the non-resonant background.

The continuous curve is a Breit–Wigner curve obtained as the best fit to the data, leaving the mass and the width as free parameters. The result is

$$m = 1530 \, \text{MeV} \quad \Gamma = 7 \pm 2 \, \text{MeV}. \tag{4.15}$$

The $\Xi^0\pi^0$ mass distribution from (4.14) shows a similar peak, about 1/2 in intensity. This result determines the isospin. Let us see how.

Since the Ξ^* decays into $\Xi\pi$, it can have isospin 1/2 or 3/2.

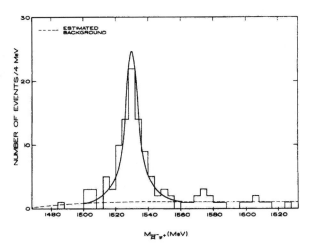

Fig. 4.5. $\Xi^-\pi^+$ mass distribution. (Schlein *et al.* 1963)

The third component of the isospin of the Ξ^{*0} is $+1/2$. It decays with isospin conservation into a Ξ, which has isospin $1/2$ and a π, which has isospin 1. We use the Clebsch–Gordan coefficients to find the weights of the charge states. If the isospin of the Ξ^* is $3/2$, we have

$$
\begin{aligned}
\left|\Xi^{*0}\right\rangle &= \left|\frac{3}{2}, +\frac{1}{2}\right\rangle = \sqrt{\frac{1}{3}}\left|\frac{1}{2}, -\frac{1}{2}\right\rangle|1, +1\rangle + \sqrt{\frac{2}{3}}\left|\frac{1}{2}, +\frac{1}{2}\right\rangle|1, 0\rangle \\
&= \sqrt{\frac{1}{3}}\left|\Xi^-\pi^+\right\rangle + \sqrt{\frac{2}{3}}\left|\Xi^0\pi^0\right\rangle.
\end{aligned}
\tag{4.16}
$$

Taking the squares of the amplitudes, we immediately find

$$
\Gamma\left(\Xi^{*0} \to \Xi^0\pi^0\right)/\Gamma\left(\Xi^{*0} \to \Xi^-\pi^+\right) = 2
$$

in contradiction with experiments. If the isospin of the Ξ^* is $1/2$, we have

$$
\begin{aligned}
\left|\Xi^{*0}\right\rangle &= \left|\frac{1}{2}, +\frac{1}{2}\right\rangle = \sqrt{\frac{2}{3}}\left|\frac{1}{2}, -\frac{1}{2}\right\rangle|1, +1\rangle - \sqrt{\frac{1}{3}}\left|\frac{1}{2}, +\frac{1}{2}\right\rangle|1, 0\rangle \\
&= \sqrt{\frac{2}{3}}\left|\Xi^-\pi^+\right\rangle - \sqrt{\frac{1}{3}}\left|\Xi^0\pi^0\right\rangle.
\end{aligned}
\tag{4.17}
$$

Hence

$$
\Gamma(\Xi^{*0} \to \Xi^0\pi^0)/\Gamma(\Xi^{*0} \to \Xi^-\pi^+) = 1/2
$$

in agreement with experiments. In conclusion, the isospin of the Ξ^* is $1/2$.

The analysis of the angular distributions determines the spin-parity as $J^P = 3/2^+$.

Question 4.1   For each value of the isospin and for every value of its third component, list the charge states of the $\Xi\pi$ system. Use the Gell-Mann and Nishijima formula (3.47).

This concludes the discussion of the $J^P = 3/2^+$ strongly decaying baryons, which are summarised in Fig. 4.6. Comparison with the $J^P = 1/2^+$ baryons (Fig. 3.1) shows

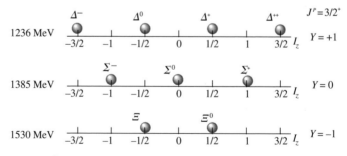

Fig. 4.6.   The $J^P = 3/2^+$ baryons with strong decay.

the deep difference between the two cases, another feature explained by the quark model.

4.3 The Dalitz plot

In Section 1.6 we calculated the phase space volume for a two-particle final state. The next more complicated case is the three-particle state. Several important processes, both decays and collisions, have such final states. We shall now discuss the kinematics and the phase space of the three-particle system with masses m_1, m_2 and m_3. Whether it is the final state of a decay or of a collision is obviously immaterial.

We treat the problem in the centre of mass reference frame, where things are simpler. We call the centre of mass energy M, the mass of the mother particle in the case of a decay, and we call $\mathbf{p}_1$, $\mathbf{p}_2$ and $\mathbf{p}_3$ the momenta and E_1, E_2 and E_3 the energies of the three particles. The constraints among these variables are the following. Since the masses are given, the energies are determined by the momenta. The nine components of the momenta must satisfy three conditions for momentum conservation, $\mathbf{p}_1 + \mathbf{p}_2 + \mathbf{p}_3 = \mathbf{0}$ and one condition for energy conservation, $E_1 + E_2 + E_3 = M$. We are left with five independent variables.

Since the three momenta add up to the null vector, they are coplanar. Let $\mathbf{n}$ be the unit vector normal to this plane. We choose two of the independent variables as the two angles that define the direction of $\mathbf{n}$. The triangle defined by the three momenta can rotate rigidly in the plane; we take the angle that defines the orientation of the triangle as the third variable.

The last two variables define the shape and the size of the triangle. If we are not interested in the polarisation of the initial state, the case to which we limit our discussion, the dependence on the angles of the matrix element is irrelevant and we can describe the final state with the last two variables only. There are a few equivalent choices: two energies, say E_1 and E_2, two kinetic energies T_1 and T_2 and the masses squared of two pairs. These variables are linked by linear relationships. Take for example m_{23}^2:

$$
\begin{aligned}
m_{23}^2 &= (E_2 + E_3)^2 - (\mathbf{p}_2 + \mathbf{p}_3)^2 = (M - E_1)^2 - \mathbf{p}_1^2 \\
&= M^2 + m_1^2 - 2ME_1.
\end{aligned}
\tag{4.18}
$$

In conclusion, a configuration of our three-particle system can be represented in a plane defined by, say, m_{12}^2 and m_{13}^2, as in Fig. 4.7. The loci of the configurations with $m_{23}^2 = $ constant are straight lines such as the one shown in Fig. 4.7.

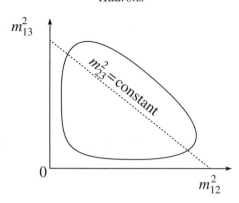

Fig. 4.7.   The Dalitz plot.

The closed line, the contour of the plot, delimits the region allowed by energy and momentum conservation. The diagram goes by the name of R. H. Dalitz (Dalitz 1956), who pointed out that the elements of its area are proportional to the phase space volume. We have already mentioned this property and we shall now prove it.

From (1.47), the phase space volume for a three-body system, ignoring constant factors that are irrelevant here, is

$$R_3 \propto \int \frac{d^3p_1}{E_1} \frac{d^3p_2}{E_2} \frac{d^3p_3}{E_3} \delta(E_1 + E_2 + E_3 - M)\delta^3(\mathbf{p}_1 + \mathbf{p}_2 + \mathbf{p}_3). \quad (4.19)$$

Integrating on $\mathbf{p}_3$, we obtain

$$R_3 \propto \int \frac{1}{E_1 E_2 E_3} d^3p_1 d^3p_2 \, \delta(E_1 + E_2 + E_3 - M). \quad (4.20)$$

We must now integrate over the angles made by the two vectors $\mathbf{p}_1$ and $\mathbf{p}_2$. The following choice is convenient. We fix the angle θ_{12} between $\mathbf{p}_1$ and $\mathbf{p}_2$ and integrate over θ_1, ϕ_1 and ϕ_2. We have

$$R_3 \propto \int \frac{1}{E_1 E_2 E_3} 4\pi p_1^2 \, dp_1 \, 2\pi p_2^2 \, dp_2 \, d(\cos\theta_{12})\delta(E_1 + E_2 + E_3 - M). \quad (4.21)$$

We now use the momentum conservation $\mathbf{p}_3 = -\mathbf{p}_1 - \mathbf{p}_2$, which gives $p_3^2 = p_1^2 + p_2^2 + 2p_1p_2 \cos\theta_{12}$. We differentiate this expression keeping p_1 and p_2 constant and obtain $2p_3 \, dp_3 = 2p_1p_2 \, d(\cos\theta_{12})$. By substituting into (4.21), we obtain

$$R_3 \propto \int \frac{p_1 \, dp_1 \, p_2 \, dp_2 \, p_3 \, dp_3}{E_1 E_2 E_3}\delta(E_1 + E_2 + E_3 - M). \quad (4.22)$$

Differentiating the relationships $E_i^2 = p_i^2 + m_i^2$ we have $p_i \, dp_i = E_i \, dE_i$, hence

$$R_3 \propto \int dE_1 \, dE_2 \, dE_3 \, \delta(E_1 + E_2 + E_3 - M). \qquad (4.23)$$

Finally, using the remaining δ-function we arrive at the conclusion

$$R_3 \propto \int dE_1 \, dE_2 \propto \int dm_{23}^2 \, dm_{13}^2 \propto \int dT_1 \, dT_2. \qquad (4.24)$$

This is what we had to prove. The expressions in the last two members are obvious consequences of the linear relationships between all these pairs of variables.

In the next section, we shall employ the Dalitz plot for the spin and parity analysis of three-pion systems. Since the three particles are equal, the plot is geometrically symmetrical. Actually, R. H. Dalitz invented the plot for his analysis of the decay of the K meson into 3π. Let us look at his reasoning.

The sum of the three centre of mass kinetic energies $T_1 + T_2 + T_3$ is the same for every 3π configuration. Now let us consider a triangle, which we take to be equilateral because the three particles have the same mass. The sum of the distances from the sides is the same for every point inside the triangle. Therefore, if the kinetic energies are measured by these distances, the energy conservation is automatically satisfied. The momentum conservation limits the allowed region inside a closed curve, which is tangent to the three sides. The diagram, shown in Fig. 4.8, is equivalent to that in Fig. 4.7. The former explicitly shows the symmetry of the problem.

This diagram is extremely useful in the study of the quantum numbers of the mesons that decay into 3π. Actually, the dependence of the decay matrix element

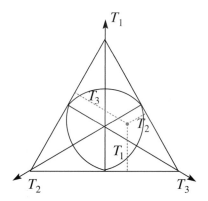

Fig. 4.8.   The 3π Dalitz plot.

on the position of the representative point in the graph is determined by the angular momentum, the parity and the isospin of the system. In practice, it is necessary to collect a sizeable number of decays of the meson to be studied. Each event is represented as a dot on the Dalitz plot. Provided there are sufficient statistics, every non-uniformity in the point density has to be ascribed to a corresponding variation of the matrix element and the quantum numbers can be determined.

4.4 Spin, parity, isospin analysis of three-pion systems

Three important mesons decay into 3π: the K that we have already met, the η and the ω, which we shall study in the following sections. In each case, the 3π final system is in a well-defined spin-parity and isospin state. The corresponding symmetry conditions of the decay matrix element lead to observable characteristics of the event distributions in the Dalitz plot and hence to the determination of the quantum numbers of the final state. It is important to note that these are not necessarily those of the decaying mesons. If the decay is strong, as for the ω, all quantum numbers are conserved; if it is electromagnetic, the isospin is not conserved, as is the case for the η; if it is weak, parity is also violated, as is the case for the K meson. Historically, these were the considerations that led Lee and Yang to the hypothesis of parity non-conservation.

A complete treatment of the methods of analysis for obtaining the quantum numbers of a 3π system was made by C. Zemach (Zemach 1964). Here, we shall limit our discussion to the simplest spin-parity assignments. We shall focus on the most conspicuous features, namely the regions of the Dalitz plot where this assignment predicts vanishing density.

We shall now see how to construct a general matrix element $\mathcal{M}$ for the 3π decay of a meson of mass M, taking into account the constraints imposed by isospin, spin and parity and by the Bose statistics. The matrix element $\mathcal{M}$ is Lorentz-invariant, but it can be shown that it reduces to a three-dimensional invariant in the centre of mass system. Therefore, we shall work in the centre of mass reference frame.

Let us begin by dealing with the isospin. The isospin of a 3π system can be any integer number between 0 and 3. As mesons with isospin larger than 1 have not been found, we shall limit our discussion to two possibilities, 0 and 1.

The three pions can have several charge states. We shall consider only those in which all pions are charged or two are charged and one is neutral. Actually, these are the states that can be observed in a bubble chamber, the instrument that made the greatest contribution to hadron spectroscopy. There are three such charge combinations: $\pi^+\pi^+\pi^-$ and its charge conjugate $\pi^-\pi^-\pi^+$, which have isospin 1, and $\pi^+\pi^-\pi^0$ that may have isospin 0 or 1.

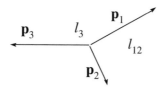

Fig. 4.9. Kinematic variables in a 3π system.

We now list the elements at our disposal to build $\mathcal{M}$. These must be covariant quantities in three dimensions (scalars, pseudoscalars, vectors and axial vectors).

We start by choosing an order π_1, π_2 and π_3 of the pions and call their momenta in the centre of mass reference frame $\mathbf{p}_1$, $\mathbf{p}_2$ and $\mathbf{p}_3$. These give us two independent vectors, taking into account the relationship

$$\mathbf{p}_1 + \mathbf{p}_2 + \mathbf{p}_3 = 0. \qquad (4.25)$$

Figure 4.9 shows the variables we shall use, in the plane of the three momenta. There is one axial vector, which is normal to the plane

$$\mathbf{q} = \mathbf{p}_1 \times \mathbf{p}_2 = \mathbf{p}_2 \times \mathbf{p}_3 = \mathbf{p}_3 \times \mathbf{p}_1. \qquad (4.26)$$

There are four scalar quantities: one is simply a constant, the others are the energies E_1, E_2 and E_3 which are linked by the relationship

$$E_1 + E_2 + E_3 = M. \qquad (4.27)$$

We choose, arbitrarily, two pions, π_1 and π_2, which we call the 'dipion'. We call I_{12} and I_3 the isospin of the dipion and of π_3 respectively. I_{12} can be 0, 1 or 2, while obviously $I_3 = 1$. The total isospin is

$$\mathbf{I} = \mathbf{I}_{12} \otimes \mathbf{I}_3. \qquad (4.28)$$

Similarly, let l_{12} be the orbital angular momentum of π_1 and π_2 (in their centre of mass frame). This is also the total angular momentum of the dipion, because the pions have no spin. Let l_3 be the orbital angular momentum between π_3 and the dipion. The total angular momentum is

$$\mathbf{J} = \mathbf{l}_{12} \otimes \mathbf{l}_3. \qquad (4.29)$$

Now, let us observe how the spin-parity of a 3π system cannot be $J^P = 0^+$. Actually, $J = 0$ implies $l_{12} = l_3$. The total parity is the product of the three intrinsic parities and the two orbital parities, namely $P = (-1)^3 (-1)^{l_{12}} (-1)^{l_3} = -1$. Hence $J^P = 0^+$ is impossible.

We shall limit this discussion to the three simplest spin-parity assignments: $J^P = 0^-$, 1^- and 1^+. This makes six cases in total, taking into account the two possible isospins $I = 0$ and $I = 1$.

For an assumed spin-parity J^P of the three-pion system, taking into account that their intrinsic parity is -1, the space part of the amplitude, which we must construct, transforms as J^{-P}. For each isospin choice, we must impose the corresponding symmetry properties under exchanges of two pions.

Firstly, let us consider $I = 0$. In this case, it must be $I_{12} = 1$. This is true for every choice of π_3. Consequently, the matrix element is antisymmetric in the exchange of every pair (completely antisymmetric). Consider now the three J^P cases.

$J^P = 0^-$ We must construct a completely antisymmetric scalar quantity. We use the energies. To obtain $1 \Leftrightarrow 2$ antisymmetry we write $E_1 - E_2$. Then we antisymmetrise completely

$$\mathcal{M} \propto (E_1 - E_2)(E_2 - E_3)(E_3 - E_1). \tag{4.30}$$

The vanishing density regions, where $\mathcal{M}$ is zero, are all the diagonals as shown in Fig. 4.10(a).

$J^P = 1^-$ We need an axial vector. We have only one of them, which, as required, is already antisymmetric. Hence

$$\mathcal{M} \propto \mathbf{q}. \tag{4.31}$$

At the periphery of the Dalitz plot, i.e. the kinematic limit, two momenta are parallel and $\mathbf{q} = 0$. The situation is shown in Fig. 4.10(b).

$J^P = 1^+$ We construct a completely antisymmetric vector. We take one of the vectors, $\mathbf{p}_1$, and make it antisymmetric in $2 \Leftrightarrow 3$ multiplying by $E_2 - E_3$. We then antisymmetrise completely

$$\mathcal{M} \propto \mathbf{p}_1(E_2 - E_3) + \mathbf{p}_2(E_3 - E_1) + \mathbf{p}_3(E_1 - E_2). \tag{4.32}$$

The three energies are equal at the centre of the plot, hence $\mathcal{M} = 0$. Consider the vertex of a diagonal, for example that corresponding to T_3. Here $\mathbf{p}_1 = \mathbf{p}_2 = -\mathbf{p}_3/2$, hence $E_1 = E_2$ and $\mathcal{M} = 0$. The result is shown in Fig. 4.10(c).

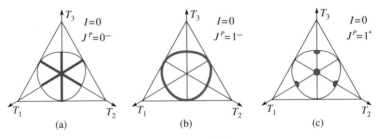

Fig. 4.10.   Vanishing density regions for $\pi^0 \pi^+ \pi^-$ with $I = 0$.

Let us now proceed to $I=1$. We have two charge states to consider. $\pi^+\pi^+\pi^-$ (and its charge conjugate). Let us take as π_3 the different one, $\pi_3 = \pi^-$. Since the other two are identical, the amplitude is symmetric under the exchange $1 \Leftrightarrow 2$. $\pi^0\pi^+\pi^-$. We take $\pi_3 = \pi^0$. First we see that $I_{12}=1$ is forbidden. If this were possible, the isospin of the dipion would be $|I_{12}, I_{12,z}\rangle = |1,0\rangle$ and for π_3, obviously, $|I_3, I_{3z}\rangle = |1,0\rangle$; these should total $|I, I_z\rangle = |1,0\rangle$. However, this cannot be, since the corresponding Clebsch–Gordan coefficient is zero. We are left with $I_{12}=0$ or $I_{12}=2$. In both cases the state is symmetric under the exchange $1 \Leftrightarrow 2$.

To sum up, the amplitude must be symmetric under the exchange $1 \Leftrightarrow 2$ in both cases.

Now, let us move on to the three spin-parities.

$J^P = 0^-$ We need a scalar, symmetric in $1 \Leftrightarrow 2$. The simplest are E_3 and a constant.

$$\mathcal{M} \propto \text{constant} \quad \mathcal{M} \propto E_3. \tag{4.33}$$

In both cases, there are no vanishing density regions. See Figure 4.11(a).

$J^P = 1^-$ We need an axial vector, symmetric in $1 \Leftrightarrow 2$. The only axial vector that we have, $\mathbf{q}$, is antisymmetric. We make it symmetric by multiplying it by an antisymmetric (in $1 \Leftrightarrow 2$) scalar quantity

$$\mathcal{M} \propto \mathbf{q}(E_1 - E_2). \tag{4.34}$$

The vanishing points are on the periphery ($\mathbf{q}$) and on the vertical diagonal ($E_1 - E_2$). See Fig. 4.11(b).

$J^P = 1^+$ We need a vector, symmetric in $1 \Leftrightarrow 2$. We take

$$\mathcal{M} \propto \mathbf{p}_1 + \mathbf{p}_2 = -\mathbf{p}_3. \tag{4.35}$$

It is zero at the foot of the vertical diagonal where $T_3 = 0$, hence $\mathbf{p}_3 = 0$. See Figure 4.11(c).

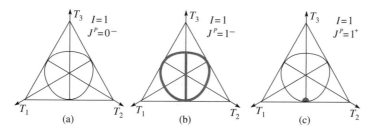

Fig. 4.11.   Vanishing density regions for the states $\pi^+\pi^+\pi^-$ and $\pi^0\pi^+\pi^-$ with $I=1$.

4.5 Pseudoscalar and vector mesons

Figure 3.2 shows the pseudoscalar mesons, i.e. those with spin-parity 0^-, that we have discussed so far. We have also discussed the measurement of the spin and parity of the charged pion. We shall now discuss the spin and parity charged K meson. Two more pseudoscalar mesons exist, called η and η', that we shall discuss shortly. Counting all their charge states, the pseudoscalar mesons are nine in number.

There are as many vector mesons, namely with spin-parity 1^-, forming isospin multiplets identical to those of the pseudoscalar mesons. We shall deal with only one of these, the ω, and simply list the quantum numbers of the others.

Let us go back, historically, to the early 1950s, when cosmic ray experiments had shown several decay topologies of strange particles of similar masses and lifetimes. This was initially interpreted as evidence of three particles. One of them, decaying into 2π, was called θ; the second, decaying into 3π, was called τ; and the third with a number of different decays was called K. The situation was clarified by the G-stack (great-stack) experiment, proposed by M. Merlin in 1953. A 15 litre emulsion stack was flown at 27 000 m height on an aerostatic balloon over the Po valley. A huge analysis and measurement effort followed, but the results were rewarding. The different decay modes were clearly identified and the masses accurately measured. It became clear that different decay modes of the same particle were being observed (Davies *et al.* 1955).

Over the same years, the Bevatron became operational at Berkeley, providing the first K^+ beams. Emulsion exposures allowed accurate measurements of the masses of the strange mesons. The masses of the θ and τ mesons were equal, to within a few parts per thousand. It was impossible to escape the conclusion that θ and τ were the same particle, now called the K meson.

Figure 4.12 shows the Dalitz plot of 220 τ events (i.e. $K^\pm \to \pi^\pm + \pi^+ + \pi^-$) collected from different experiments. The spin-parity analysis was originally done by R. H. Dalitz (Dalitz 1956). We profit from the conclusions of Section 4.4, in particular with regard to Figs. 4.10 and 4.11, and compare the vanishing density regions shown in those figures with the data in Fig. 4.12. The distribution of the data being uniform without any depletion, we conclude that the 3π state has $I = 1$ and $J^P = 0^-$.

However, the K^+ (the θ events) also decays into two pions $K^+ \to \pi^+ + \pi^0$. In this case if $J = 0$ the parity must be positive. The problem became known as the $\theta-\tau$ puzzle. The puzzle was solved by T. D. Lee and C. N. Yang (Lee & Yang 1956) who made the revolutionary hypothesis that the weak interactions violate parity conservation and by the experiments that immediately confirmed this hypothesis. We shall discuss this issue in Chapter 7.

We conclude by observing that the K^+ decay also violates the isospin conservation, by $\Delta I = 1/2$.

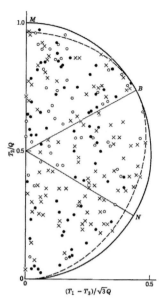

Fig. 4.12. Dalitz plot for 'τ events'. The kinetic energies are divided by the Q-value of the decay. The diagram is folded about the vertical axis, because of two equal pions. (Orear *et al.* 1956)

We now turn to the η meson. It was discovered by M. Bloch, A. Pevsner and collaborators (Pevsner *et al.* 1961) in the reaction

$$\pi^+ + d \rightarrow \pi^+ + \pi^- + \pi^0 + p + p \tag{4.36}$$

with the 72'' Alvarez bubble chamber filled with liquid deuterium and exposed to a π^+ beam with 1.23 GeV momentum. Figure 4.13(a) shows the $\pi^+\pi^-\pi^0$ mass distribution. Two resonances are clearly seen: the η at a lower mass, the ω, which we shall discuss shortly, at a higher mass.

No charged state of the η has ever been observed, hence its isospin $I = 0$.

The mass and the width are

$$m_\eta = 548\,\text{MeV} \qquad \Gamma_\eta = 1.3\,\text{MeV}. \tag{4.37}$$

Very soon after, the typically electromagnetic decay $\eta \rightarrow 2\gamma$ was observed to happen with a probability similar to that of the decay into $\pi^+\pi^-\pi^0$. The present values of the branching ratios are

$$\Gamma\left(\eta \rightarrow \pi^+ + \pi^- + \pi^0\right)/\Gamma_{\text{tot}} = 28\% \qquad \Gamma(\eta \rightarrow 2\gamma)/\Gamma_{\text{tot}} = 39.4\%. \tag{4.38}$$

The 2γ decay establishes that the charge conjugation of the η is $C = +1$. Since $I = 0$, the G-parity is $G = C = +1$. However, the 3π final state has $G = -1$, hence $I = 1$. Therefore, this decay violates the isospin and cannot be strong.

Turning to the spin-parity, Fig. 4.13(b) shows the Dalitz plot. We compare it with Fig. 4.11, which is relevant for $I = 1$. Again, the only case that has no zeros, in agreement with the uniform experimental distribution, is $J^P = 0^-$. We understand why the G-conserving decay into 2π is forbidden, namely because 2π cannot have $J^P = 0^-$. On the other hand, the decay into 4π is forbidden by energy conservation.

The η' meson. Without entering into any detail, we simply state that another pseudoscalar, zero-isospin, meson exists (Kalbfleish *et al.* 1964, Goldberg *et al.* 1964). It is called η' and has the same quantum numbers as the η. Its mass and its width are

$$m_{\eta'} = 958 \,\text{MeV} \qquad \Gamma_{\eta'} = 0.2 \,\text{MeV}. \qquad (4.39)$$

The η' meson has like the η, important electromagnetic decays (2γ, $\omega\gamma$, $\rho\gamma$) and a small width. Surprisingly, its mass is enormous, when compared to the pion.

The ω meson was discovered in 1961 in the reaction

$$\bar{p} + p \rightarrow \pi^+ + \pi^+ + \pi^0 + \pi^- + \pi^- \qquad (4.40)$$

in the 72″ Alvarez hydrogen bubble chamber exposed to the antiproton beam of the Bevatron (Maglic *et al.* 1961).

Note that, for every event, four pion triplets with zero charge exist, four with unit (both signs) charge and two of double charge. Figure 4.14(a) shows the distributions of these three charge state triplets. A narrow resonance clearly appears in the neutral combination, not in the charged ones. This fixes the isospin, $I = 0$. The ω mass and width are

$$m_\omega = 782 \,\text{MeV} \qquad \Gamma_\omega = 8 \,\text{MeV}. \qquad (4.41)$$

Figure 4.14(b) shows the Dalitz plot of the $\pi^+\pi^-\pi^0$ combinations, chosen so as to have a mass in the peak region. Notice that this sample not only includes ω decays, but also background events. The same figure shows the radial distribution of the events, in radial zones of equal areas. The curve is the square of the (4.31) matrix element summed to a background, assumed to be constant. The agreement establishes $J^P = 1^-$.

We shall now consider the other vector mesons ($J^P = 1^-$) without entering into any detail, but only summarising their properties. The isospin multiplets of the vector mesons are equal to those of the pseudoscalar mesons.

The ρ meson has three charge states, ρ^+, ρ^0 and ρ^- and isospin $I = 1$. It decays mainly into 2π, hence $G = +1$. It follows that the neutral state, ρ^0 has $C = -1$. Its mass is $m \approx 770 \,\text{MeV}$ and its width is $\Gamma \approx 150 \,\text{MeV}$ (Erwin *et al.* 1961).

The K^* and $\bar{K}^*$ mesons. The charge states of the former are $+1$ and 0, those of the latter -1 and 0. Therefore, both have isospin $I = 1/2$. They decay into $K\pi$ and

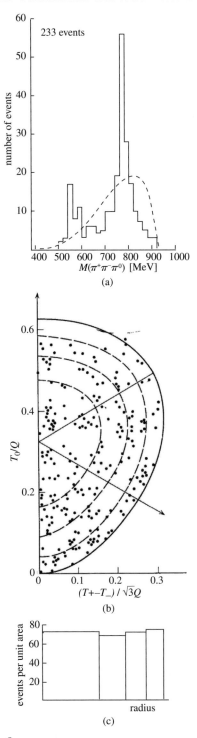

Fig. 4.13. (a) $\pi^+\pi^-\pi^0$ mass distribution. The dotted curve is the phase space (Pevsner *et al.* 1961); (b) Dalitz plot of the decay $\eta \rightarrow \pi^+\pi^-\pi^0$; (c) events in equal area zones. (Alff *et al.* 1962)

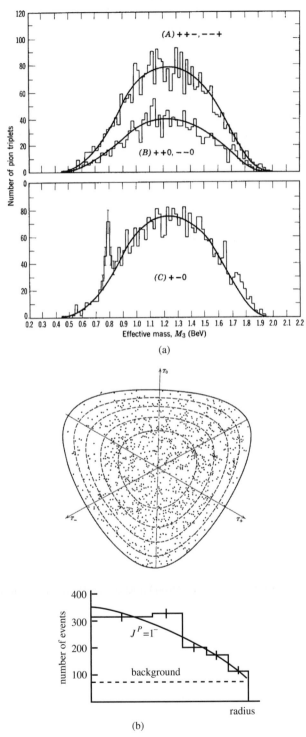

Fig. 4.14.   (a) Mass distributions of the triplets with charge 0, 1 and 2. (Maglic *et al.* 1961) (b) Dalitz plot of the decay $\omega \to \pi^+\pi^-\pi^0$ and numbers of events in equal area zones. (Alff *et al.* 1962 © Nobel Foundation 1968)

$\bar{K}\pi$ respectively. The mass of both is $m \approx 892$ MeV and the width is $\Gamma \approx 50$ MeV (Alston *et al.* 1961).

The ϕ meson has the same quantum numbers as the ω meson, $J^{PC} = 1^{--}$, $I^G = 0^-$. Its mass is $m = 1019.5$ MeV. Its width is very small, considering its large mass, $\Gamma = 4.3$ MeV (Connoly *et al.* 1963).

An important related feature is the following. Clearly, the energetically favourite decay channel is the same as that of the ω, namely $\phi \to 3\pi$. However, in this case the branching ratio is small

$$\Gamma(\phi \to 3\pi)/\Gamma_{\text{tot}} = 15.6\% \qquad (4.42)$$

while

$$\Gamma(\phi \to K\bar{K})/\Gamma_{\text{tot}} = 83\%. \qquad (4.43)$$

We experimentally observe that the ϕ meson 'prefers' to decay into K mesons, even if its mass is barely large enough to allow these decays, $Q = 32$ MeV. This fact hinders the decay and makes the ϕ lifetime long by strong interaction standards. There are two reasons for this behaviour: the quark content of the ϕ, which is mainly $s\bar{s}$, and a dynamical QCD property, as we shall see in Chapter 6.

4.6 The quark model

We shall now summarise the hadronic states we have met.

- Nine pseudoscalar mesons in two $SU(2)$ singlets, two doublets, one triplet.
- Nine vector mesons in the same multiplets.
- Eight $J^P = 1/2^+$ baryons (and as many antibaryons) in two doublets, with hypercharge $Y = +1$ and $Y = -1$ respectively and a triplet and a singlet with $Y = 0$.
- Nine $J^P = 3/2^+$ baryons in the multiplets: a quartet with $Y = +1$, a triplet with $Y = 0$ and a doublet with $Y = -1$. As we shall see immediately, we are still missing one hyperon, a singlet with $Y = -2$.

In 1964, G. Zweig at CERN (Zweig 1964) and M. Gell-Mann in the USA (Gell-Mann 1964) independently proposed that the hadrons are made up of constituents, called quarks by Gell-Mann. Baryons are made of three quarks, the mesons of a quark and an antiquark. To be precise, we are dealing with the so-called 'valence' quarks. Actually, the internal structure of the hadrons also has other elements, as we shall see in Chapter 6.

This scheme extends the isospin internal symmetry, which is based on the group $SU(2)$ to $SU(3)$, a larger unitary group. We immediately stress that the $SU(3)$ symmetry has two very different roles in subnuclear physics: (1) the classification

Table 4.1 *Quantum numbers and masses of the three lowest-mass quarks*

	Q	I	I_z	S	C	B	T	$\mathcal{B}$	Y	mass
d	$-1/3$	$1/2$	$-1/2$	0	0	0	0	$1/3$	$1/3$	3–7 MeV
u	$+2/3$	$1/2$	$+1/2$	0	0	0	0	$1/3$	$1/3$	1.5–3.0 MeV
s	$-1/3$	0	0	-1	0	0	0	$1/3$	$-2/3$	95 ± 25 MeV

of the hadrons, or rather the hadrons with up (*u*), down (*d*) and strange (*s*) valence quarks, which we are considering here; (2) the symmetry of the charges of one of the fundamental forces, the strong force, as we shall discuss in Chapter 6. The two roles are completely different and, to avoid confusion, we shall call the former $SU(3)_f$, where the suffix stands for 'flavour', even if mathematically the group is the same in both cases.

We now know six different quarks, each with a different flavour. The two quarks present in normal matter, *u* and *d*, are an isospin doublet, the former with third component $I_z = +1/2$, the latter with $I_z = -1/2$. We can say that I_z is the flavour of each of them. The flavour of *s* is the strangeness, which is *negative*, with the value $S = -1$. The other three quarks that we shall meet in the following sections are the charm quark (*c*), with flavour which is also called charm, with $C = +1$; the beauty quark (*b*), with flavour called beauty, $B = -1$; and the top quark (*t*) with flavour $T = +1$. The sign convention is that the sign of the quark flavour is the same as that of its electric charge.

Table 4.1 gives the quantum numbers of the first three quarks and their masses. Their antiquarks have opposite values of all the quantum numbers, but isospin *I*.

The characteristics of the quarks are surprising. Their electric charges and their baryon number are fractional. More surprisingly, nobody has ever succeeded in observing a free quark. They have been sought in violent collisions of every type of beam, hadrons, muons, neutrinos, photons, etc. on different targets; they have been looked for in cosmic rays; they were searched for with Millikan-type experiments on ordinary matter and even on rocks brought by the astronauts from the Moon. No quark was ever found. We know today that the reason for this is in the very nature of the colour force that binds the hadrons, as we shall see in Chapter 6.

The $SU(3)_f$ representations, namely the multiplets in which the hadrons are grouped, are more complicated than in $SU(2)$. The latter are represented in one dimension, labelling the members by the third component of the isospin. See for example Figs. 3.1, 3.2 and 4.6. Note that $SU(2)$ has as a subgroup $U(1)$, to which the charges of the particles correspond.

Two variables are needed to represent an $SU(3)_f$ multiplet. Therefore, we draw them in a plane, taking as axes the third isospin component and the hypercharge.

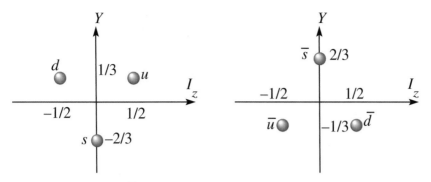

Fig. 4.15.   The **3** and **3̄** representations.

$SU(3)$ has $SU(2)$ as a subgroup and the $SU(3)_f$ multiplets have a substructure made up of $SU(2)$ multiplets.

There are two different fundamental representations, both with dimensions equal to three. They are called **3** and **3̄**. We use them to classify the quarks and the antiquarks respectively, as shown in Fig. 4.15. Both **3** and **3̄** contain an $SU(2)$ doublet and an $SU(2)$ singlet.

4.7 Mesons

Mesons are made of a quark and an antiquark. Therefore, they must be members of the multiplets belonging to the product of **3** by **3̄**. Group theory tells us that

$$\mathbf{3} \otimes \mathbf{\bar{3}} = \mathbf{1} \oplus \mathbf{8} \tag{4.44}$$

which gives us nine places, exactly the number we need. Let us look at the isospin and hypercharge structure, shown in Fig. 4.16. Notice that in the 'centre' of the octet there are two states with $I_z = Y = 0$, one with $I = 1$, one with $I = 0$. We see that the octet and the singlet provide us with exactly the isospin multiplets we need to classify the pseudoscalar and the vector mesons. If the $SU(3)_f$ were exact, all particles in a multiplet would have the same mass. In practice, there are sizeable differences. Unlike the $SU(2)$, which is good for the strong interactions and broken by the electromagnetic interactions, $SU(3)_f$ is already broken by the former.

Now consider the spin-parity of the mesons. We expect the quark–antiquark pair to be, for the ground-level mesons we are considering, an S wave, as is usual for ground states. In this hypothesis the configurations should be $^1S_0$ with $J^{PC} = 0^{-+}$ and $^3S_1$ with $J^{PC} = 1^{--}$. This is just what we observe.

Starting with the pseudoscalar mesons, we now have to lodge the nine mesons in the octet and in the singlet. We immediately recognise that $\pi^+ = u\bar{d}$, $\pi^- = d\bar{u}$,

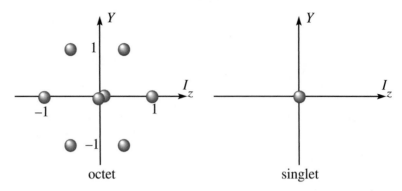

Fig. 4.16.   The octet and the singlet.

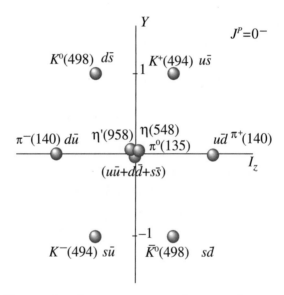

Fig. 4.17.   The pseudoscalar mesons nonet; the approximate values of their masses are in MeV.

$K^+ = u\bar{s}$, $K^0 = d\bar{s}$, $K^- = s\bar{u}$ and $\bar{K}^0 = s\bar{d}$, as shown in Fig. 4.17. We still have the three neutral mesons: π^0, η and η'. Since the π^0 is in an isospin triplet it must be in the octet. As such, it does not contain $\bar{s}s$, which has zero isospin. Moreover, it is antisymmetric in the exchange $\bar{u}u \leftrightarrow \bar{d}d$. Labelling the state with its $SU(3)_f$ and $SU(2)$ representations, **8** and **3**, we have

$$\left|\pi^0\right\rangle \equiv |\mathbf{8}, \mathbf{3}\rangle = \frac{1}{\sqrt{2}}\left|-\bar{u}u + \bar{d}d\right\rangle. \tag{4.45}$$

The $SU(3)_f$ singlet, which is also an isotopic singlet, is the following completely symmetric state

$$|\eta_1\rangle \equiv |\mathbf{1}, \mathbf{1}\rangle = \frac{1}{\sqrt{3}}|u\bar{u} + d\bar{d} + s\bar{s}\rangle \qquad (4.46)$$

which we have not yet identified with a meson. The third combination, namely the iso-singlet of the octet, must be orthogonal to the other two. Imposing this condition, one finds

$$|\eta_8\rangle \equiv |\mathbf{8}, \mathbf{1}\rangle = \frac{1}{\sqrt{6}}|\bar{u}u + \bar{d}d - 2\bar{s}s\rangle. \qquad (4.47)$$

We have two iso-singlets, η_1 and η_8, and two physical states, states η and η'. However, we cannot identify the latter with the former. Given that $SU(3)_f$ is broken already by strong interactions, the η and η' are not pure octet and pure singlet. Indeed, the apparently simple question of the flavour composition of the pseudo-scalar completely neutral mesons, η and η', turns out to be highly non-trivial. The reason is that for these states the colour interaction induces continuous transitions between quark–antiquark ($u\bar{u}$, $d\bar{d}$, $s\bar{s}$). A consequence is the large values of the η and η' masses. The correct description of these complicated mixtures of quark–antiquark pairs and gluons can be achieved only within QCD theory and cannot be given at an elementary level. We notice, however, that the experimental and theoretical work on this issue was extremely important in the historical develop-ment of QCD. As we shall see in Section 6.8, the QCD vacuum is indeed a very active medium, in which energy fluctuations continuously take place. Particularly important are the pseudoscalar fluctuations, which couple with η and η', strongly contributing to their mass.

The structure of the vector mesons is similar and is shown in Fig. 4.18. Here too, the symmetry is broken, as seen from the relevant differences between the masses. Differently from the pseudoscalar case, the two physical isospin singlets, the ω and the ϕ, are two orthogonal linear superpositions of the $SU(3)_f$ octet and singlet states. These superpositions turn out to be that one of the mesons, the ω, almost does not contain a valence s quark, while the other, the ϕ, is made up almost entirely of strange quarks, i.e.

$$\begin{aligned}|\omega\rangle &= \frac{1}{\sqrt{2}}|u\bar{u} + d\bar{d}\rangle \\ |\phi\rangle &= |s\bar{s}\rangle. \end{aligned} \qquad (4.48)$$

This explains, firstly, why the masses of the ρ and the ω are almost equal, secondly, why the 'hidden strangeness' ϕ meson decays preferentially into $K\bar{K}$, i.e. into final

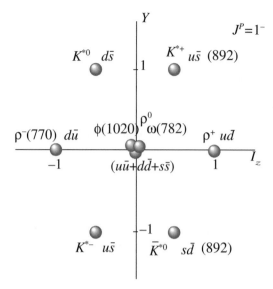

Fig. 4.18.   The vector mesons; the approximate values of the masses are in MeV.

states in which the s and $\bar{s}$ are still present. If the final state is non-strange, s and $\bar{s}$ must first annihilate into pure energy of the colour field; subsequently this energy must give rise to non-strange quark–antiquark pairs to produce, in turn, the observed hadronic final state, 3π for example. As we shall discuss in Chapter 6, QCD foresees that this process is severely suppressed.

Many other mesons exist beyond the ground-level pseudoscalar and vector mesons; their quantum numbers are compatible with those of a spin 1/2 particle–antiparticle pair in excited states with non-zero orbital momenta. In particular their spin-parities are amongst those in Table 3.1 but not, for example, $J^{PC} = 0^{+-}$, 0^{--}, 1^{-+}.

4.8 Baryons

Baryons are made of three quarks. Therefore, their classification in $SU(3)_f$ multiplets is less simple than that of mesons. The correspondence between a baryon and the quarks has no ambiguity if the three quarks are identical, for example uuu can only be Δ^{++} and ddd can only be Δ^{-}. However, uud can be p or Δ^{+}, uds can be Σ^0, Λ^0, $\Sigma^0(1385)$, or $\Lambda(1405)$. The physical states correspond to different exchange symmetries of the corresponding three quarks.

Let us start by dealing with three equal quarks. We are still missing a case, namely the baryon with three strange quarks sss. According to the quark model,

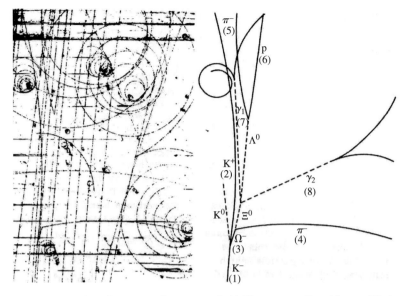

Fig. 4.19.   The first Ω^- event. (Barnes *et al.* 1964, courtesy Brookhaven National Laboratory)

a hyperon, called Ω^-, with strangeness $S = -3$ must exist. Since the isospin of s is zero, the Ω^- must be an isospin singlet. We can predict its mass by looking at Fig. 4.6. We see that the $SU(3)_f$ breaking is such that the masses of the decimet members increase by about 145 MeV for a decrease of strangeness of one unit. Therefore the Ω^- mass should be about 1675 MeV. But if this is true, it cannot decay strongly! Actually the lowest mass final state with $S = -3$ and $\mathcal{B} = 1$ is $\Xi^0 K^-$, which has a mass of 1809 MeV. In conclusion, the quark model was making a precise prediction: a metastable hyperon with the mentioned quantum numbers and mass should exist and must have escaped detection.

The search started immediately and the particle was discovered in a bubble chamber at the Brookhaven National Laboratory in 1964 (Barnes *et al.* 1964). The reaction was

$$K^- + p \to \Omega^- + K^+ + K^0 \tag{4.49}$$

which conserves the strangeness. The information provided by the bubble chamber is so complete that a single event was sufficient for the discovery. It is shown in Fig. 4.19, which we now analyse.

The track of the Ω^- terminates where the hyperon decays as

$$\Omega^- \to \Xi^0 + \pi^-. \tag{4.50}$$

This weak decay violates the strangeness by one unit, $\Delta S = -1$. Since the Ξ^0 is neutral it does not leave a track, decaying, again weakly with $\Delta S = -1$, as

$$\Xi^0 \rightarrow \Lambda + \pi^0. \tag{4.51}$$

Two other decays follow: a weak one, again with $\Delta S = -1$

$$\Lambda \rightarrow p + \pi^- \tag{4.52}$$

and an electromagnetic one

$$\pi^0 \rightarrow 2\gamma. \tag{4.53}$$

Finally, both γs materialise into two electron–positron pairs (a rather lucky circumstance in a hydrogen bubble chamber). Having determined the momenta of all the charged particles by measuring their tracks, a kinematic fitting procedure is performed by imposing the energy-momentum conservation at each decay vertex. In this way, the event is completely reconstructed. The resulting mass and lifetime are

$$m = 1672 \, \text{MeV} \qquad \tau = 82 \, \text{ps} \tag{4.54}$$

in perfect agreement with the prediction.

The discovery of the Ω^- marked a triumph of the quark model, but at the same time aggravated an already existing problem. Three baryons, Δ^{++}, Δ^- and Ω^- are composed of three identical quarks. Since they are the ground states, the orbital momenta of the quarks must be zero. Therefore, the spatial part of their wave function is symmetric. Their spin wave function is also symmetric, the total spin being 3/2. In conclusion, we have states of three equal fermions that are completely symmetric, in contradiction with the Pauli principle. The solution of this puzzle led to the discovery of colour: there are three quarks for each flavour, each with a different colour charge, called red, green and blue.

With the colour degree of freedom, the wave function is the product of four factors

$$\Psi = \psi_{\text{space}} \psi_{\text{spin}} \psi_{SU(3)_f} \psi_{\text{colour}}. \tag{4.55}$$

The Pauli principle requires this product to be antisymmetric. We must now anticipate that the baryon colour wave function ψ_{colour} is *antisymmetric* in the exchange of every quark pair. As we have already said, the baryons we are considering, both those with $1/2^+$ and those with $3/2^+$, are the ground states and their quarks are in an S wave. Therefore, ψ_{space} is symmetric. It follows that $\psi_{\text{spin}} \psi_{SU(3)_f}$ must be symmetric

$$\psi_{\text{spin}} \psi_{SU(3)_f} = \text{symmetric}. \tag{4.56}$$

Table 4.2 *The symmetries of the states resulting from the combination of three spin 1/2*

$J=3/2$	S	$\uparrow\uparrow\uparrow$	$\frac{1}{\sqrt{3}}(\uparrow\uparrow\downarrow+\uparrow\downarrow\uparrow+\downarrow\uparrow\uparrow)$	$\frac{1}{\sqrt{3}}(\downarrow\downarrow\uparrow+\downarrow\uparrow\downarrow+\uparrow\downarrow\downarrow)$	$\downarrow\downarrow\downarrow$
$J=1/2$	M,A		$\frac{1}{\sqrt{2}}(\uparrow\downarrow-\downarrow\uparrow)\uparrow$	$\frac{1}{\sqrt{2}}(\uparrow\downarrow-\downarrow\uparrow)\downarrow$	
$J=1/2$	M,S		$\frac{1}{\sqrt{6}}(\uparrow\downarrow+\downarrow\uparrow)\downarrow-\sqrt{\frac{2}{3}}\uparrow\uparrow\downarrow$	$-\frac{1}{\sqrt{6}}(\uparrow\downarrow+\downarrow\uparrow)\uparrow-\sqrt{\frac{2}{3}}\downarrow\downarrow\uparrow$	
J_z		$+\frac{3}{2}$	$+\frac{1}{2}$	$-\frac{1}{2}$	$-\frac{3}{2}$

This not only solves the paradox of both ψ_{spin} and $\psi_{SU(3)f}$ being symmetric for three equal quarks, but also tells us much more. Let us see.

We start by examining the symmetries of the combinations of three $SU(2)$ doublets, **2** (namely spin 1/2). We first take the product of two of them

$$\mathbf{2} \otimes \mathbf{2} = \mathbf{1_A} \oplus \mathbf{3_S} \tag{4.57}$$

where the suffixes indicate the symmetry. In other words, combining two 1/2 spins we obtain a state of total spin 0 that is antisymmetric, and one of total spin 1 that is symmetric. We proceed by taking the product of the result with the third doublet

$$\mathbf{2} \otimes \mathbf{2} \otimes \mathbf{2} = (\mathbf{1_A} \otimes \mathbf{2}) \oplus (\mathbf{3_S} \otimes \mathbf{2}) = \mathbf{2_{MA}} \oplus \mathbf{2_{MS}} \oplus \mathbf{4_S}. \tag{4.58}$$

Here **MA** means mixed-antisymmetric, namely antisymmetric in the exchange of two quarks, and similarly, **MS** stands for mixed-symmetric, which is symmetric in a two-quark exchange. Again, in other words, the spin 0 combined with the third spin 1/2 gives a spin 1/2 that is antisymmetric in the exchange of the first two, (**MA**). The spin 1 with the spin 1/2 gives a spin 1/2 that is symmetric in the exchange of the first two, (**MS**), and a spin 3/2 (**S**). The situation is spelled out in Table 4.2.

We should now proceed in a similar manner to combine three $SU(3)_f$ triplets, namely three representations **3**. We do not assume the reader to have enough knowledge of the group, and give only the result

$$\mathbf{3} \otimes \mathbf{3} \otimes \mathbf{3} = \mathbf{10_S} \oplus \mathbf{8_{MS}} \oplus \mathbf{8_{MA}} \oplus \mathbf{1_A}. \tag{4.59}$$

We can understand the structure by the following simple arguments.

- A three-quark system can always be made symmetric in the exchange of each pair (completely symmetric) (**S**). There are three different possibilities with equal quarks (*ddd*, *uuu*, *sss*), six with two equal and one different quarks (*ddu*, *dds*, *uud*, *uus*, *ssd*, *ssu*) and one with all the quarks different (*dus*); in total **10** states.
- Both the mixed-symmetric (**MS**) and mixed-antisymmetric (**MA**) combinations are possible only if at least one of the quarks is different. There are six

combinations with two equal and one different quarks. Each of these can be **MS**, as for example $(ud + du)u/\sqrt{2}$, or **MA** as $(ud - du)u/\sqrt{2}$. There are six different orderings of three all-different quarks (u, d, s). We have already used one of these combinations for the **S** states and we shall use a second one for the **A** states immediately. With the remaining four, two **MS** and two **MA** combinations can be arranged. Summing up, we have in total **8 MS** states and **8 MA** states.

- The three-quark system can be made completely antisymmetric (**A**) only if all of them are different; **1** state

$$[(sdu - sud + dus - uds + usd - dsu)/6].$$

We now use the symmetry properties we have found and combine the $SU(3)_f$ and spin multiplets (**SU3, Spin**) in order to fulfil the Pauli principle, namely Eq. (4.56). We have only two possibilities, i.e.
one spin 3/2 decimet $\psi^S_{SU(3)_f} \psi_{\text{spin}} \equiv (\mathbf{10_S}, \mathbf{4_S})$
one spin 1/2 octet

$$\frac{1}{\sqrt{2}} \left(\psi^{M,S}_{SU(3)_f} \psi^{M,S}_{\text{spin}} + \psi^{M,A}_{SU(3)_f} \psi^{M,A}_{\text{spin}} \right) \equiv (\mathbf{8_{MS}}, \mathbf{2_{MS}}) + (\mathbf{8_{MA}}, \mathbf{2_{MA}}).$$

These are precisely the multiplets observed in Nature! This result is far from being insignificant and goes beyond the quark model. Actually, the quark model alone would foresee the existence of all the multiplets (4.59). The observed restriction to an octet and a decimet is a consequence of the dynamics, namely the antisymmetry of ψ_{colour}, a fundamental characteristic of QCD. Notice, in particular, that an $SU(3)_f$ hyperon singlet, say a Λ_1, does not exist. The Λ is pure octet.

The octet of $1/2^+$ baryons is shown in Fig. 4.20, the $3/2^+$ decimet in Fig. 4.21.

We conclude this section with an important observation, looking back at the last column of Table 4.1 with the quark masses. As the quarks are never free, we cannot define their mass precisely, but we need to extend the concept of mass itself. This is possible only within a well-defined theoretical scheme. The Standard Model provides this scheme, as we shall see in Chapter 6.

Actually, one issue is already clear. For composite systems like atoms and nuclei the difference between the mass of the system and the sum of the masses of its constituents is small. On the contrary, the mass of the hadrons is enormously larger than the sum of the masses of their quarks. Notably, u and d have extremely small masses in comparison to the nucleon mass. The mass of the non-strange hadrons is predominantly energy of the colour field. As we shall see in Chapter 6, the colour force is independent of the flavour. This explains why, even if the d quark mass is twice as large as that of the u, isospin is a good symmetry. The $SU(3)_f$ case is similar, considering that the s-quark mass is small but not negligible in comparison

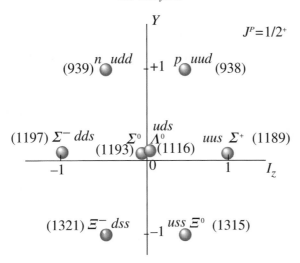

Fig. 4.20.   The baryon octet; in parentheses the masses in MeV.

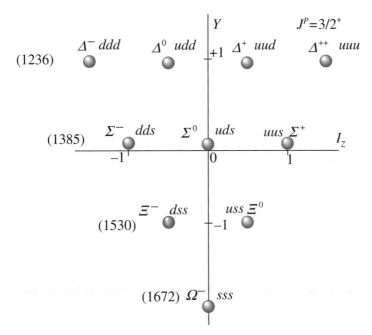

Fig. 4.21.   The baryon decimet; in parentheses the masses in MeV.

to the hadron masses. Therefore the latter symmetry is already broken at the level of the strong interactions. In conclusion, both isospin and $SU(3)_f$ symmetries, when considered at a fundamental level, appear to be accidental. This does not mean that they are less important.

4.9 Charm

Unlike for strangeness, the existence of hadrons with a fourth flavour, called 'charm' was theoretically predicted. However, charm was established by the observation of totally unexpected phenomena. S. Glashow, I. Iliopoulos and L. Maiani introduced in 1970 (Glashow *et al.* 1970) a mechanism, which became known as GIM, to explain the experimentally observed suppression of the 'neutral current' weak processes between quarks of different flavour that should have been faster by several orders of magnitude. We shall discuss the issue in Chapter 7. Here we shall consider the essential aspect of the problem in the following example. The decay $K^+ \to \pi^0 + v_e + e^+$ is a weak process, as is evident from the presence of a neutrino, and is called a 'charged current' process because the weak interaction, transforming a positive hadron into a neutral one, carries electrical charge. On the other hand, the decay $K^+ \to \pi^+ + v + \bar{v}$ is, for similar reasons, a weak 'neutral current' process. The two processes are very similar and should proceed with similar partial widths. However, the former decay is strongly suppressed, the measured branching ratios (Yao *et al.* 2006) being

$$
\begin{aligned}
\mathrm{BR}(K^+ \to \pi^+ v \bar{v}) &= \left(1.5^{+1.3}_{-0.9}\right) \times 10^{-10} \\
\mathrm{BR}(K^+ \to \pi^0 e^+ v_e) &= (4.98 \pm 0.07) \times 10^{-2}.
\end{aligned}
\tag{4.60}
$$

The GIM mechanism accounted for the suppression of such neutral current processes, introducing a fourth quark c carrying a new flavour, charm. The electric charge of c is $+2/3$.

A rough evaluation of the masses of the hypothetical charmed hadrons led to values around 2 GeV. Like strangeness, charm is conserved by the strong and electromagnetic interactions and is violated by the weak interactions. The lowest mass charmed mesons, the ground levels, have spin-parity 0^-. There are three such states, two non-strange mesons called D^+ and D^0 with valence quarks $c\bar{u}$ and $c\bar{d}$ respectively and one strange meson called D_s, that contains $c\bar{s}$. They have positive charm, $C = +1$. Their antiparticles, D^-, $\bar{D}^0$ and $\bar{D}_s$ have negative charm. These mesons decay weakly with lifetimes of the order of the picosecond. Their lifetimes are roughly an order of magnitude shorter than those of the K mesons, due to their higher mass.

As we shall see in Section 7.9, another clear prediction of the four-quark GIM model was that the mesons with positive charm would decay preferentially into negative strangeness final states, i.e. at the quark level, by $c \to s + \cdots$. In particular, the favoured decays of D^0 and D^+ are

$$
D^0 \to K^- \pi^+ \qquad D^+ \to K^- \pi^+ \pi^+
\tag{4.61}
$$

and similarly for their charge conjugates. While

$$D^0 \to K^+\pi^- \qquad D^0 \to \pi^+\pi^- \qquad D^+ \to K^+\pi^+\pi^- \qquad D^+ \to \pi^+\pi^+\pi^- \quad (4.62)$$

are strongly suppressed.

In 1974 S. Ting and collaborators at the AGS proton accelerator at Brookhaven (Aubert *et al.* 1974) built a spectrometer designed for the search for heavy particles with the same quantum numbers as the photon, $J^{PC} = 1^{--}$. They were not looking for charm. To search for particles of unknown mass one needs to explore a mass spectrum as wide as possible. With the assumed quantum numbers, the particle, which was to be called J, would decay into e^+ and e^-. The idea was to search for this decay in hadronic collisions. The overall reaction to search for, calling the non-detected part of the final state X, is

$$p + N \to J + X \to e^+ + e^- + X. \quad (4.63)$$

The spectrometer must then detect two particles of opposite sign. The main difficulty is that the two charged particles are almost always pions. Only once in a million or so might electrons be produced. Consequently, the spectrometer must provide a rejection power against hadrons of at least 10^8, and must measure the two momenta with high accuracy, both in absolute value (p_1 and p_2) and in direction (θ_1 and θ_1). The energies are known, assuming the particles to be electrons and the mass of the e^+e^- system is calculated using

$$m(e^+e^-) = \sqrt{2m_e^2 + 2E_1E_2 - 2p_1p_2 \cos(\theta_1 + \theta_2)}.$$

The particles sought should appear as peaks in this mass distribution.

Figure 4.22 shows a schematic of the spectrometer, which has two arms, one for the positive, one for the negative particles. In each arm, the measurement of the angle is decoupled from that of the momentum by having the dipole magnets bend

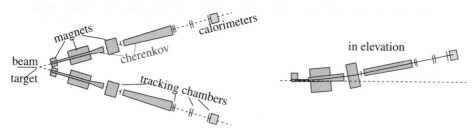

Fig. 4.22. The Brookhaven double-arm spectrometer. (Aubert *et al.* 1974 © Nobel Foundation 1976)

in the vertical plane. The range in m can be varied, changing the current intensity in the magnets and thus varying the acceptance for p_1 and p_2.

The necessary rejection power against pions is obtained with the following elements in each arm: (1) two threshold Cherenkov counters, designed to see the electrons, but not pions and heavier hadrons; (2) a calorimeter that measures the longitudinal profile of the shower, thus distinguishing electrons from hadrons. The spectrometer was designed to be able to handle very high particle fluxes, up to 10^{12} protons hitting the target per second.

Figure 4.23 shows the $m(e^+e^-)$ distribution, showing a spectacular resonance. Its mass is $m_J = 3100\,\text{MeV}$ and has the outstanding feature of being extremely narrow. Actually, its width is smaller than the experimental resolution and only an upper limit could be determined, $\Gamma < 5\,\text{MeV}$.

At the same times, the SPEAR e^+e^- collider, built by B. Richter and collaborators, was operational at the SLAC Laboratory. Its maximum energy was $\sqrt{s} = 8$ GeV. A general-purpose detector, called Mark I, had been completed with tracking chambers in a magnetic field and shower counters. After a period of collecting data at high energy, the decision was made to return to $\sqrt{s} \approx 3$ GeV, to check some anomalies that had previously been observed. The energy of the collider was varied in small steps, while measuring the cross sections of different processes. A huge resonance appeared with all the cross sections jumping up by more than two orders of magnitude (Augustin *et al.* 1974).

Figure 4.24 shows the cross sections of the processes

$$e^+e^- \rightarrow \text{hadrons} \qquad e^+e^- \rightarrow \mu^+\mu^- \qquad e^+e^- \rightarrow e^+e^- \qquad (4.64)$$

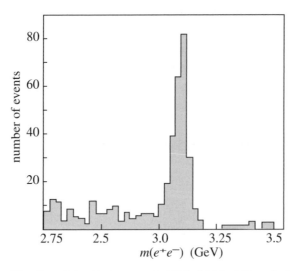

Fig. 4.23.   The *J* particle. (Aubert *et al.* 1974 © Nobel Foundation 1976)

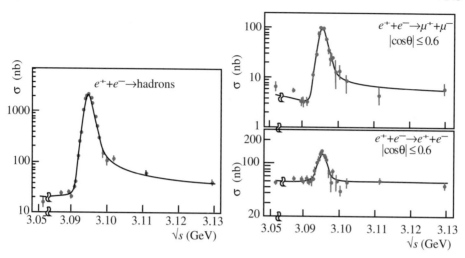

Fig. 4.24.   The ψ particle. (Boyarski 1975 © Nobel Foundation 1976)

as functions of the centre of mass energy. Note the logarithmic vertical scales and the much expanded energy scale. The resonance is very high and extremely narrow. Since the discovery at SLAC was independent, a different name was given to the new particle, ψ. It is now called J/ψ.

The quantum numbers of the J/ψ are expected to be $J^{PC} = 1^{--}$, because it decays into e^+e^- in the Ting experiment and because it is produced in e^+e^- collisions, both processes being mediated by a photon. However, the characteristics of the J/ψ were so surprising that these assignments had to be confirmed.

Firstly, the J/ψ might have $J^P = 1^+$ and decay with parity violation. To test this possibility, consider the electron–positron annihilation into a muon pair. In general the reaction proceeds in two steps, which outside the resonance are the annihilation of the initial electron pair into a photon and the materialisation of the photon into the final muon pair

$$e^+e^- \to \gamma \to \mu^+\mu^-. \tag{4.65a}$$

In resonance an additional contribution is given by the annihilation into a J/ψ that then decays into the final muon pair

$$e^+e^- \to J/\psi \to \mu^+\mu^-. \tag{4.65b}$$

Now, if the J/ψ has $J^P = 1^-$, the two amplitudes interfere because their intermediate states have the same quantum numbers, while, if the J/ψ has $J^P = 1^+$, they do not. Figure 4.25 shows the ratio between the cross sections of $e^+e^- \to \mu^+\mu^-$ and $e^+e^- \to e^+e^-$. The expected effect of the interference is a dip below the

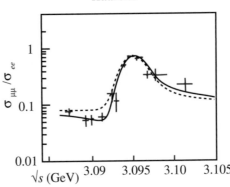

Fig. 4.25.　Ratio between $e^+e^- \to \mu^+\mu^-$ and $e^+e^- \to e^+e^-$ cross sections. (Boyarski 1975 © Nobel Foundation 1976)

resonance, shown as a continuous line. The dotted line is the expectation for $J^P = 1^+$. Comparison with the data establishes that $J^P = 1^-$.

Secondly, consider the hadronic decays. Given the very small width of the J/ψ, they might be electromagnetic, as for the η meson. By measuring the cross sections for $e^+e^- \to n\pi$ as a function of energy, the resonance was found for $n = 3$ and 5, but not for $n = 2$ or 4. Therefore the G-parity is conserved, i.e. the J/ψ decays strongly. Finally, the isospin of the J/ψ was established to be $I = 0$, observing that its branching ratio into $\rho^0\pi^0$ is equal to that into $\rho^+\pi^-$ (or $\rho^-\pi^+$).

After the discovery of the ψ, a systematic search started at SPEAR, scanning in energy at very small steps. Ten days later, a second narrow resonance was found, which was called ψ' (Abrams *et al.* 1974). The quantum numbers of the ψ' were established, in similar manner as for the ψ, to be those of the photon, $J^{PC} = 1^{--}$.

At Frascati the ADONE e^+e^- collider had been designed with a maximum energy $\sqrt{s} = 3$ GeV, just a little too small to detect the resonance. Nevertheless, when S. Ting communicated the discovery, the machine was brought above its nominal maximum energy, up to 3100 MeV (Bacci *et al.* 1974). The resonance appeared immediately. A fine-step scanning at lower energies did not show any narrow resonance.

From the line shape of the resonance, one can extract accurate values of the mass and the width. The values obtained at SLAC were

$$m(\psi) = 3097 \text{ MeV} \qquad \Gamma(\psi) = 91 \text{ keV}$$
$$m(\psi') = 3686 \text{ MeV} \qquad \Gamma(\psi') = 281 \text{ keV}. \tag{4.66}$$

Again note the width values that are surprisingly small for strongly decaying particles. Let us see how the widths are determined. One might think of taking simply the half maximum width. However, this is the convolution of the natural width and the experimental resolution and the method does not work if the latter is much wider than the former, as in this case.

We can use Eq. (4.6) to represent the behaviour of the cross sections of the processes (4.65) around the resonance. That expression is only an approximation, but it is good enough to understand the logic of the procedure. In the present case, we have $s_a = s_b = 1/2$ and $J = 1$. Calling Γ_e the initial state (e^+e^-) partial width, we have

$$\sigma(E) = \frac{3\pi}{s} \frac{\Gamma_e \Gamma_f}{(E - M_R)^2 + (\Gamma/2)^2}. \tag{4.67}$$

A quantity that can easily be measured is the 'peak area', which is not altered by the experimental resolution. The integration of (4.67) gives (see Problem 4.24)

$$\int \sigma(E) \, dE = \frac{6\pi^2}{M^2} \frac{\Gamma_e \Gamma_f}{\Gamma}. \tag{4.68}$$

The partial widths that appear in the numerator are determined by the measurements of the ratios between the peak areas in the corresponding channels. Then (4.68) gives the total width.

The extremely small widths show that J/ψ and ψ' are hadronic states of a completely new type, they are 'hidden charm' states. Both are made of a $c\bar{c}$ pair in a $^3S_1$ configuration, as follows from the quantum numbers $J^{PC} = 1^{--}$. The former is the fundamental level (1^3S_1 in spectroscopic notation), the latter is the first radial excited level (2^3S_1). However, further experimental work was needed before the extremely small values of the widths could be understood.

The Mark I detector started the search for charmed pseudoscalar mesons at $\sqrt{s} = 4.02$ GeV in 1976, after having improved its K to π discrimination ability, in the channels

$$e^+ + e^- \to D^0 + \bar{D}^0 + X \qquad e^+ + e^- \to D^+ + D^- + X \tag{4.69}$$

where X means anything else. The mesons are expected to have very short lifetimes. Consequently they appear as resonances in the final state. A first resonance was observed in the $K^\pm \pi^\mp$ mass distributions in multiparticle events, corresponding to the $D^0 \to K^- \pi^+$ and $\bar{D}^0 \to K^+ \pi^-$ decays. Its mass was 1865 MeV and its width smaller than the experimental resolution (Goldhaber *et al.* 1976).

Soon after (Peruzzi *et al.* 1976), as shown in Fig. 4.26, the charged D mesons were observed at the slightly larger mass of 1875 MeV (now 1869 MeV) in the channels

$$D^+ \to K^- \pi^+ \pi^+ \qquad D^- \to K^+ \pi^- \pi^-. \tag{4.70}$$

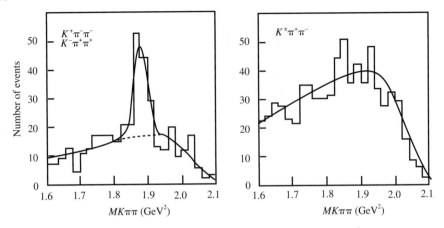

Fig. 4.26. $K\pi\pi$ mass distributions in multiparticle events at $\sqrt{s} = 4.02\,\text{GeV}$. (Peruzzi *et al.* 1976 © Nobel Foundation 1976)

However, no resonance was present in the channels

$$D^+ \to K^+\pi^+\pi^- \qquad D^- \to K^-\pi^+\pi^-. \tag{4.71}$$

This is precisely what the four-quark model requires.

We can now explain the reason for the narrow widths of J/ψ and ψ'. The reason is the same as for the ϕ. However, while the ϕ can decay into non-QCD-suppressed channels

$$\phi \to K^0\bar{K}^0; K^+K^-$$

even if slowly, due to the small Q, in the cases of J/ψ and ψ' the corresponding decay channels

$$J/\psi \to D^0\bar{D}^0; D^+D^- \qquad \psi' \to D^0\bar{D}^0; D^+D^- \tag{4.72}$$

are closed because $m_{D^+} + m_{D^-}$ and $m_{D^0} + m_{\bar{D}^0}$ are both larger than $m_{J/\psi}$ and $m_{\psi'}$.

This conclusion is confirmed by the observation, again at SLAC, of the third 3^3S_1 level, called ψ'', with mass and width

$$m(\psi'') = 3770\,\text{MeV} \qquad \Gamma(\psi'') = 24\,\text{MeV}. \tag{4.73}$$

In this case, the decay channels (4.72) are open and consequently the width is 'normal'.

Before leaving the subject, let us take another look at Fig. 4.24: the resonance curves are not at all symmetric around the maximum as the Breit–Wigner formula would predict, their high-energy side is higher than the low-energy one. This feature is general for resonances in the e^+e^- colliders. Actually, the energy on the horizontal axis is the energy we know, namely the energy of the colliding beams. This is not always the nominal electron–positron collision energy. It may happen that one or both initial particles radiate a photon (bremsstrahlung) before colliding. In this case, the collision energy is smaller than the collider energy. If the latter is above the resonance energy, the process returns the collision to resonance, increasing the cross section. On the other hand, if the machine energy is below the resonance, the initial state radiation takes the collision energy even farther from resonance, decreasing the cross section. This explains the high-energy tails.

A further element in the history of the discovery of charm must be mentioned, namely the precursor observations made in Japan before 1974, which were substantially ignored in the West. In Japan the technology of nuclear emulsions exposed to cosmic rays at high altitudes by airplane and balloon flights had progressed considerably. This was true, in particular, for K. Niu and collaborators, who had developed the so-called 'emulsion chamber'.

An emulsion chamber is made up of two main components. The first is a sandwich of several emulsion sheets; the second is another sandwich of lead plates, about 1 mm thick, alternated with emulsion sheets. The former gives an accurate tracking of the charged particles; the latter provides the gamma conversion (and the detection of the π^0s), the identification of the electrons and the measurement of their energy. The momenta of the charged particles are determined by an accurate measurement of the multiple scattering of the tracks.

The first example of associated production of charm, published in 1971, is shown in Fig. 4.27 (Niu *et al.* 1971). The short horizontal bars in the figure show the points

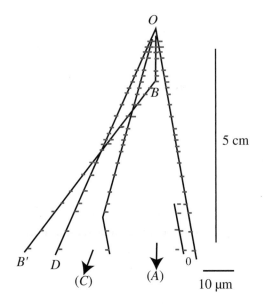

Fig. 4.27.   The first associated production of charm particles. (Adapted from Niu *et al.* 1971)

Table 4.3 *The lowest-mass hidden and open charm mesons*

State	Quark	M (MeV)	Γ/τ	J^{PC}	I	Principal decays
$J/\psi\,(1^3S_1)$	$c\bar{c}$	3097	93 keV	1^{--}	0	hadrons (88%), e^+e^- (6%), $\mu^+\mu^-$ (6%)
$\psi'\,(2^3S_1)$	$c\bar{c}$	3686	281 keV	1^{--}	0	$\psi + 2\pi\,(50\%)$
$\psi''\,(3^3S_1)$	$c\bar{c}$	3770	24 MeV	1^{--}	0	$D\bar{D}$ dominant
η_c	$c\bar{c}$	2980	26 MeV	0^{-+}	0	hadrons
D^+	$c\bar{d}$	1869	1 ps	0^-	1/2	K^-+ others; $\bar{K}^0 +$ others
D^0	$c\bar{u}$	1865	0.4 ps	0^-	1/2	$K^- +$ others; $\bar{K}^0 +$ others
D_s^+	$c\bar{s}$	1968	0.5 ps	0^-	0	$K^\pm +$ others; $K^0/\bar{K}^0 +$ others

where the tracks crossed the different emulsion sheets (the measurement accuracy is much better than the length of the bars). The straight lines are the interpolated tracks.

We now analyse the picture, making the following observations. The primary interaction has all the features of a strong interaction. Two particles decay, after 1.38 cm and 4.88 cm respectively, corresponding to proper times of the order of several 10^{-14} s. Therefore, the two particles are produced in association and decay weakly. The primary particle had several TeV energy, as evaluated from the measured energies of the secondary particles. Note that at the time no accelerator at this energy scale existed. Tracks OB and BB' and the π^0, shown by the photons that materialise in the lower part, are coplanar.

Niu dubbed as X the particle decaying in B and evaluated its mass to be $m_X = 1.8$ GeV if it was a meson, $m_X = 2.9$ GeV if it was a baryon. Consequently, it could not be a strange particle: a new type of hadron had been discovered. In the following years the Japanese groups observed other examples of the X particles, neutral and charged, in emulsion chambers exposed both to cosmic rays and to the proton beam at Fermilab. The new particles had all the characteristics foreseen for charmed hadrons.

Many hadrons containing the c quark, the charmed hadrons, are known today. Table 4.3 summarises the characteristics of the lowest-mass mesons. The charmed vector mesons, not shown, have larger masses and can decay strongly, without violating charm conservation.

Several 'hidden charm' mesons exist. Indeed, we can think of the $c\bar{c}$ system as an 'atom', which is called charmonium, in which quark and antiquark are bound by strong interaction. The charmonium bound states have the same quantum numbers as the hydrogen atom. The ψs are amongst these. We mention in particular the 1^1S_0 level, which is called η_c. It has the same quantum numbers as the η and the η', $J^{PC} = 0^{-+}$, $I = 0$, and in principle could mix with them. This does not happen in practice, due to the large mass difference, and η_c is a pure $c\bar{c}$ state.

Obviously the charmed hyperons exist too, containing one, two or three c quarks and consequently with $C = 1, 2$ or 3, with any combination of the other quarks. We shall not enter into any detail. However, we report the characteristics of the principal charmed hadrons in Appendix 3.

Question 4.2 Write down the possible charm values for a meson of any charge and those for a meson with charge $Q = +1$.

Question 4.3 Why is the radiative decay $\psi' \rightarrow \psi + \gamma$ forbidden? Why is $\psi \rightarrow \pi^0 + \pi^0$ forbidden?

4.10 The third family

The basic constituents of ordinary matter are electrons and, inside the nucleons, the up and down quarks with charges $-1/3$ and 2/3 respectively. In the nuclear beta decays a fourth particle appears, the electron neutrino. In total, we have two quarks and two leptons.

However, there is more in Nature. As we have seen, in cosmic rays a second charged lepton is present, the μ, identical to but heavier than the electron, and its associated neutrino. In cosmic radiation too, hadrons containing two further quarks, s with charge $-1/3$ and c with charge 2/3, have been discovered. Nature appears to have repeated the same scheme twice. The two groups of elementary particles are called first and second 'families'.

We have already seen that a third lepton and its neutrino exist. This suggests the existence of two more quarks, in three colours each, with charges $-1/3$ and $+2/3$. These are the 'beauty' (also called 'bottom'), b, and 'top', t, quarks. Their flavours are, respectively, $B = -1$ and $T = +1$. Like the other flavours, B and T are conserved by strong and electromagnetic interactions and are violated by weak ones.

In 1977, L. Lederman and collaborators (Herb *et al.* 1977) built a two-arm spectrometer at Fermilab designed to study $\mu^+ \mu^-$ pairs produced by high-energy hadronic collisions. The reaction studied was

$$p + (\text{Cu}, \text{Pt}) \rightarrow \mu^+ + \mu^- + X. \tag{4.74}$$

The 400 GeV proton beam extracted from the Tevatron was aimed at a metal target made of copper or platinum. The two arms measure the momenta of the positive and negative particles respectively. Since the events looked for are extremely rare, the spectrometer must accept very intense fluxes and must provide a high rejection power against charged pions and other hadrons. The rejection is obtained by using a sophisticated 'hadron filter' located on the path of the secondary particles, before they enter the arms of the spectrometer. A block of beryllium 18 radiation lengths thick stops the hadrons while letting the muons through. The price to pay is some degradation in the momentum measurement. The corresponding resolution on the mass of the two-muon system was $\Delta m_{\mu\mu}/m_{\mu\mu} \approx 2\%$.

To have an idea of the orders of magnitude, we mention that at every extraction, namely at every accelerator cycle, 10^{11} protons hit the target. With a total exposure of 1.6×10^{16} protons on target, a sample of 9000 $\mu^+\mu^-$ events with $m_{\mu\mu} > 5\,\text{GeV}$ was obtained. The $m(\mu^+\mu^-)$ mass distribution is shown in Fig. 4.28(a) and, after subtracting a non-resonating, i.e. continuum, background, in Fig. 4.28(b). Three barely resolved resonances are visible, which were generically called Υ.

The precision study of the new resonances was made at the e^+e^- colliders at DESY (Hamburg) and at Cornell in the USA. Figure 4.29, with the data from the latter laboratory, shows that the peaks are extremely narrow. The measurement of the masses and the widths of the Υs, made with the method we discussed for ψ, gave the results

$$
\begin{aligned}
m(1\,^3S_1\Upsilon) &= 9460\,\text{MeV} & \Gamma(1\,^3S_1\Upsilon) &= 53\,\text{keV} \\
m(2\,^3S_1\Upsilon) &= 10023\,\text{MeV} & \Gamma(2\,^3S_1\Upsilon) &= 43\,\text{keV} \\
m(3\,^3S_1\Upsilon) &= 10352\,\text{MeV} & \Gamma(3\,^3S_1\Upsilon) &= 26\,\text{keV}.
\end{aligned}
\tag{4.75}
$$

The situation is very similar to that of the ψs, now with three very narrow resonances, all with $J^{PC} = 1^{--}$ and $I = 0$: they are interpreted as the states $^3S_1$ of

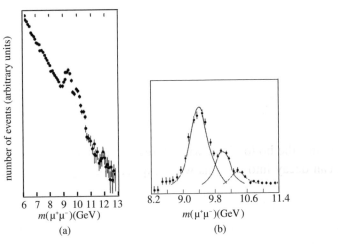

Fig. 4.28. The $\mu^+\mu^-$ mass spectrum: (a) full; (b) after continuum background subtraction. (Herb *et al.* 1977 © Nobel Foundation 1988)

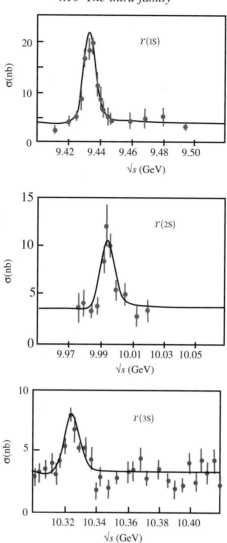

Fig. 4.29. The hadronic cross section measured by the CLEO experiment at the CESR e^+e^- collider, showing the $\Upsilon(1^1S_3)$, $\Upsilon(2^1S_3)$, $\Upsilon(3^1S_3)$ states. (From Andrews *et al.* 1980)

the $b\bar{b}$ 'atom', the bottomium, with increasing principal quantum number. None of them can decay into hadrons with 'explicit' beauty, because their masses are below threshold.

The lowest-mass beauty hadrons are the pseudoscalar mesons with a $\bar{b}$ anti-quark and a *d*, *u*, *s* or *c* quark. Therefore there are two charged, $B^+ = u\bar{b}$ and $B_c^+ = c\bar{b}$, and two neutral, $B^0 = d\bar{b}$ and $B_s^0 = s\bar{b}$, mesons and their antiparticles. The masses of B^0 and B^+ are practically equal, the mass of the B_s^0 is about one

Table 4.4 *The principal hidden and open beauty hadrons*

State	Quark	M (MeV)	Γ/τ	J^{PC}	I
$\Upsilon(1^1S_3)$	$b\bar{b}$	9460	54 keV	1^{--}	0
$\Upsilon(2^1S_3)$	$b\bar{b}$	10023	32 keV	1^{--}	0
$\Upsilon(3^1S_3)$	$b\bar{b}$	10355	20 keV	1^{--}	0
$\Upsilon(4^1S_3)$	$b\bar{b}$	10580	20 MeV	1^{--}	0
B^+	$u\bar{b}$	5279	1.6 ps	0^-	1/2
B^0	$d\bar{b}$	5279	1.5 ps	0^-	1/2
B^0_s	$s\bar{b}$	5368	1.5 ps	0^-	0
B^+_c	$c\bar{b}$	6286	0.5 ps	0^-	0

hundred MeV higher, due to the presence of the s and that of the B^+_c one thousand MeV higher due to the c. Table 4.4 gives a summary of the beauty particles we are discussing.

The pseudoscalar beauty mesons, as the lowest-mass beauty states, must decay weakly. Their lifetimes, shown in Table 4.4, are, surprisingly, of the order of a picosecond, larger than those of the charmed mesons, notwithstanding their much larger masses. As we shall see in Chapter 7, in the weak decay of every quark, not only of the strange one, both the electric charge and the flavour change. In the case of charm, there are two possibilities, $c \to s + \cdots$ and $c \to d + \cdots$. The former, as we saw, is favoured, the second is suppressed. Notice that in the former case the initial and final quarks are in the same family, in the latter they are not. In the case of beauty, the 'inside family' decay $b \to t + \cdots$ cannot take place because the t mass is larger than the b mass. The beauty must decay as $b \to c + \cdots$, i.e. with change of one family (from the third to the second), or as $b \to u + \cdots$, i.e. with change of two families (from the third to the first). We shall come back to this hierarchy in Section 7.9.

The non-QCD-suppressed decays of the Υs are those into a beauty–antibeauty pair. The smaller masses of these pairs are $m_{B^+} + m_{B^-} = 2m_{B^0} = 10\,558$ MeV and $2m_{B^0_s} = 10\,740$ MeV.

Therefore $\Upsilon(1^1S_3)$, $\Upsilon(2^1S_3)$ and $\Upsilon(3^1S_3)$ are narrow. The next excited level, the $\Upsilon(4^1S_3)$, is noticeable. Since it has a mass of $10\,580$ MeV, the decay channels

$$\Upsilon(4^3S_1) \to B^0 + \bar{B}^0 \quad \text{or} \quad \to B^+ + B^- \tag{4.76}$$

are open. The width of the $\Upsilon(4^1S_3)$ is consequently larger, namely 20 MeV.

A consideration is in order here. The production experiments, such as the Ting and Lederman experiments, have the best chance of discovering new particles, because they can explore a wide range of masses. After the discovery, when one knows where to look, the e^+e^- colliders are the ideal instruments for the accurate determination of their properties.

The third family still needed an up-type quark, but it took 20 years from the discovery of the τ and 18 from that of beauty to find it. This was because the top is very heavy, more than 170 GeV in mass. Taking into account that it must be produced in a pair, a very high centre of mass energy is necessary. Finally, in 1995, the CDF experiment at the Tevatron collider at Fermilab at $\sqrt{s}=2$ TeV reported 27 top events with an estimated background of 6.7 ± 2.1 events. More statistics were collected over the following years thanks to a substantial increase in the collider luminosity.

Let us see now how the top was discovered. We must anticipate a few concepts that we shall develop in Chapter 6. Consider a quark, or an antiquark, immediately after its production in a hadronic collision. It moves rapidly in a very intense colour field, which it contributes to produce. The energy density is so high that the field materialises in a number of quark–antiquark pairs. Quarks and antiquarks, including the original one, then join to form hadrons. This process, which traps the quark into a hadron, is called 'hadronisation'. In this process, the energy-momentum that initially belonged to the quark is distributed amongst several hadrons. In the reference frame of the quark, their momenta are typically of half a GeV. In the reference frame of the collision, the centre of mass of the group moves with the original quark momentum, which is typically several dozen GeV. Once hadronised, the quark appears to our detectors as a 'jet' of particles in a rather narrow cone.

Top is different from the other flavours in that there are no top hadrons. Actually, the hadronisation, even if extremely fast, takes a non-zero time, of the order of 10^{-23} s. The top decays by weak interactions, but, being very heavy, its lifetime is shorter than the hadronisation time. Unlike the other quarks, the top lives freely, but very briefly.

At Tevatron top production is a very rare event; it happens once in 10^{10} collisions. Experimentally one detects the top by observing its decay products. To distinguish these from the background of non-top events one must look at the channels in which the top 'signature' is as different from the background as possible. The top decays most probably into final states containing a W boson and a b quark or antiquark. Therefore one searches for the processes

$$p + \bar{p} \rightarrow t + \bar{t} + X \qquad t \rightarrow W^+ + b \qquad \bar{t} \rightarrow W^- + \bar{b}. \qquad (4.77)$$

The W boson, the mediator of the weak interactions, has a mass of 80 GeV and a very short lifetime. It does not leave an observable track and must be detected by observing its daughters. The W decays most frequently into a quark–antiquark pair, but these decays are difficult to distinguish from the much more common events with quarks directly produced by the proton–antiproton annihilation. We must search for rare but cleaner cases, such as those in which both *W*s decay into leptons

$$W \to e\nu_e \text{ or } \to \mu\nu_\mu. \tag{4.78}$$

Another clean channel occurs when one *W* decays into a lepton and the other into a quark–antiquark pair, requiring the presence of a *b* and a *b̄* from the *t* and *t̄* decays. Namely, one searches for the following sequence of processes

$$p + \bar{p} \to t + \bar{t} + X \qquad t \to W^+ + b \to W^+ + \text{jet}(b)$$
$$\bar{t} \to W^- + \bar{b} \to W^- + \text{jet}(\bar{b}) \qquad W \to e\nu_e \text{ or } \to \mu\nu_\mu \tag{4.79}$$
$$W \to q\bar{q}' \to \text{jet} + \text{jet}.$$

The requested 'topology' must have: one electron or one muon; one neutrino; four hadronic jets, two of which contain a beauty particle. Figure 4.30 shows this topology pictorially. Figure 4.31 shows one of the first top events observed by CDF in 1995 (Abe *et al.* 1995). The right part of the figure is an enlarged view of the tracks near the primary vertex showing the presence of two secondary vertices. They flag the decays of two short-lived particles, such as the beauties. The high-resolution picture is obtained thanks to a silicon-microstrip vertex detector (see Section 1.11). The calorimeters of CDF surround the interaction point in a 4π solid angle, as completely as possible. This makes it possible to check if the sum of the momenta of the detected particles is compatible with zero. If this is not the case, the 'missing momentum' must be the momentum of the undetectable particles, the vector sum of the neutrinos' momentum. The missing momentum is also shown in Fig. 4.31.

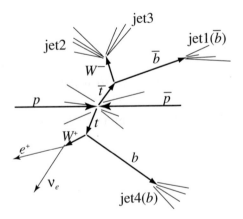

Fig. 4.30. Schematic view of reactions (4.79); the flight lengths of the *W*s and the *t*s are exaggerated.

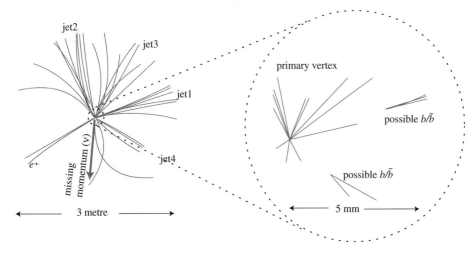

Fig. 4.31. An example of reaction (4.79) from CDF (Abe *et al.* 1995). One sees the four hadronic jets, the track of an electron, certified as such by the calorimeter, and the direction of the reconstructed missing momentum. The enlargement shows the *b* candidates in jets 1 and 4. (Courtesy Fermi National Laboratory)

As the top decays before hadronising, we can measure its mass from the energies and momenta of its decay products, as for any free particle. The result is

$$m_t = 174.2 \pm 3.3 \, \text{GeV}. \tag{4.80}$$

4.11 The elements of the Standard Model

Let us now summarise the hadronic spectroscopy we have studied. The hadrons have six additive quantum numbers, called flavours, which are: two values of the third component of the isospin (I_z), the strangeness (S), the charm (C), the beauty (B) and the top (T). All the flavours are conserved by the strong and the electromagnetic interactions and are violated by the weak interactions. There is a quark for each flavour. Quarks do not exist as free particles (with the exception of top), rather they live inside the hadrons, to which they give flavour, baryonic number and electric charge. They have spin 1/2 and, by definition, positive parity.

With a generalisation of (3.46), we define as flavour hypercharge

$$Y = B + S + C + B + T. \tag{4.81}$$

Its relationship to the electric charge is given by the Gell-Mann and Nishijima equation

$$Q = I_z + \frac{Y}{2}. \tag{4.82}$$

Table 4.5 *Quantum numbers and masses of the quarks*

	Q	I	I_z	S	C	B	T	$\mathcal{B}$	Y	Mass
d	$-1/3$	$1/2$	$-1/2$	0	0	0	0	$1/3$	$1/3$	3–7 MeV
u	$+2/3$	$1/2$	$+1/2$	0	0	0	0	$1/3$	$1/3$	1.5–3.0 MeV
s	$-1/3$	0	0	-1	0	0	0	$1/3$	$-2/3$	95 ± 25 MeV
c	$+2/3$	0	0	0	$+1$	0	0	$1/3$	$4/3$	1.25 ± 0.09 GeV
b	$-1/3$	0	0	0	0	-1	0	$1/3$	$-2/3$	4.20 ± 0.07 GeV
t	$+2/3$	0	0	0	0	0	$+1$	$1/3$	$4/3$	174.2 ± 3.3 GeV

By convention, the flavour of a particle has the same sign as its electric charge. Therefore the strangeness of K^+ is $+1$, the beauty of B^+ is $+1$, the charm of D^+ is $+1$, both strangeness and charm of D_s^- are -1, etc.

Table 4.5, a complete version of Table 4.1, gives the quantum numbers of the quarks, and their masses.

In Nature there are three families of quarks and leptons, each with the same structure: an up-type quark, with charge $+2/3$, a down-type quark with charge $-1/3$, a charged lepton with charge -1 and a neutrino. We shall see an experimental proof of the number of families in Chapter 9.

In the following chapters we shall study, even if at an elementary level, the fundamental properties of the interactions between quarks and leptons, namely their subnuclear dynamics. For each of the three fundamental interactions different from gravitation there are 'charges', which are the sources and receptors of the corresponding force, and vector mesons that mediate them. The fundamental characteristics of the charges and the mediators are very different in the three cases, as we shall study in the following chapters. We anticipate a summary of the main properties.

1. The electromagnetic interaction has the simplest structure. There is only one charge, the electric charge, with two different types. Charges of the same type repel each other, charges of different types attract each other. The two types are called positive and negative. Note that these are arbitrary names. The mediator is the photon, which is massless and has no electric charge. In Chapter 5, we shall study the fundamental aspects of quantum electrodynamics (QED) and we shall introduce instruments that we shall use for all the interactions.

2. The strong interaction sources and receptors are the 'colour' charges, where the name colour has nothing to do with everyday colours. The structure of the colour charges is more complex than that of the electric charge, as we shall study in Chapter 6. There are three charges of different colours, instead of the one of QED, called red R, green G and blue B. The quarks have one colour charge, and only one; the leptons, which have no strong interaction, have no

colour charge. The colour force between quarks is independent of their flavours. For example, the force between a red up quark and a green strange quark is equal to the force between a red down and a green beauty, provided the states are the same. There are 18 quarks in total, with six flavours and three colours. As for the electric charge, one might define a positive and negative 'redness', a positive and negative 'greenness' and a positive and negative 'blueness'. However, positive and negative colour charges are called 'colour' and 'anticolour' respectively. This is simply a matter of names. The repulsive or attractive character of the colour force between quarks cannot be established simply by looking at the signs of their charges, a fundamental difference compared to the electromagnetic force. The colour force mediators are the gluons, which are massless. The limited range of the strong force is due not to the mass of the mediators, but to a more complex mechanism, which we shall see. There are eight different gluons, which have colour charges, hence they also interact strongly amongst themselves. We shall study quantum chromodynamics (QCD) in Chapter 6.

3. Weak interactions have a still different structure. All the fundamental fermions, quarks, charged leptons and neutrinos have weak charges. The weak charge of a fermion depends on its 'chirality'. This term was created from the Greek word 'cheir', which means 'hand', to indicate handedness, but this meaning is misleading. Actually, chirality is the eigenvalue of the Dirac γ_5 matrix. It can be equal to $+1$ or -1. A state is often called 'right' if its chirality is positive, 'left' if it is negative. Again, these commonly used terms induce confusion with circular polarisation states, which are not the states of positive and negative chirality. Electrons and positrons can have both positive and negative chirality, while, strangely enough, only negative chirality neutrinos exist. The mediators of the weak interactions are three, two charged, W^+, W^- and one neutral, Z^0. All of them are massive, the mass of the former being about 80 GeV and of the latter about 90 GeV. The mediators have weak charges and, consequently, interact between themselves, as the gluons do. The phenomenology of weak interactions is extremely rich. We have space here to discuss only a part of it, in Chapters 7, 8 and 9.

Table 4.6 contains all the known fundamental fermions, particles and antiparticles, with their interaction charges. The colour is the apex at the left of the particle symbol.

Two observations are in order, both on neutrinos. Neutrinos are the most difficult particles to study, due to their extremely small interaction probability. They are also amongst the most interesting. Their study has always provided surprises.

Table 4.6 *The 24 fundamental fermions and their antiparticles.*
Each column is a family.

Fermions			Antifermions		
R_d	R_s	R_b	$^R\bar{d}$	$^R\bar{s}$	$^R\bar{b}$
G_d	G_s	G_b	$^{\bar{G}}\bar{d}$	$^{\bar{G}}\bar{s}$	$^{\bar{G}}\bar{b}$
B_d	B_s	B_b	$^B\bar{d}$	$^B\bar{s}$	$^B\bar{b}$
R_u	R_c	R_t	$^R\bar{u}$	$^R\bar{c}$	$^R\bar{t}$
G_u	G_c	G_t	$^{\bar{G}}\bar{u}$	$^{\bar{G}}\bar{c}$	$^{\bar{G}}\bar{t}$
B_u	B_c	B_t	$^B\bar{u}$	$^B\bar{c}$	$^B\bar{t}$
ν_e	ν_μ	ν_τ	$\bar{\nu}_e$	$\bar{\nu}_\mu$	$\bar{\nu}_\tau$
e^-	μ^-	τ^-	e^+	μ^+	τ^+

- The neutrino states in the table are the states of defined lepton flavour. These are the states in which neutrinos are produced by the weak interactions and the states that we can detect, again by weak interactions. Nevertheless, unlike for the other particles in the table, these *are not the stationary states*. The stationary states, called ν_1, ν_2 and ν_3 are quantum superpositions of ν_e, ν_μ and ν_τ. The stationary states are the states of definite mass, but do not have definite flavour and, therefore, cannot be classified in a family.
- What we have just said implies that the lepton flavour numbers are not conserved. Moreover, even if never observed so far, we cannot completely exclude a very small violation of the total lepton number. Actually, the lepton number is the only quantum number that distinguishes the neutrino from the antineutrino. If it is violated, neutrino and antineutrino may well be two states of the same particle. This is not, of course, the assumption of the Standard Model.

Problems

4.1. Consider the following three states: π^0, $\pi^+\pi^+\pi^-$ and ρ^+. Define which of them is a G-parity eigenstate and, for this case, give the eigenvalue.

4.2. Consider the particles ω, ϕ, K and η. Define which of them is a G-parity eigenstate and, for this case, give the eigenvalue.

4.3. From the observation that the strong decay $\rho^0 \rightarrow \pi^+\pi^-$ exists but $\rho^0 \rightarrow \pi^0\pi^0$ does not, what information can be extracted about the ρ quantum numbers: J, P, C, G, I?

4.4. Find the distance travelled by a K^* with momentum $p = 90\,\text{GeV}$ in a lifetime.

4.5. In a bubble chamber experiment on a K^- beam, a sample of events of the reaction $K^- + p \to \Lambda^0 + \pi^+ + \pi^-$ is selected. A resonance is detected both in the $\Lambda^0 \pi^+$ and in the $\Lambda^0 \pi^-$ mass distributions. In both, the mass of the resonance is $M = 1385\,\text{MeV}$ and its width $\Gamma = 50\,\text{MeV}$. It is called $\Sigma(1385)$. (a) What are the strangeness, the hypercharge, the isospin and its third component of the resonance $\Lambda^0 \pi^+$? (b) If the study of the angular distributions establishes that the orbital angular momentum of the $\Lambda^0 \pi$ systems is $L = 1$, what are the possible spin-parity values J^P?

4.6. The $\Sigma(1385)$ hyperon is produced in the reaction $K^- + p \to \pi^- + \Sigma^+(1385)$, but is not observed in $K^+ + p \to \pi^+ + \Sigma^+(1385)$. Its width is $\Gamma = 50\,\text{MeV}$; its main decay channel is $\pi^+ \Lambda$. (a) Is the decay strong or weak? (b) What are the strangeness and the isospin of the hyperon?

4.7. State the three reasons forbidding the decay $\rho^0 \to \pi^0 \pi^0$.

4.8. The ρ^0 has spin 1; the f^0 meson has spin 2. Both decay into $\pi^+ \pi^-$. Is the $\pi^0 \gamma$ decay forbidden for one of them, for both, or for none?

4.9. Calculate the branching ratio $\Gamma(K^{*+} \to K^0 + \pi^+)/\Gamma(K^{*+} \to K^+ + \pi^0)$ assuming, in turn, that the isospin of the K^* is $I_{K^*} = 1/2$ or $I_{K^*} = 3/2$.

4.10. Calculate the ratios $\Gamma(K^- p)/\Gamma(\bar{K}^0 n)$ and $\Gamma(\pi^- \pi^+)/\Gamma(\bar{K}^0 n)$ for the $\Sigma(1915)$ that has $I = 1$.

4.11. A low-energy antiproton beam is introduced into a bubble chamber. Two exposures are made, one with the chamber full of liquid hydrogen (to study the interactions on protons) and one with the chamber full of liquid deuterium (to study the interactions on neutrons). The beam energy is such that the antiprotons come to rest in the chamber. We know that the stopped antiprotons are captured in an 'antiproton' atom and, when they reach an S wave, annihilate. The $\bar{p}p$ and $\bar{p}n$ in an S wave are, in spectroscopic notation, the triplet $^3S_1$ and the singlet $^1S_0$.

List the possible values of the total angular momentum and parity J^P and isospin I.

Establish the eigenstates of C and of G and give the eigenvalues.

What are the quantum numbers of the possible initial states of the process $\bar{p}p \to \pi^- \pi^- \pi^+$?

Consider the following three groups of processes. Compute for each the ratios between the processes:

a. $\bar{p}n \to \rho^0 \pi^-; \bar{p}n \to \rho^- \pi^0$

b. $\bar{p}p(I = 1) \to \rho^+ \pi^-; \bar{p}p(I = 1) \to \rho^0 \pi^0; \bar{p}p(I = 1) \to \rho^- \pi^+$

c. $\bar{p}p(I = 0) \to \rho^+ \pi^-; \bar{p}p(I = 0) \to \rho^0 \pi^0; \bar{p}p(I = 0) \to \rho^- \pi^+$

4.12. Establish the possible total isospin values of the $2\pi^0$ system.

4.13. Find the Dalitz plot zeros for the $3\pi^0$ states with $J^P = 0^-$, 1^- and 1^+.

4.14. Knowing that the spin and parity of the deuteron are $J^P = 1^+$, give its possible states in spectroscopic notation.

4.15. What are the possible charm (C) values of a baryon, in general? What is the charm value if the charge is $Q = 1$, and what is it if $Q = 0$?

4.16. A particle has baryon number $B = 1$, charge $Q = +1$, charm $C = 1$, strangeness $S = 0$, beauty $B = 0$, top $T = 0$. Define its valence quark content.

4.17. Consider the following quantum number combinations, with in each case $B = 1$ and $T = 0$: $Q, C, S, B = -1, 0, -3, 0$; $Q, C, S, B = 2, 1, 0, 0$; $Q, C, S, B = 1, 1, -1, 0$; $Q, C, S, B = 0, 1, -2, 0$; $Q, C, S, B = 0, 0, 0, -1$. Define their valence quark contents.

4.18. Consider the following quantum number combinations, with in each case $B = 0$ and $T = 0$: $Q, S, C, B = 1, 0, 1, 0$; $Q, S, C, B = 0, 0, -1, 0$; $Q, S, C, B = 1, 0, 0, 1$; $Q, S, C, B = 1, 0, 1, 1$. Define their valence quark contents.

4.19. Explain why each of the following particles cannot exist according to the quark model: a positive strangeness and negative charm meson; a spin 0 baryon; an antibaryon with charge $+2$; a positive meson with strangeness -1.

4.20. Suppose you do not know the electric charges of the quarks. Find them using the other columns of Table 4.5.

4.21. What are the possible electric charges in the quark model of (a) a meson, (b) a baryon?

4.22. The mass of the J/ψ is $m_J = 3.097$ GeV and its width is $\Gamma = 91$ keV. What is its lifetime? If it is produced with $p_J = 5$ GeV in the L reference frame, what is the distance travelled in a lifetime? Consider the case of a symmetric $J/\psi \to e^+ e^-$ decay, i.e. with the electron and the positron at equal and opposite angles $\pm\theta_e$ to the direction of the J/ψ. Find this angle and the electron energy in the L reference frame. Find θ_e if $p_J = 50$ GeV.

4.23. Consider a D^0 meson produced with energy $E = 20$ GeV. We wish to resolve its production and the decay vertices in at least 90% of cases. What spatial resolution will we need? Mention adequate detectors.

4.24. Consider the cross section of the process $e^+ e^- \to f^+ f^-$ as a function of the centre of mass energy $\sqrt{s}$ near a resonance of mass M_R and total width Γ. Assuming that the Breit–Wigner formula correctly describes its line shape, calculate its integral over energy (the 'peak area'). Assume $\Gamma/2 \ll M_R$.

4.25. A 'beauty factory' is (in particle physics) a high-luminosity electron–positron collider dedicated to the study of the $e^+ e^- \to B^0 \bar{B}^0$ process. Its centre of mass energy is at the $\Upsilon(4^1S_3)$ resonance, namely at 10 580 MeV. This is only 20 MeV above the sum of the masses of the two Bs. Usually, in a collider the energies of the two beams are equal. However, in such a configuration the two *B*s are produced with very low energies. They travel distances that are too small to be measured. Therefore, the beauty factories are asymmetric. Consider PEP2 at SLAC, where the electron momentum is $p_{e-} = 9$ GeV and the positron momentum is $p_{e+} = 3$ GeV. Consider the case in which the two *B*s are produced with the same energy. Find the distance travelled by the *B*s in a lifetime and the angles of their directions to the beams.

4.26. A baryon decays strongly into $\Sigma^+ \pi^-$ and $\Sigma^- \pi^+$, but not into $\Sigma^0 \pi^0$ or $\Sigma^+ \pi^+$, even if all are energetically possible. (1) What can you tell about its isospin? (2) You should check your conclusion by looking at the ratio between the widths in the two observed channels. Neglecting phase space differences, which is the value you expect?

4.27. Write the diffusion amplitudes of the following processes in terms of the total isospin amplitudes: (1) $K^- p \to \pi^- \Sigma^+$, (2) $K^- p \to \pi^0 \Sigma^0$, (3) $K^- p \to \pi^+ \Sigma^-$, (4) $\bar{K}^0 p \to \pi^0 \Sigma^+$, (5) $\bar{K}^0 p \to \pi^+ \Sigma^0$.

Further reading

Alvarez, L. (1968); Nobel Lecture, *Recent Developments in Particle Physics* http://nobelprize.org/nobel_prizes/physics/laureates/1968/alvarez-lecture.pdf

Fowler, W. B. & Samios, N. P. (1964); *The Omega minus experiment. Sci. Am.* October 36

Lederman, L. M. (1988); Nobel Lecture, *Observations in Particle Physics from Two Neutrinos to the Standard Model* http://nobelprize.org/nobel_prizes/physics/laureates/1988/lederman-lecture.pdf

Richter, B. (1976); Nobel Lecture, *From the Psi to Charm: The Experiments of 1975 and 1976* http://nobelprize.org/nobel_prizes/physics/laureates/1976/richter-lecture.html

Ting, S. B. (1976); Nobel Lecture, *The Discovery of the J Particle: A Personal Recollection* http://nobelprize.org/nobel_prizes/physics/laureates/1976/ting-lecture.html

5

Quantum electrodynamics

5.1 Charge conservation and gauge symmetry

The coupling constant of the electromagnetic interaction is the **fine structure constant**

$$a = \frac{1}{4\pi\varepsilon_0} \frac{q_e^2}{\hbar c} \approx \frac{1}{137}, \tag{5.1}$$

where q_e is the elementary charge. Note that a has no physical dimensions; it is a pure number, which is small. It is one of the fundamental constants in physics and one of the most accurately measured.

We assume that the electric charge is conserved absolutely. The best experimental limit is obtained by searching for the decay of the electron, which, since it is the lightest charged particle, can decay only by violating charge conservation. The present limit is

$$\tau_e > 4 \times 10^{26} \text{ yr.} \tag{5.2}$$

Notice that this limit is much weaker than that of the proton decay.

The theoretical motivations for charge conservation are extremely strong, since they are a consequence of the 'gauge' invariance of the theory.

Let us start by recalling how the same property already appears in classical electromagnetism. As the reader will remember, charge conservation

$$\nabla \cdot \mathbf{j} - \frac{\partial \rho}{\partial t} = 0 \tag{5.3}$$

is a consequence of the Maxwell equations, i.e. it is deeply built into the theory ($\mathbf{j}$ is the current density and ρ the charge density). Furthermore, the Maxwell equations are invariant under the gauge transformations of the potentials $\mathbf{A}$ and ϕ

$$\mathbf{A} \Rightarrow \mathbf{A}' = \mathbf{A} + \nabla\chi \qquad \phi \Rightarrow \phi' = \phi - \frac{\partial\chi}{\partial t} \tag{5.4}$$

where $\chi(\mathbf{r},t)$ is called the 'gauge function'.

V. Fock discovered in 1929 that in quantum mechanics this invariance can be obtained only if the wave function of the charged particles is transformed at the same time as the potentials. If, for example, the source of the field is the electron of wave function ψ, the transformation is

$$\psi \Rightarrow \psi' = e^{i\chi(\mathbf{r},t)}\psi. \tag{5.5}$$

Note that the phase is just the gauge function. As we shall see in Section 5.3, in relativistic quantum mechanics ψ becomes itself an operator, the field of the electrons. More generally, the sources of the electromagnetic field are the matter fields. Therefore, the field equations determining the time evolution of the matter fields and of the electromagnetic field are not independent, but closely coupled. Hence the gauge invariance of the theory determines the interaction.

Gauge invariance is a basic principle of the Standard Model. All the fundamental interactions, not only the electromagnetic one, are gauge invariant. The gauge transformations of each of the three interactions form a Lie group. Equation (5.5) corresponds to the simplest possibility, the unitary group $U(1)$, which is the symmetry of QED. The symmetry groups of the other interactions are more complex: $SU(3)$ for QCD and $SU(2) \otimes U(1)$ for the electroweak interaction.

We have already used $SU(2)$ and $SU(3)$ to classify the hadrons and to correlate the cross sections and the decay rates of different hadronic processes. We have observed that these symmetries are only approximate due to the fact that two of the six quarks have negligible masses, compared to the hadrons, and that the mass of the third, even if not completely negligible, is still small. We now meet the same symmetry groups. However, their role is now much deeper because they determine the very structure of the fundamental interactions.

We conclude by observing that other 'charges' that might look similar at first sight, namely the baryonic and the leptonic numbers, do not correspond to a gauge invariance. Therefore, from a purely theoretical point of view, their conservation is not as fundamental as that of the gauge charges.

5.2 The Lamb and Retherford experiment

In 1947, W. Lamb and R. Retherford performed a crucial atomic physics experiment on the simplest atom, hydrogen (Lamb & Retherford 1947). The result showed that the motion of the atomic electron could not be described simply by the Dirac equation in an external, classically given field. The theoretical developments that followed led to a novel description of the interaction between charged particles and the electromagnetic field, and to the construction of the first quantum field theory, quantum electrodynamics, QED.

Let us start by recalling the aspects of the hydrogen atom relevant for this discussion. We shall use the spectroscopic notation, nL_j, where n is the principal quantum number, L is the orbital angular momentum and j is the total electronic angular momentum (i.e. it does not include the nuclear angular momentum, as we shall not need the hyperfine structure). We have not included the spin multiplicity $2s + 1$ in the notation since this, being $s = 1/2$, is always equal to 2. Since the spin is $s = 1/2$, there are two values of j for every L, $j = L + 1/2$ and $j = L - 1/2$, with the exception of the S wave, for which it is only $j = 1/2$.

A consequence of the $-1/r$ dependence of the potential on the radius r is a large degree of degeneracy in the hydrogen levels. In a first approximation the electron motion is non-relativistic ($\beta \approx 10^{-2}$) and we can describe it by the Schrödinger equation. As is well known, the energy eigenvalues in a $V \propto -1/r$ potential depend only on the principal quantum number

$$E_n = -\frac{Rhc}{n^2} = -\frac{13.6}{n^2}\,\text{eV} \tag{5.6}$$

where R is the Rydberg constant.

However, the high-resolution experimental observation of the spectrum, for example with a Lummer plate or a Fabry–Perot interferometer, resolves the spectral lines into multiplets. This is called the 'fine structure' of the spectrum.

We are interested here in the $n = 2$ levels. Their energy above the fundamental level is

$$E_2 - E_1 = Rhc\left(1 - \frac{1}{4}\right) = \frac{3}{4}Rhc = 10.2\,\text{eV}. \tag{5.7}$$

We recall that the fine structure is a relativistic effect. It is theoretically interpreted by describing the electron motion with the Dirac equation. The equation is solved by expanding in a power series of the fine structure constant, which is much smaller than one. We give the result at order α^2 ($= (1/137)^2$)

$$E_{n,j} = -\frac{Rhc}{n^2}\left[1 + \frac{\alpha^2}{n}\left(\frac{1}{j + 1/2} - \frac{3}{4n}\right)\right]. \tag{5.8}$$

We see that all levels, apart from the S level, split into two. This is the well-known spin-orbit interaction due to the orbital and the spin magnetic moments of the electron.

However, the degeneracy is not completely eliminated: states with the same values of the principal quantum number n and of the angular momentum j with a different orbital momentum L have the same energy. In particular, the levels $2S_{1/2}$ and $2P_{1/2}$ are still degenerate. The aim of the Lamb experiment was to check this

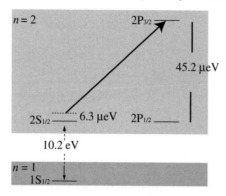

Fig. 5.1. Sketch of the levels relevant to the Lamb experiment.

crucial prediction, namely whether it really is $E(2S_{1/2}) - E(2P_{1/2}) = 0$, or, in other words, whether there is a shift between these levels. We can expect this shift, even if it exists, to be small in comparison to the energy splits of the fine structure, which, as shown in Fig. 5.1, are tens of μeV.

The energy of a level cannot be measured in absolute value, but only in relative value. Lamb and Retherford measured the energy differences between three (for redundancy) $2P_{3/2}$ levels, taken as references, and the $2S_{1/2}$ level, searching for a possible shift (now called the Lamb shift) of the latter. The method consisted in forcing transitions between these states with an electromagnetic field and measuring the resonance frequency (order of tens of GHz). One of these transitions is shown as an arrow in Fig. 5.1.

Figure 5.1 shows the levels relevant to the experiment; the solid line for the $2S_{1/2}$ level is drawn according to Eq. (5.8), the dotted line includes the Lamb shift.

Let us assume that $E(2S_{1/2}) > E(2P_{1/2})$. This is the actual case; the discussion for the opposite case would be completely similar, inverting the roles of the levels. In our hypothesis, $2S_{1/2}$ is metastable, meaning that its lifetime is of the order of 100 μs, much longer than the usual atomic lifetimes, which are of the order of 10 ns. Indeed, one of the a-priori possible transitions, the $2S_{1/2} \Rightarrow 1S_{1/2}$, is forbidden by the $\Delta l = \pm 1$ selection rule and the second, $2S_{1/2} \Rightarrow 1P_{1/2}$, would be extremely slow, because the transition probability is proportional to the cube of the shift.

Now consider the energy levels in a magnetic field. All the energy levels split depending on the projection of the angular momentum in the direction of **B** (Zeeman effect). Figure 5.2 gives the energies, in frequency units, of $2S_{1/2}$ and $2P_{1/2}$ as functions of the field. We have let the $2S_{1/2}$ and $2P_{1/2}$ energies be slightly different at zero field, because this possible difference is precisely the sought-after Lamb shift.

Note that when the field increases, the level $(2S_{1/2}, m = -1/2)$ approaches the $2P_{1/2}$ levels and even crosses some of them. Therefore, it mixes with these levels,

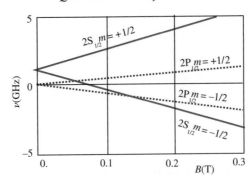

Fig. 5.2.   Sketch of the dependence of the energy levels on the magnetic field.

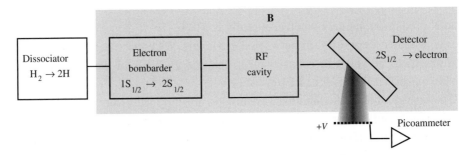

Fig. 5.3.   Schematic block diagram of Lamb and Retherford apparatus.

loses its metastability and decays in times of the order of 10^{-8} s. On the other hand, the level $(2S_{1/2},\ m=+1/2)$ moves farther from the $2P_{1/2}$ levels and remains metastable.

Let us discuss at this point the logic of the experiment with the help of Fig. 5.3. The principal elements of the apparatus are:

1. The oven, where at 2500 K, 65% of the H_2 molecules dissociate into atoms. The atoms and the remaining molecules exit from an aperture with a Maxwellian velocity distribution with an average speed $\langle v \rangle \approx 8000$ m/s.

2. The $1S_{1/2}$ to $2S_{1/2}$ excitation stage. This cannot be done with light because the transition is forbidden, as already mentioned. Instead, the atoms are bombarded with electrons of approximately 10 eV energy. In this way, one succeeds in exciting to the $2S_{1/2}$ level only a few atoms, about one in 10^8.

3. The separation of the Zeeman levels. The rest of the apparatus is in a magnetic field of adjustable intensity perpendicular to the plane of the figure. The atoms in the metastable level $(2S_{1/2},\ m=+1/2)$ fly in a lifetime over distance $d = 10^{-4}$ (s)$\times 8 \times 10^3$ (m/s)$= 0.8$ m, for enough to cross the apparatus.

The non-metastable atoms, those in the level $(2S_{1/2}, m = -1/2)$ in particular, can travel only $d \approx 10^{-8}$ (s) $\times 8 \times 10^3$ (m/s) $= 0.08$ mm.

4. The pumping stage. The beam, still in the magnetic field, enters a cavity in which the radiofrequency field is produced. Its frequency can be adjusted to induce a transition from the $(2S_{1/2}, m = +1/2)$ level to one of the Zeeman $2P_{3/2}$ levels. There are four of these, but one of them, $(2P_{3/2}, m = -3/2)$, cannot be reached because this would require $\Delta m = -2$. The other three $(2P_{3/2}, m = -1/2)$, $(2P_{3/2}, m = +1/2)$, $(2P_{3/2}, m = +3/2)$, however, can be reached. Therefore, for a fixed magnetic field value, there are three resonance frequencies for transitions from $(2S_{1/2}, m = +1/2)$ to a $2P_{3/2}$ level. The atoms pumped into one of these levels, which are unstable, decay immediately. Therefore, the resonance conditions are detected by measuring the disappearance, or a strong decrease, of the intensity of the metastable $(2S_{1/2}, m = +1/2)$ atoms after the cavity.

5. The excited atoms detector: a tungsten electrode. The big problem is that the atoms to be detected, i.e. those in the $(2S_{1/2}, m = +1/2)$ level, are a very small fraction of the total, a few in a billion as we have seen, when they are present. However, they are the only excited ones that reach the detector; the others have already decayed. To build a detector sensitive to the excited atoms only, Lamb used their capability of extracting electrons from a metal. The atoms in the $n = 2$ level, which are 10.2 eV above the fundamental level, when in contact with a metal surface de-excite and a conduction electron is freed. This is energetically favoured because the work function of tungsten is $W_W \approx 6$ eV < 10.2 eV. Obviously, atoms in the fundamental level cannot do that.

6. Electron detection. This latter operation is relatively easy: an electrode, at a positive potential relative to the tungsten (which is earthed) collects the electron flux, measured as an electric current with a picoammeter.

The results are given in Fig. 5.4. The measuring procedure was the following: the value of the radiofrequency in the cavity, v, was fixed; the magnetic field intensity was then varied and the detector current measured in search of the resonance conditions, which appeared as minima in the current intensity. The points in Fig. 5.4 were obtained.

The resonance frequencies correspond to the energy differences ΔE between the levels according to

$$hv = \Delta E. \tag{5.9}$$

One can see that the experimental points fall into three groups, each with a linear correlation. Clearly each group corresponds to a transition. The three lines extrapolate to a unique value at zero field, as expected, but they are shifted from positions expected according to Dirac's theory, the dotted lines. The experiment

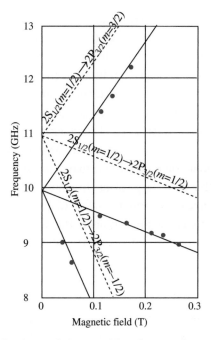

Fig. 5.4. Measured values of the transition frequencies at different magnetic field intensities (dots). Linear interpolations of the data (continuous lines) and behaviour expected in the absence of the shift (dotted lines). (Adapted from Lamb & Retherford 1947)

shows that the $S_{1/2}$ level is shifted by about 1 GHz. More precisely, the Lamb-shift value as measured in 1952 was

$$\Delta E\left(2S_{1/2} - 2P_{1/2}\right) = 1057.8 \pm 0.1 \,\text{MHz}. \qquad (5.10)$$

In the same year as Lamb's discovery, 1947, P. Kusch (Kusch & Foley 1947) made an accurate measurement of the electron gyromagnetic ratio g, or better of its difference from the expected value 2. The result was

$$(g - 2)/2 = +1.19 \times 10^{-3}. \qquad (5.11)$$

We shall see the consequences of both observations in the following sections.

5.3 Quantum field theory

The theoretical developments originated by the discoveries in the previous section led to the creation of the fundamental description of the basic forces, the quantum field theories. To interpret the Lamb experiment we must not think of the electric field of the proton seen by the electron as an external field classically given once and forever, as for example in the Bohr description of the atom. On the contrary, the field itself is a quantum system, made of photons that interact with the charges. Moreover, while the Dirac equation remains valid, its interpretation changes, its argument becoming itself a field, the quantum field of the electrons. We shall proceed in our description by successive approximations.

Let us use for the first time, with the help of intuition, a Feynman diagram. It is shown in Fig. 5.5 and represents an electron interacting with a nucleus. We must think of a time coordinate on a horizontal axis running from left to right and of a vertical axis giving the particle position in space. The thin lines represent the electron, which exchanges a photon, the wavy line, with the nucleus of charge *Ze*. The nucleus is represented by a line parallel to the time axis because, having a mass much larger than the electron, it does not move during the interaction. The Feynman diagram, and Fig. 5.5 in particular, represents a well-defined physical quantity, the probability amplitude of a process.

Now consider a free electron in vacuum. The quantum vacuum is not really empty, because processes such as that shown in Fig. 5.6 continually take place. The diagram shows the electron emitting and immediately reabsorbing a photon.

In a similar way, a photon in vacuum is not simply a photon. Figure 5.7 shows a photon that materialises into an e^+e^- pair followed by their re-annihilation into a photon. This process is called 'vacuum polarisation'.

The e^+e^- pair production and annihilation also occur for the virtual photon mediating the electron–nucleus interaction as shown by the diagram in Fig. 5.8.

The careful reader will have noticed that the processes we have just described do not conserve the energy. Indeed they are only possible on one condition. Namely, a measurement capable of detecting the energy violation ΔE must have energy

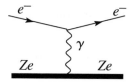

Fig. 5.5.   Diagram of an electron interacting with a nucleus.

Fig. 5.6.   Diagram of an electron emitting and reabsorbing a photon.

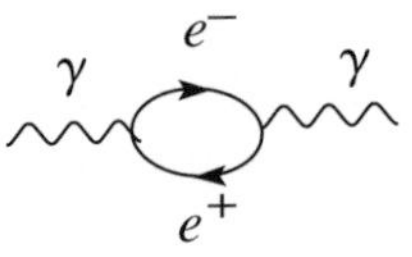

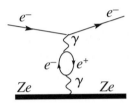

Fig. 5.7.   Vacuum polarisation by a photon.

Fig. 5.8.   An electron interacting with a nucleus with vacuum polarisation.

resolution better than ΔE. However, according to the uncertainty principle, this requires some time. Therefore, if the duration Δt of the violation is very short, namely if

$$\Delta E \Delta t \leq \hbar \tag{5.12}$$

the violation is not detectable, and may occur.

In conclusion, the atomic electron interacts both with the external field and with its own field. As in classical electromagnetism, this self-interaction implies an infinite value of the electron mass-energy. H. Bethe made a fundamental theoretical contribution in 1947, a month after the Lamb and Retherford experiment (Bethe 1947). He observed that the problem of the infinite value of the auto-interaction term could be avoided because such a term is not observable. One could 'renormalise' the mass of the electron by subtracting an infinite term.

After this subtraction, if the electron is in vacuum the contribution of the self-interaction is zero (by construction). However, this does not happen for a bound electron. Indeed, we can imagine the electron as moving randomly around its unperturbed position, due to the above-mentioned quantum fluctuations. The electron appears as a small charged sphere (the radius is of the order of a femto-metre) and, consequently, its binding energy is a little less than that of a point particle. This small increase in energy is a little larger for the zero orbital momentum states such as $2S_{1/2}$, compared to that of the $2P_{1/2}$. This is because, in the latter case, the electron has a smaller probability of being close to the nucleus.

Now consider the new interpretation of the Dirac equation mentioned above. If the electron field is not quantised $|\psi|^2$ is the probability of finding the electron.

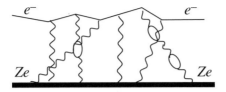

Fig. 5.9.   An electron bound to a nucleus.

However, as we have seen, the hydrogen atom does not always contain only one electron. Sometimes two electrons are present, together with a positron; or even three electrons and two positrons can be there. As long as the system is bound, the electron moves in the neighbourhood of the nucleus, continuously exchanging photons, as in the diagram in Fig. 5.9. In QED the number of particles is not a constant. We must describe by a quantum field not only the interaction – the electromagnetic field – but also the particles, such as the electron, that are the sources of that field. The electron field contains operators that 'create' and 'destroy' the electrons. Consider the simple diagram of Fig. 5.5. It shows two oriented electron-lines, one entering the 'vertex' and one leaving it. The correct meaning of this is that the initial electron disappears at the vertex, it is destroyed by an 'annihilation operator'; at the same time, a 'creation operator' creates the final electron. Asking whether the initial and final electrons are the same or different particles is meaningless because all the electrons are identical.

5.4 The interaction as an exchange of quanta

Now consider, in general, a particle a interacting through the field mediated by the boson V. When moving in vacuum it continually emits and reabsorbs V bosons, as shown in Fig. 5.10(a).

Now suppose that another particle b, with the same interaction as a, comes close to a. Then, sometimes, a mediator emitted by a can be absorbed not by a but by b, as shown in Fig. 5.10(b).

We say that particles a and b interact by exchanging a field quantum V.

The V boson in general has a mass m different from zero, and consequently the emission process $a \rightarrow a + V$ violates energy conservation by $\Delta E = m$. The violation is equal and opposite in the absorption process. The net violation lasts only for a short time, Δt, that must satisfy the relationship $\Delta E \Delta t \leq \hbar$. As the V boson can reach a maximum distance $R = c\Delta t$ in this time, the range of the force is finite

$$R = c\Delta t = c\hbar/m. \tag{5.13}$$

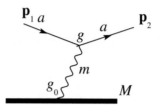

Fig. 5.10. Diagram showing the world-lines of: (a) particle a emitting and reabsorbing a V boson; (b) particles a and b exchanging a V boson.

Fig. 5.11. Diagram for the scattering of particle a in the potential of an infinite-mass centre M.

This is a well-known result: the range of the force is inversely proportional to the mass of its mediator.

The diagram in Fig. 5.10(b) gives the amplitude for the elastic scattering process $a + b \rightarrow a + b$. It contains three factors, namely the probability amplitudes for the emission of V, its propagation from a to b and the absorption of V. The internal line is called the 'propagator' of V. We shall now find the mathematical expression of the propagator using a simple argument.

We start with the non-relativistic scattering of a particle a of mass m from the central potential $\phi(\mathbf{r})$. The potential is due to a centre of forces of mass M, much larger than m. Let g be the 'charge' of a, which therefore has energy $g\phi(\mathbf{r})$, and let g_0 be the charge of the central body. Note that, since it is in a non-relativistic situation, the use of the concepts of potential and potential energy is justified.

The scattering amplitude is given by the diagram in Fig. 5.11, where $\mathbf{p}_1$ and $\mathbf{p}_2$ are the momenta of a before and after the collision. The central body does not move, assuming its mass to be infinite.

The momentum

$$\mathbf{q} = \mathbf{p}_2 - \mathbf{p}_1 \tag{5.14}$$

transferred from the centre to a is called 'three-momentum transfer'. Obviously, a transfers the momentum $-\mathbf{q}$ to the centre of forces.

Let us calculate the transition matrix element. In the initial and final states the particle a is free, hence its wave functions are plane waves. Neglecting uninteresting constants, we have

$$\langle \psi_f | g\phi(r) | \psi_i \rangle \propto g \int \exp(i\mathbf{p}_2 \cdot \mathbf{r})\, \phi(r) \exp(-i\mathbf{p}_1 \cdot \mathbf{r})\, dV$$

$$= g \int \exp[i\mathbf{q} \cdot \mathbf{r}]\, \phi(r)\, dV. \tag{5.15}$$

Notice how the scattering amplitude does not depend separately on the initial and final momenta, but only on their difference, the three-momentum transfer. Calling this amplitude $f(\mathbf{q})$, we have

$$f(\mathbf{q}) \propto \int \exp[i\mathbf{q} \cdot \mathbf{r}]\, \phi(r)\, dV. \tag{5.16}$$

We see that the scattering amplitude is proportional to the Fourier transform of the potential. The momentum transfer is the variable conjugate to the distance from the centre.

We can now assume the potential corresponding to a meson of mass m to be the Yukawa potential of range $R = 1/m$

$$\phi(r) = \frac{g_0}{4\pi r} \exp\left(-\frac{r}{R}\right) = \frac{g_0}{4\pi r} \exp(-rm). \tag{5.17}$$

Let us calculate the scattering amplitude

$$f(\mathbf{q}) = g \int_{\text{space}} \phi(r)\, e^{i\mathbf{q}\cdot\mathbf{r}}\, dV = g \int_{\text{space}} \phi(r)\, e^{iqr\cos\theta} d\varphi \sin\theta\, d\theta\, r^2\, dr$$

$$= g2\pi \int_0^\infty \phi(r) r^2\, dr \int_0^\pi e^{iqr\cos\theta}\, d\cos\theta = g4\pi \int_0^\infty \phi(r) \frac{\sin qr}{qr} r^2\, dr$$

that, with the potential (5.17) becomes

$$f(\mathbf{q}) = g g_0 \int_0^\infty e^{-rm} \frac{\sin qr}{q}\, dr = g_0 g \int_0^\infty e^{-mr} \left(\frac{e^{iqr} - e^{-iqr}}{2iq}\right) r^2\, dr.$$

Finally, calculating the above integral, we obtain the very important equation

$$f(\mathbf{q}) = \frac{g_0 g}{|\mathbf{q}|^2 + m^2}. \tag{5.18}$$

As anticipated, the amplitude is the product of the two 'charges' and the propagator, for which we now have the expression.

We now consider the relativistic situation, no longer assuming an infinite mass of the diffusion centre. Therefore, the particle a and the particle of mass M exchange both momentum and energy. The kinematic quantities are defined in Fig. 5.12.

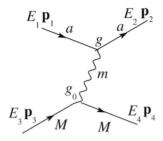

Fig. 5.12.   Basic diagram for the elastic scattering of two particles.

The relevant quantity is now the four-momentum transfer. Its norm is

$$t \equiv (E_2 - E_1)^2 - (\mathbf{p}_2 - \mathbf{p}_1)^2 = (E_4 - E_3)^2 - (\mathbf{p}_4 - \mathbf{p}_3)^2 \tag{5.19}$$

which, we recall, is negative or zero.

We noted above that the emission and absorption processes at the vertices do not conserve energy and, we may add, momentum. When using the Feynman diagrams we take a different point of view, assuming that at every vertex energy and momentum are conserved. The price to pay is the following. Since the energy of the exchanged particle is $E_2 - E_1$ and its momentum is $\mathbf{p}_2 - \mathbf{p}_1$, the square of its mass is given by Eq. (5.19). This is not the physical mass of the particle on the propagator. We call it a 'virtual particle'.

We do not calculate, but simply give the relativistic expression of the scattering amplitude, i.e.

$$f(t) = \frac{g_0 g}{m^2 - t} \tag{5.20}$$

very similar to (5.18). The 'vertex factors' are the probability amplitudes for emission and absorption of the mediator, i.e. the charges of the interacting particles. The propagator, namely the probability amplitude for the mediator to move from one particle to the other is

$$\Pi(t) = \frac{1}{m^2 - t}. \tag{5.21}$$

The probabilities of the physical processes, cross sections or decay speeds, are proportional to $|\Pi(t)|^2$, to the coupling constants and to the phase space volume.

5.5 The Feynman diagrams and QED

From the historical point of view, quantum electrodynamics (QED) was the first quantum field theory to be developed. It was created independently by Sin-Itiro

Tomonaga (Tomonaga 1946), Richard Feynman (Feynman 1948) and Julian Schwinger (Schwinger 1948). Feynman, in particular, developed the rules for evaluating the transition matrix elements. In QED, and in general in all quantum field theories, the probability of a physical process is expressed as a series of diagrams that become more and more complex as the order of the expansion increases. These 'Feynman diagrams' represent mathematical expressions, defined by a set of precise rules, which we shall not discuss here. However, the Feynman diagrams are also pictorial representations that clearly suggest interaction mechanisms to our intuition, and we shall use them as such.

Consider the initial and final states of a scattering or a decay process. They are defined by specifying the initial and final particles and the values of the momenta of each of them. We must now consider that there is an infinite number of possibilities for the system to go from the initial to the final state. Each of these has a certain probability amplitude, a complex number with an amplitude and a phase. The probability amplitude of the process is the sum, or rather the integral, of all these partial amplitudes. The probability of the process, the quantity we measure, is the absolute square of the sum.

The diagrams are drawn on a sheet of paper, on which we imagine two axes, one for time, the other for space (we have only one dimension for the three spatial dimensions), as in Fig. 5.13. The particles, both real and virtual, are represented by lines, which are their world-lines. A solid line with an arrow is a fermion; it does not move in Fig. 5.14(a), it moves upwards in Fig. 5.14(b). The arrow shows the direction of the flux of the charges relative to time. For example, if the fermion is an

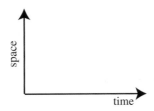

Fig. 5.13. Space-time reference frame used for Feynman diagrams.

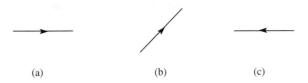

(a) (b) (c)

Fig. 5.14. Representation of the fermions, world-lines in the Feynman diagrams.

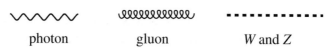

photon   gluon   *W* and *Z*

Fig. 5.15.   Representations of the world-lines of the vector mesons mediating the interactions in the Feynman diagrams.

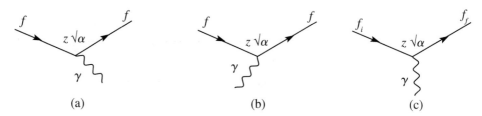

(a)   (b)   (c)

Fig. 5.16.   The electromagnetic vertex.

electron, its electric charge and electron flavour advance with it in time. In Fig. 5.14(c) all the charges go back in time: it is a positron moving forward in time. We shall soon return to this point.

We shall use the symbols in Fig. 5.15 for the vector mesons mediating the fundamental interactions, i.e. the 'gauge bosons'.

An important element of the diagrams is the vertex, shown in Fig. 5.16 for the electromagnetic interaction. The particles f are fermions, of the same type on the two sides of the vertex, of electric charge z. In Fig. 5.16(a) the initial f disappears in the vertex, while two particles appear in the final state: a fermion f and a photon. The initial state in Fig. 5.16(b) contains a fermion f and a photon that disappear at the vertex; in the final state there is only one fermion f. The two cases represent the emission and the absorption of a photon. Actually the mathematical expression of the two diagrams is the same, evaluated at different values of the kinematic variables, namely the four-momenta of the photon. Therefore, we can draw the diagram in a neutral manner, as in Fig. 5.16(c) (where we have explicitly written the indices i and f for 'initial' and 'final').

The vertex corresponds to the interaction Hamiltonian

$$z\sqrt{a}A_{\mu}\bar{f}\gamma^{\mu}f. \tag{5.22}$$

The operators f and $\bar{f}$ are Dirac bi-spinors. Their actions in the vertex are: f destroying the initial fermion (f_i in the figure), $\bar{f}$ creating the final fermion (f_f). The combination $\bar{f}\gamma^{\mu}f$ is called 'electromagnetic current' and interacts with A_{μ}, the quantum analogue of the classical four-potential. The four-potential is due to a second charged particle that does not appear in the figure, because the vertex it shows is only a part of the diagram. Figure 5.17 shows an example of a complete diagram, the diagram of the elastic scattering

$$e^- + \mu^- \rightarrow e^- + \mu^-. \tag{5.23}$$

It contains two electromagnetic vertices.

The lines representing the initial and final particles are called 'external legs'. The four-momenta of the initial and final particles, which are given quantities, define the external legs completely. On the other hand, there are infinite possible values of the virtual photon four-momenta, corresponding to different directions of its line. The scattering amplitude is the sum of these infinite possibilities. The diagram represents this sum. Therefore, we can draw the propagator in any direction. For example, the two parts of Fig. 5.17 are the same diagram whether the photon is emitted by the electron and absorbed by the muon or, vice versa, it is emitted by the muon and absorbed by the electron.

The probability amplitude is given by the product of two vertex factors (5.22)

$$\left(\sqrt{a} A_\mu \bar{e} \gamma^\mu e \right) \left(\sqrt{a} A_\mu \bar{\mu} \gamma^\mu \mu \right). \tag{5.24}$$

Note that, since the emission and absorption probability amplitudes are proportional to the charge of the particle, namely to $\sqrt{a}$, the scattering amplitude is proportional to a ($= 1/137$) and the cross section to a^2.

Summarising, the internal lines of a Feynman diagram represent virtual particles, which exist only for short times, since they are emitted and absorbed very soon after. The relationship between their energy and their momentum is not that of real particles. We shall see that, although they live for such a short time, the virtual particles are extremely important.

The amplitudes of the electromagnetic processes, such as (5.23), are calculated by performing an expansion in a series of terms of increasing powers of a, called a perturbative series. The diagram of Fig. 5.17 is the lowest term of the series, called at 'tree-level'. Figure 5.18 shows two of the next-order diagrams. They contain

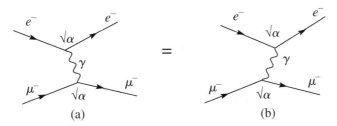

Fig. 5.17.   Feynman diagram for the electron–muon scattering.

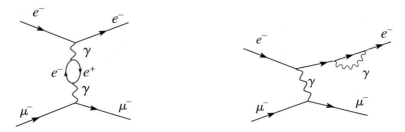

Fig. 5.18. Two diagrams at next to the tree-level.

four virtual particles and are proportional to α^2 ($=1/137^2$). One can understand that the perturbative series rapidly converges, due to the smallness of the coupling constant. In practice, if a high accuracy is needed, the calculations may be lengthy because the number of different diagrams grows enormously with increasing order. In the higher-order diagrams, closed patterns of virtual particles are always present. They are called 'loops'.

5.6 Analyticity and the need for antiparticles

Consider the two-body scattering

$$a + b \rightarrow c + d. \tag{5.25}$$

Let us consider the two invariant quantities: the centre of mass energy squared

$$s = (E_a + E_b)^2 - (\mathbf{p}_a + \mathbf{p}_b)^2 = (E_c + E_d)^2 - (\mathbf{p}_c + \mathbf{p}_d)^2 \tag{5.26}$$

where the meaning of the variables should be obvious, and the norm of the four-momentum transfer

$$t = (E_b - E_a)^2 - (\mathbf{p}_b - \mathbf{p}_a)^2 = (E_d - E_c)^2 - (\mathbf{p}_d - \mathbf{p}_c)^2. \tag{5.27}$$

We recall that $s \geq 0$ and $t \leq 0$.

The amplitude corresponding to a Feynman diagram is an analytical function of these two variables, representing different physical processes for different values of the variables, joined by analytical continuation. Consider for example the following processes: electron–muon scattering and electron–positron annihilation into a muon pair

$$e^- + \mu^- \rightarrow e^- + \mu^- \quad \text{and} \quad e^- + e^+ \rightarrow \mu^- + \mu^+. \tag{5.28}$$

Figure 5.19 shows the Feynman diagrams. They are drawn differently, but they represent the same function. They are called the '*s* channel' and the '*t* channel' respectively.

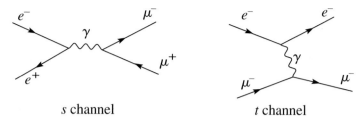

Fig. 5.19. Photon exchange in *s* and *t* channels.

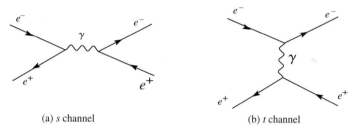

Fig. 5.20. Feynman diagrams for $e^- + e^+ \rightarrow e^- + e^+$ showing the photon exchange in the *s* and the *t* channels.

In the special case $a = c$ and $b = d$ the particles in the initial and final states are the same for the two channels. Therefore, as shown in an example in Fig. 5.20, the two channels contribute to the same physical process. Its cross section is the absolute square of their sum, namely the sum of the two absolute squares and of their cross product, the interference term.

Returning to the general case, we recall that $\sqrt{s}$ and $\sqrt{t}$ are the masses of the virtual particles exchanged in the corresponding channel. In the *t* channel the mass is imaginary, while it is real in the *s* channel. In the latter, something spectacular may happen. When $\sqrt{s}$ is equal, or nearly equal, to the mass of a real particle, such as the J/ψ for example, the cross section shows a resonance. Notice that the difference between virtual and real particles is quantitative, not qualitative.

Up to now we have discussed boson propagators, but fermion propagators also exist. Figure 5.21 shows the *t* channel and the *s* channel diagrams for Compton scattering.

Let us focus on the *t* channel in order to make a very important observation. As we know, all the diagrams, differing only by the direction of the propagator, are the same diagram. In Fig. 5.22(a) the emission of the final photon, event *A*, happens before the absorption of the initial photon, event *B*. The shaded area is the light cone of *A*. In Fig. 5.22(a) the virtual electron-line is inside the cone. The *AB* interval

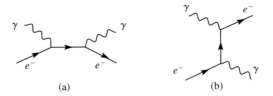

Fig. 5.21. A fermion propagator. Compton scattering.

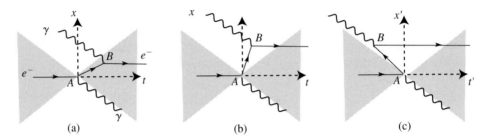

Fig. 5.22. Compton scattering Feynman diagram. The grey region is the light cone. (a) The virtual electron world-line is inside the cone (time-like); (b) the virtual electron world-line is outside the light cone (space-like); (c) as in (b), as seen by an observer in motion relative to the first one.

is time-like, the electron speed is less than the speed of light. In Fig. 5.22(b) the *AB* interval is outside the light cone, it is space-like. We state without proof that the diagram is not zero in these conditions, in other words, virtual particles can travel faster than light. This is a consequence of the analyticity of the scattering amplitude that follows, in turn, from the uncertainty of the measurement of the speeds intrinsic to quantum mechanics.

This observation has a very important consequence. If two events, *A* and *B*, are separated by a space-like interval, the order of their sequence in time is reference-frame dependent. We can always find a frame in which event *B* precedes event *A*, as shown in Fig. 5.22(c). An observer in this frame sees the photon disappearing in *B* and two electrons appearing, one advancing and one going back in time. He interprets the latter as an antielectron, with positive charge, moving forward in time. Event *B* is the materialisation of a photon in an electron–positron pair. Event *A* coming later in time is the annihilation of the positron of the pair with the initial electron.

We must conclude that the virtual particle of one observer is the virtual anti-particle of the other. However, the sum of all the configurations, which is what the diagram is for, is Lorentz-invariant. Lorentz invariance and quantum mechanics, once joined together, necessarily imply the existence of antiparticles.

Every particle has an amplitude to go back in time, and therefore has an antiparticle. This is true for both fermions and bosons. Consider for example Fig. 5.17(b). We can read it thinking that the photon is emitted at the upper vertex, moves backward in time, and is absorbed at the lower vertex, or that it is emitted at the lower vertex, moves forward in time and is absorbed at the upper vertex. The two interpretations are equivalent because the photon is completely neutral, i.e. photon and antiphoton are the same particle. This is the reason why there is no arrow in the wavy line representing the photon in Fig. 5.15.

We now consider the gauge bosons of the weak interactions. The Z is, like the photon, completely neutral, it is its own antiparticle. On the other hand, W^+ and W^- are each the antiparticle of the other. A W^+ moving back in time is a W^- and vice versa. To be rigorous this would require including an arrow in the graphic symbol of the Ws in Fig. 5.15, but this is not really needed in practice.

The situation for the gluons is similar. The gluons are eight in total, two completely neutral and three particle–antiparticle pairs. We shall study them in Chapter 6.

5.7 Electron–positron annihilation into a muon pair

When an electron and a positron annihilate they produce a pure quantum state, with the quantum numbers of the photon, $J^{PC} = 1^{--}$. We have already seen how resonances appear when $\sqrt{s}$ is equal to the mass of a vector meson.

Actually, the contributions of the e^+e^- colliders to elementary particle physics were also extremely important outside the resonances. In the next chapter we shall see what they have taught us about strong interaction dynamics, namely QCD. Now consider the process

$$e^+ + e^- \rightarrow \mu^+ + \mu^- \tag{5.29}$$

at energies high compared to the masses of the particles. This process is easily described by theory, because it involves only leptons that have no strong interactions. It is also easy to measure because the muons can be unambiguously identified.

Figure 5.23 shows the lowest-order diagram for reaction (5.29), the photon exchange in the s channel. The t channel does not contribute. The differential cross section of (5.29) is given by Eq. (1.53). Neglecting the electron and muon masses, we have $p_f = p_i$ and

$$\frac{d\sigma}{d\Omega_f} = \frac{1}{(8\pi)^2} \frac{1}{E^2} \frac{p_f}{p_i} \overline{\sum_{\text{initial}} \sum_{\text{final}} |M_{fi}|^2} = \frac{1}{(8\pi)^2} \frac{1}{s} \frac{1}{4} \sum_{\text{spin}} |M_{fi}|^2. \tag{5.30}$$

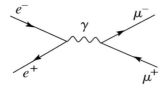

Fig. 5.23.   Lowest-order diagram for $e^+ + e^- \rightarrow \mu^+ + \mu^-$.

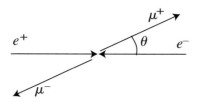

Fig. 5.24.   Initial and final momenta in the scattering $e^+ + e^- \rightarrow \mu^+ + \mu^-$, defining the scattering angle θ.

We do not perform the calculation; we give the result directly. Defining the scattering angle λ as the angle between the μ^- and the e^- (Fig. 5.24), we have

$$\frac{1}{4}\sum_{\text{spin}} |M_{fi}|^2 = (4\pi a)^2 (1 + \cos^2 \theta). \qquad (5.31)$$

We observe here that the cross section in (5.30) is proportional to $1/s$. This important feature is common to the cross sections of the collisions between point-like particles at energies much larger than all the implied masses, both of the initial and final particles and of the mediator. This can be understood by a simple dimensional argument. The cross section has the physical dimensions of a surface, or, in NU, of the reciprocal of an energy squared. Under our hypothesis, the only available dimensional quantity is the centre of mass energy. Therefore the cross section must be inversely proportional to its square. This argument fails if the mediator is massive at energies not very high compared to its mass. We shall consider this case in Section 7.2.

Let us discuss the origin of the angular dependence (5.31). Since reaction (5.29) proceeds through a virtual photon the total angular momentum is defined to be $J = 1$. We take the angular momenta quantisation axis z along the positron line of flight. As we shall show in Section 7.4 the third components of the spins of the electron and the positron can be either both $+1/2$ or both $-1/2$, but not one $+1/2$ and one $-1/2$.

In the final state we choose as quantisation axis z', the line of flight of one of the muons, say the μ^+. The third component of the orbital momentum is zero

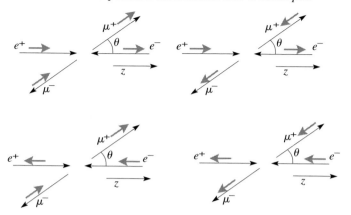

Fig. 5.25.   Four polarisation states for $e^+ + e^- \rightarrow \mu^+ + \mu^-$.

and therefore the third component of the total angular momentum can be, again, $m' = +1$ or $m' = -1$. The components of the final spins must again be either both $+1/2$ or both $-1/2$. In total, we have four cases, as shown in Fig. 5.25.

The matrix element for each $J = 1$, m, m' case is proportional to the rotation matrix from the axis z to the axis z', namely to $d^1_{m,m'}(\theta)$, i.e. the four contributions are proportional to

$$d^1_{1,1}(\theta) = d^1_{-1,-1}(\theta) = \frac{1}{2}(1 + \cos\theta) \quad d^1_{1,-1}(\theta) = d^1_{-1,1}(\theta) = \frac{1}{2}(1 - \cos\theta). \quad (5.32)$$

The contributions are distinguishable and we must sum their absolute squares. We obtain the angular dependence $(1 + \cos^2\theta)$ that we see in Eq. (5.31). This result is valid for all the spin 1/2 particles.

The arguments we have made give the correct dependence on energy and on the angle, but cannot give the proportionality constant. The complete calculation gives for the total cross section

$$\sigma = \frac{4}{3}\pi\frac{a^2}{s} = \frac{86.8\,\text{nb}}{s(\text{GeV}^2)}. \quad (5.33)$$

We introduce now a very important quantity called the 'hadronic cross section'. It is the sum of the cross sections of the electron–positron annihilations in all the hadronic final states

$$e^+ + e^- \rightarrow \text{hadrons}. \quad (5.34)$$

Figure 5.26 shows the hadronic cross section as a function of $\sqrt{s}$ from a few hundred MeV to about 200 GeV. Notice the logarithmic scales. The dotted line is the 'point-like' cross section, which does not include resonances. We see a very

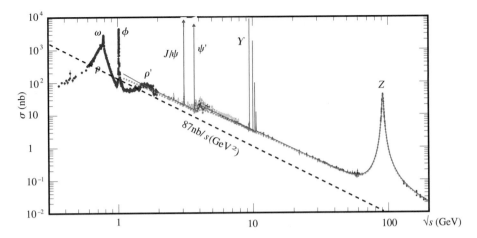

Fig. 5.26. The hadronic cross section. (Adapted from Yao *et al.* 2006 by permission of Particle Data Group and Institute of Physics)

rich spectrum of resonances, the ω, the ρ (and the ρ', which we have not mentioned), the ϕ, the ψs, the Υs and finally the Z.

Before leaving this figure, we observe another feature. While the hadronic cross section generically follows the $1/s$ behaviour, it shows a step every so often. These steps correspond to the thresholds for the production of quark–antiquark pairs of flavours of increasing mass.

5.8 The evolution of α

We have already mentioned that infinite quantities are met in quantum field theories and that the problem is solved by the theoretical process called 'renormalisation'. In QED two quantities are renormalized, the charge and the mass. We are interested in the charge, namely the coupling constant. One starts by defining a 'naked' charge that is infinite, but not observable, and an 'effective' charge that we measure. Then one introduces counter terms in the Lagrangian, which are subtracted cancelling the divergences. The counter terms are infinite.

The situation is illustrated pictorially in Fig. 5.27. The coupling constant at each vertex is the naked constant. However, when we measure, all the terms of the series contribute, reducing the naked charge to the effective charge. Note that the importance of the higher-order terms grows as the energy of the virtual photon increases. Therefore, the effective charge depends on the distance at which we measure it. We understand that if we go closer to the charge we include diagrams of higher order.

We proceed by analogy considering a small sphere with a negative charge immersed in a dielectric medium. The charge polarises the molecules of the

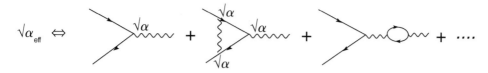

Fig. 5.27. The lowest-order diagrams contributing to the electromagnetic vertex, illustrating the relationship between the 'naked' coupling constant and the 'effective' (measured) one.

Fig. 5.28. A charge in a dielectric medium.

medium which tend to become oriented toward the sphere, as shown in Fig. 5.28. This causes the well-known screening action that macroscopically appears as the dielectric constant. Imagine measuring the charge from the deflection of a charged probe particle. In such a scattering experiment the distance of closest approach of the probe to the target is a decreasing function of the energy of the probe. Consequently, higher-energy probes will 'see' a larger charge on the sphere.

In quantum physics the vacuum becomes, spontaneously, polarised at microscopic level. Actually, e^+e^- pairs appear continuously, live for a short time, and recombine. If a charged body is present the pairs become oriented. If its charge is, for example, negative the positrons tend to be closer to the body, the electrons somewhat farther away, as schematically shown in Fig. 5.29. The virtual particle cloud that forms around the charged body reduces the power of its charge at a distance by its screening action.

If we repeat the scattering experiment with the probe particle, we find an effective charge that is larger and larger at smaller and smaller distances.

The fine structure constant, which we shall call simply a without the suffix 'eff', is not, as a consequence of the above discussion, constant, rather it 'evolves' with the four-momentum transfer or, in other cases, with the centre of mass energy at which we perform the measurement. Let us call Q^2 the relevant Lorentz-invariant variable, namely s or t depending on the situation. The coupling constants of all the

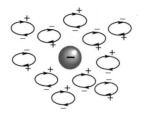

Fig. 5.29.   A charge in a vacuum.

fundamental forces are functions of Q^2. These functions are almost completely specified by renormalisation theory, which, however, is not able to fix an overall scale constant, which must be determined experimentally.

Suppose for a moment that only one type of charged fermions exists, the electron. Then only e^+e^- pairs fluctuate in the vacuum. The expression of a is

$$a(Q^2) = \frac{a(\mu^2)}{1 - \frac{a(\mu^2)}{3\pi} \ln\left(|Q|^2/\mu^2\right)} \tag{5.35}$$

where μ is the scale constant that the theory is unable to fix. Note that it has the dimension of the energy. Note also that in (5.35) the dependence is on the absolute value of Q^2 not on its sign.

Equation (5.35) is valid at small values of $|Q|$ when only e^+e^- pairs are effectively excited. At higher values more and more particle–antiparticle pairs are resolved, $\mu^+\mu^-, \tau^+\tau^-, u\bar{u}, d\bar{d}, \ldots$

Every pair contributes proportionally to the square of its charge. The complete expression is

$$a(Q^2) = \frac{a(\mu^2)}{1 - z_f \frac{a(\mu^2)}{3\pi} \ln\left(|Q|^2/\mu^2\right)} \tag{5.36}$$

where z_f is the sum of the squares of the charges (in units of the electron charge) of the fermions that effectively contribute at the considered value of $|Q|^2$, in practice with mass $m < |Q|$.

For example, in the range $10\text{ GeV} < Q < 100\text{ GeV}$, three charged leptons, two up-type quarks, u and c (charge 2/3) and three down-type quarks, d, s and b (charge 1/3) contribute, and we obtain

$$z_f = 3(\text{leptons}) + 3(\text{colours}) \times \frac{4}{9} \times 2(u, c) + 3 \times \frac{1}{9} \times 3(d, s, b) = 6.67$$

hence

$$a(Q^2) = \frac{a(\mu^2)}{1 - 6.67\frac{a(\mu^2)}{3\pi}\ln\left(|Q|^2/\mu^2\right)} \qquad \text{for } 10\,\text{GeV} < |Q| < 100\,\text{GeV}. \quad (5.37)$$

The dependence on Q^2 of the reciprocal of a is particularly simple, namely

$$a^{-1}(Q^2) = a^{-1}(\mu^2) - \frac{z_f}{3\pi}\ln\left(|Q|^2/\mu^2\right). \qquad (5.38)$$

We see that a^{-1} is a linear function of $\ln\left(|Q|^2/\mu^2\right)$, as long as thresholds for more virtual particles are not crossed. The crossing of thresholds is an important aspect of the evolution of the coupling constants, as we shall see.

The fine structure constant cannot be measured directly, rather its value at a certain Q^2 is extracted from a measured quantity, for example a cross section. The relationship between the former and the latter is obtained by a theoretical calculation in the framework of QED.

The fine structure constant has been determined at $Q^2 = 0$ with an accuracy of 0.7 ppb (parts per billion, 1 billion $= 10^9$), by measuring the electron magnetic moment with an accuracy of 0.7 ppt (parts per trillion, 1 trillion $= 10^{12}$). On the theoretical side, the QED relationship between the magnetic moment and the fine structure constant has been calculated to the eighth order by computing 891 Feynman diagrams. The result is (Gabrielse *et al.* 2006)

$$a^{-1}(0) = 137.035\,999\,710 \pm 0.000\,000\,096. \qquad (5.39)$$

The evolution, or 'running', of a has been determined both for $Q^2 > 0$ and for $Q^2 < 0$ at the e^+e^- colliders.

To work at $Q^2 > 0$ we use an s channel process, measuring the cross section of the electron–positron annihilations into fermion–antifermion pairs (for example $\mu^+\mu^-$)

$$e^+ + e^- \rightarrow f^+ + f^-.$$

Figure 5.30 shows the first three diagrams of the series contributing to the process.

The measured quantities are the cross sections as functions of $Q^2 = s$, from which the function $a(s)$ is extracted with a QED calculation. The result is shown in Fig. 5.31 in which $1/a$ is given at different energies. The data show that, indeed, a is not a constant and that its behaviour perfectly agrees with the prediction of quantum field theory.

A high-precision determination of a at the Z mass was made by the LEP experiments, with a combined resolution of 35 ppm (parts per million).

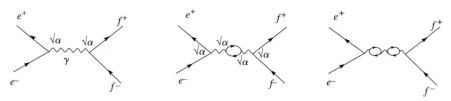

Fig. 5.30. Three diagrams for $e^+ + e^- \rightarrow f^+ + f^-$.

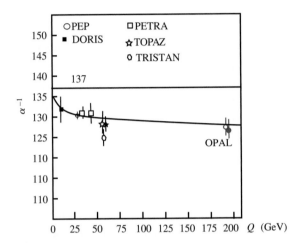

Fig. 5.31. $1/a$ vs. energy. (From Abbiendi *et al.* 2004)

The value is

$$a^{-1}\left(M_Z^2\right) = 128.936 \pm 0.046. \qquad (5.40)$$

To verify the prediction of the theory for space-like momenta, namely for $Q^2 < 0$, we measure the differential cross section of the elastic scattering (called Bhabha scattering)

$$e^+ + e^- \rightarrow e^+ + e^-. \qquad (5.41)$$

The four-momentum transfer depends on the centre of mass energy and on the diffusion angle θ (see Fig. 5.32) according to the relationship

$$|Q|^2 = -t = \frac{s}{2}(1 - \cos\theta). \qquad (5.42)$$

Figure 5.33 shows the lowest-order diagrams contributing to the Bhabha scattering in the t channel. We see that $|Q|^2$ varies from zero in the forward direction ($\theta = 0$) to s at $\theta = 180°$ and that to have a large $|Q|^2$ range one must work at high energies.

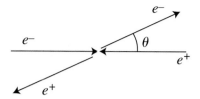

Fig. 5.32.   Bhabha scattering.

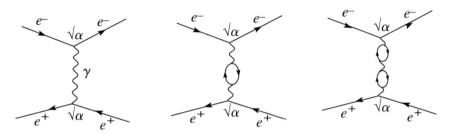

Fig. 5.33.   Three diagrams for the Bhabha scattering.

Another condition is set by the consideration that we wish to study a t channel process. As a consequence, we should be far from the Z peak where the s channel is dominant. The highest energy reached by LEP, $\sqrt{s} = 198$ GeV, satisfies both conditions. The L3 experiment measured the differential cross section at this energy between almost $0°$ and $90°$, corresponding to 1800 GeV$^2 < |Q|^2 < 21\,600$ GeV^2.

Let $d\sigma^{(0)}/dt$ be the differential cross section calculated with a constant value of α and $d\sigma/dt$ the cross section calculated with α as in (5.37). The relationship between them is

$$\frac{d\sigma}{dt} = \frac{d\sigma^{(0)}}{dt} \left[\frac{a(t)}{a(0)}\right]^2.$$ (5.43)

To be precise, things are a little more complicated, due mainly to the s channel diagrams. However, these contributions can be calculated and subtracted.

Figure 5.34(a) shows the measurement of the Bhabha differential cross section. The dotted curve is $d\sigma^{(0)}/dt$ and is clearly incompatible with the data. The solid curve is $d\sigma/dt$ with $a(t)$ given by Eq. (5.37), in perfect agreement with the data. Figure 5.34(b) shows a number of measurements of $1/\alpha$ at different values of $-Q^2$. In particular, the trapezoidal band is the result of the measurement just discussed. The solid curve is Eq. (5.37), the dotted line is the constant as measured at $Q^2 = 0$.

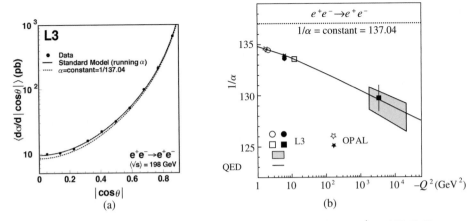

Fig. 5.34.   (a) Differential cross section of Bhabha scattering at $\sqrt{s}=198$ GeV as measured by L3 (Achard *et al.* 2005); (b) $1/a$ in the space-like region from the L3 and OPAL experiments (Abbiendi *et al.* 2006, as in Mele 2005).

Problems

5.1. Estimate the speeds of an atomic electron, a proton in a nucleus, a quark in a nucleon.

5.2. Evaluate the order of magnitude of the radius of the hydrogen atom.

5.3. Calculate the energy difference due to the spin-orbit coupling between the levels $P_{3/2}$ and $P_{1/2}$ for $n=2$ and $n=3$ for the hydrogen atom ($Rhc=13.6$ eV).

5.4. Consider the process $e^+ + e^- \rightarrow \mu^+ + \mu^-$. Evaluate the spatial distance between the two vertices of the diagram Fig. 5.19 in the CM reference frame and in the reference frame in which the electron is at rest.

5.5. Draw the tree-level diagrams for Compton scattering $\gamma + e^- \rightarrow \gamma + e^-$.

5.6. Draw the diagrams at the next to tree order for Compton scattering. [In total 17]

5.7. Give the values that the cross section of $e^+ e^- \rightarrow \mu^+ \mu^-$ would have in the absence of resonance at the ρ, the ψ, the Υ and the Z. What is the fraction of the angular cross section $\theta > 90°$?

5.8. Calculate the cross sections of the processes $e^+ e^- \rightarrow \mu^+ \mu^-$ and $e^+ e^- \rightarrow$ hadrons at the J/ψ peak ($m_\psi = 3.097$ GeV) and for the ratio of the former to its value in the absence of resonance. Neglect the masses and use the Breit–Wigner approximation. $\Gamma_e/\Gamma = 5.9\%$, $\Gamma_h/\Gamma = 87.7\%$.

5.9. Consider the narrow resonance Υ ($m_\Upsilon = 9.460$ GeV) that was observed at the $e^+ e^-$ colliders in the channels $e^+ e^- \rightarrow \mu^+ \mu^-$ and in $e^+ e^- \rightarrow$ hadrons. Its width is $\Gamma_\gamma = 54$ keV. The measured 'peak areas' are $\int \sigma_{\mu\mu}(E)\, dE = 8$ nb MeV and $\int \sigma_h(E)\, dE = 310$ nb MeV. In the Breit–Wigner approximation calculate the partial widths Γ_μ and Γ_h. Assume all the leptonic widths to be equal.

5.10. Two photons flying in opposite directions collide. Let E_1 and E_2 be their energies. (1) Find the minimum value of E_1 needed to allow the process $\gamma_1 + \gamma_2 \rightarrow e^+ + e^-$ to occur if $E_2 = 10$ eV. (2) Answer the same question if $E_1 = 2E_2$. (3) Find the CM speed in the latter case. (4) Draw the lowest-order Feynman diagram of the process.

5.11. Calculate the reciprocal of the fine structure constant at $Q^2 = 1$ TeV^2, knowing that $a^{-1}\left(M_Z^2\right) = 129$ and that $M_Z = 91$ GeV. Assume that no particles beyond the known ones exist.

5.12. If no threshold is crossed $a^{-1}(Q^2)$ is a linear function of $\ln(|Q|^2/\mu^2)$. What is the ratio between the quark and lepton contributions to the slope of this linear dependence for $4 < Q^2 < 10$ GeV^2?

Further reading

Feynman, R. P. (1985); *QED*. Princeton University Press

Feynman, R. P. (1987); *The reason for antiparticles*. In *Elementary Particles and the Laws of Physics*. Cambridge University Press

Jackson, J. D. & Okun, L. B. (2001); *Historical roots of gauge invariance. Rev. Mod. Phys.* **73** 663

Kusch, P. (1955); Nobel Lecture, *The Magnetic Moment of the Electron* http://nobelprize.org/nobel_prizes/physics/laureates/1955/kusch-lecture.pdf

Lamb, W. E. (1955); Nobel Lecture, *Fine Structure of the Hydrogen Atom* http://nobelprize.org/nobel_prizes/physics/laureates/1955/lamb-lecture.pdf

6

Chromodynamics

6.1 Hadron production at electron–positron colliders

We have already anticipated the importance of the experimental study of the process

$$e^+ + e^- \rightarrow \text{hadrons} \tag{6.1}$$

at the electron–positron colliders. We shall now see why.

We interpret the process as a sequence of two stages. In the first stage a quark–antiquark pair is produced

$$e^+ + e^- \rightarrow q + \bar{q}. \tag{6.2}$$

Here q and $\bar{q}$ can be any quark above threshold, namely with mass m such that $2m < \sqrt{s}$. The second stage is called hadronisation, the process in which the quark and the antiquark produce hadronic jets, as shown in Fig. 6.1.

The energies of the quarks are of the order of $\sqrt{s}$. Their momenta are of the same order of magnitude, at sufficiently high energy that we can neglect their masses, and are directed in equal and opposite directions, because we are in the centre of mass frame. The quark immediately radiates a gluon, similarly to an electron radiating a photon, but with a higher probability due to the larger coupling constant. The gluons, in turn, produce quark–antiquark pairs and quarks and antiquarks radiate more gluons, etc. During this process, quarks and antiquarks join to form hadrons. The radiation is most likely soft, the hadrons having typical momenta of $0.5-1$ GeV. In the collider frame, the typical hadron momentum component in the direction of the original quark is a few times smaller than the quark momentum. Its transverse component p_T (which is the same in both frames) is between about 0.5 and 1 GeV.

Therefore, the opening angle of the group of hadrons is of the order

$$\frac{p_T}{p} \approx \frac{0.5}{\sqrt{s}/2} = \frac{1}{\sqrt{s}} \tag{6.3}$$

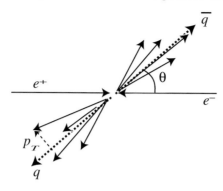

Fig. 6.1.   Hadronisation of two quarks into jets.

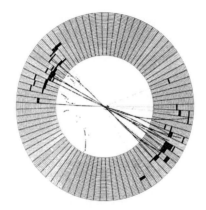

Fig. 6.2.   Two-jet event in the JADE detector of the PETRA collider at DESY.
(Naroska 1987)

with $\sqrt{s}$ in GeV. If, for example, $\sqrt{s} = 30$ GeV the group opening angle is of several degrees and it appears as a rather narrow 'jet'. If the energy is low the opening angle is so wide that the jets overlap and are not distinguishable.

Figure 6.2 shows the transverse (to the beams) projection of a typical hadronic event in the JADE detector of the PETRA collider at the DESY laboratory at Hamburg, with centre of mass energy $\sqrt{s} = 30$ GeV. The final state quark pairs appear clearly as two back-to-back jets.

Nobody has ever seen a quark by trying to extract it from a proton. To see the quarks we must change our point of view, as we have just done, and focus our attention on the energy and momentum flux rather than on the single hadrons. The quark then appears as such a flux in a narrow solid angle with the shape of a jet.

The total hadronic cross section (6.1) can be measured both at high energies, when the quarks appear as well-separated jets, and at lower energies, where the hadrons are distributed over all the solid angle and the jets cannot be identified. It is useful to express this cross section in units of the point-like cross section, i.e. the one for $\mu^+\mu^-$ that we have studied in Section 5.7, namely

$$R = \frac{\sigma(e^+ + e^- \rightarrow \text{hadrons})}{\sigma(e^+ + e^- \rightarrow \mu^+ + \mu^-)}. \tag{6.4}$$

If the quarks are point-like, without any structure, this ratio is simply given by the ratio of the sum of the electric charges

$$R = \sum_i q_i^2/1 \tag{6.5}$$

where the sum is over the quark flavours with production above threshold.

In 1969 the experiments at ADONE first observed that the hadronic production was substantially larger than expected. However, at the time quarks had not yet been accepted as physical entities and a correct theoretical interpretation was impossible. In retrospect, since the u, d and s quarks are produced at the ADONE energies ($1.6 < \sqrt{s} < 3$ GeV), we expect $R = 2/3$, whilst the experiments indicated values between 1 and 3. This was the first, not understood, evidence for colour.

Actually, the quarks of every flavour come in three types, each with a different colour. Consequently R is three times larger

$$R = 3 \sum_{\text{flavour}} q_i^2. \tag{6.6}$$

Figure 6.3 shows the R measurements in the range $2 \text{ GeV} < \sqrt{s} < \sqrt{40} \text{ GeV}$. In the energy region $2 \text{ GeV} < \sqrt{s} < 3 \text{ GeV}$ quark–antiquark pairs of three flavours, u, d and s can be produced; between 5 GeV and 10 GeV $c\bar{c}$ pairs are also produced; and finally between 20 GeV and 40 GeV also $b\bar{b}$ pairs are produced. In each case R

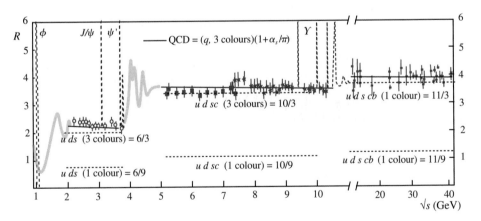

Fig. 6.3. Ratio R of hadronic to point-like cross section in e^+e^- annihilation as a function of $\sqrt{s}$. (Yao *et al.* 2006)

is about three times larger than foreseen in the absence of colour. To be precise, QCD also interprets well the small residual difference above the prediction of Eq. (6.6). This is due to the gluons, which themselves have colour charges. QCD predicts that (6.6) must be multiplied by the factor $(1 + a_s/\pi)$, where a_s is the QCD coupling constant, corresponding to the QED a, as we shall see shortly.

Question 6.1 Evaluate a_s at $\sqrt{s} = 40$ GeV from Fig. 6.3. Compare your result with Fig. 6.25.

In Section 5.7 we studied the differential cross section for the electron–positron annihilation into two point-like particles of spin 1/2. If the spin of the quarks is 1/2, the cross section of the process

$$e^+ + e^- \rightarrow q + \bar{q} \rightarrow \text{jet} + \text{jet} \tag{6.7}$$

should be

$$\frac{d\sigma}{d\Omega} = \frac{z^2 a^2}{s}(1 + \cos^2 \theta) \tag{6.8}$$

where z is the quark charge.

The scattering angle θ is the angle between, say, the electron and the quark. As we cannot measure the direction of the quarks, we take the common direction of the total momenta of the two jets. We know only the absolute value $|\cos \theta|$ because we cannot tell the quark from the antiquark jet. Figure 6.4 shows the measured angular cross section of (6.7) at $\sqrt{s} = 35$ GeV. It shows that quark spin is 1/2.

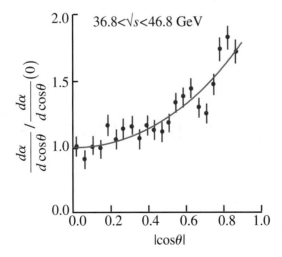

Fig. 6.4. Two-jet differential cross section as measured by CELLO at DESY. The curve is $1 + \cos^2 \theta$. (Adapted from Beherend *et al.* 1987)

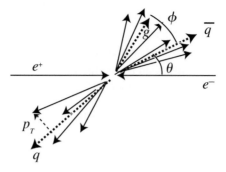

Fig. 6.5.   Sketch of the gluon radiation.

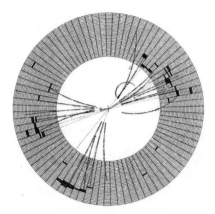

Fig. 6.6.   Three-jet event at the JADE detector at the PETRA collider at DESY.
(Naroska 1987)

We have seen that soft gluon radiation by quarks gives rise to the hadronisation process. More rarely the quarks radiate 'hard' gluons, meaning with a large relative momentum. The hard gluons hadronise, much like the quarks do, becoming visible as a hadronic jet. At the typical collider energies, $\sqrt{s} = 30$–100 GeV, a third jet appears in the detector about 10% of the time. We are then observing a gluon. Figure 6.5 shows the schematics; Fig. 6.6 shows a three-jet event observed by the JADE detector at the PETRA collider at $\sqrt{s} = 30$ GeV. In conclusion, at large enough centre of mass energy the gluons are clearly detectable as hadronic jets.

Unfortunately, the physical characteristics of the gluon jets are very similar to those of the quark jets. Therefore, it is only possible to establish which is the gluon jet on a statistical basis. The following criterion is adequate: in every three-jet event we classify the jets in decreasing centre of mass energy order, E_1, E_2, E_3, and we define jet 3 as the gluon. We then transform to the jet 1–jet 2 centre of mass frame and compute the angle ϕ between the common direction of the pair and jet 3.

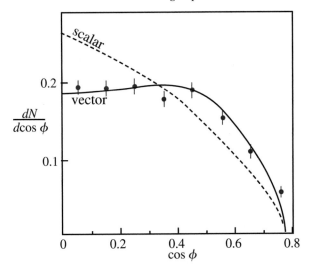

Fig. 6.7. Angular distribution of the gluon jet. (Adapted from Brandelik & Wu 1984)

The distribution of ϕ depends on the gluon spin. Figure 6.7 shows the measurement of the $\cos \phi$ distribution made by the TASSO experiment at PETRA. The two curves are calculated assuming the spin-parity of the gluon to be 0^+ and 1^-. The data clearly show that the gluon is a vector particle.

In conclusion, we have seen that the experiments at the e^+e^- colliders at centre of mass energies of several tens of GeV have given the following fundamental pieces of information:

1. The hadronic cross section cannot be understood without the colour charges.
2. The quarks are observed as hadronic jets with the angular distribution of the elementary spin 1/2 particles.
3. The gluons are seen as a third jet. Its angular distribution relative to the quark jet is that foreseen for a vector particle. Sometimes more gluons are radiated and are detected as further jets.
4. The R value shows that not only the quarks, but also the gluons are coloured.

6.2 Scattering experiments

We start by considering a classical experiment in optics. Suppose we have a film with transparency which varies as a function of the position on its surface. We want to measure this function. We prepare an almost monochromatic collimated light beam and measure the diffraction pattern of the target in the focal plane of a lens located behind the target (Fraunhofer conditions). This figure is the square of the

Fourier transform of the amplitude transparency of the target. The transverse component of the wave vector is the conjugate variable of the position vector. Therefore the minimum resolvable detail is of the order of the wavelength.

In a similar manner, we use probes of adequate resolving power to study the structure of microscopic objects, such as a nucleus or a nucleon.

The probes are particle beams. The situation is simple if these are point-like particles such as electrons and neutrinos. The two are complementary: the electrons 'see' the electric charges inside the nucleon, the neutrinos the weak charges.

We must now study the kinematics of the collisions. We start with the elastic scattering of small-mass spinless particles, say electrons neglecting their spin, of mass m, with a large-mass particle, say a nucleus, of mass M. Figure 6.8 defines the kinematic variables in the laboratory frame.

For elastic scattering, knowledge of the momentum and the energy of the incident particle and measurement of the energy and the direction of the scattered particle completely determine the event. Let us see.

The four-momenta and their norms are

$$p_\mu = (E, p) \qquad p'_\mu = (E', p') \qquad p_\mu p^\mu = p'_\mu p'^\mu = m_e^2$$
$$P_\mu = (M, 0) \qquad P'_\mu = (E, P_r) \qquad P_\mu P^\mu = p'_\mu p'^\mu = M^2. \tag{6.9}$$

The energy and momentum conservation gives

$$p_\mu + P_\mu = p'_\mu + P'_\mu \Rightarrow p_\mu p^\mu + P_\mu P^\mu + 2p_\mu P^\mu = p'_\mu p'^\mu + P'_\mu P'^\mu + 2p'_\mu P'^\mu. \tag{6.10}$$

Taking into account that $p_\mu p^\mu = p'_\mu p'^\mu = m_e^2$ and that $P_\mu P^\mu = P'_\mu P'^\mu = M^2$, this gives $p_\mu P^\mu = p'_\mu P'^\mu$, which is $EM - 0 = E'E_r - \mathbf{p}' \cdot \mathbf{p}_r$. Considering that $E_r = E + M - E'$ and that $\mathbf{p}_r = \mathbf{p} - \mathbf{p}'$ we have

$$EM = E'(E + M - E') - \mathbf{p}' \cdot (\mathbf{p} - \mathbf{p}') = E'E + E'M - pp' \cos\theta - m_e^2.$$

That is the looked-for relationship. It becomes very simple if the electron energy is high enough. We then neglect the term m_e^2 and take the momenta equal to

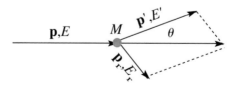

Fig. 6.8. Kinematic variables for the elastic scattering of a particle of mass m by a particle of mass M in the laboratory frame.

energies, obtaining

$$E' = \frac{E}{1 + \dfrac{E}{M}(1 - \cos\theta)} = \frac{E}{1 + \dfrac{2E}{M}\sin^2\dfrac{\theta}{2}}. \tag{6.11}$$

This important relationship shows, in particular, that the energy transferred to the target $E - E'$ becomes negligible for large target mass, namely if $E/M \ll 1$. However, the momentum transfer is not negligible.

Now consider the scattering of an electron by the electrostatic potential of a nucleus $\phi(r)$. We have seen in Section 5.4 that the scattering matrix element is the Fourier transform of the potential. Equation (5.15) becomes

$$\left\langle \psi_f | q_e\phi(r) | \psi_i \right\rangle \propto q_e \int \exp[i\mathbf{q}\cdot\mathbf{r}]\phi(r)\, dV \tag{6.12}$$

where $\mathbf{q} = \mathbf{p} - \mathbf{p}'$ is the three-momentum transfer.

We want to find the dependence of the matrix element on the charge density (the electric charge in this case) $\rho(r)$. This is the source of the potential and, according to the electrostatic equation

$$\nabla^2\phi = -\frac{\rho}{\varepsilon_0}. \tag{6.13}$$

We now use the relationship $\nabla^2 \exp(i\mathbf{q}\cdot\mathbf{r}) = -q^2 \exp(i\mathbf{q}\cdot\mathbf{r})$ and the identity $\int \phi\nabla^2[\exp(i\mathbf{q}\cdot\mathbf{r})]dV = \int [\exp(i\mathbf{q}\cdot\mathbf{r})]\nabla^2\phi\, dV$, obtaining

$$\left\langle \psi_f | q_e\phi(r) | \psi_i \right\rangle \propto \frac{q_e}{\varepsilon_0}\frac{1}{q^2} \int \rho(r)\exp[i\mathbf{q}\cdot\mathbf{r}]\, dV. \tag{6.14}$$

It is now easy to calculate the cross section, but since we are not interested in the proof, we simply give the result

$$\frac{d\sigma}{d\Omega} = \frac{q_e^2}{(2\pi)^2\varepsilon_0^2}\frac{E'^2}{|\mathbf{q}|^4}\left| \int \rho(r)\exp[i\mathbf{q}\cdot\mathbf{r}]\, dV \right|^2. \tag{6.15}$$

Let us now call $f(\mathbf{r})$ the target charge density normalised to one, namely

$$f(\mathbf{r}) = \frac{1}{Zq_e}\rho(\mathbf{r}) \tag{6.16}$$

and $F(\mathbf{q})$ its Fourier transform

$$F(\mathbf{q}) = \int f(\mathbf{r})\exp(i\mathbf{q}\cdot\mathbf{r})dV. \tag{6.17}$$

Then we can write (6.15) as

$$\frac{d\sigma}{d\Omega} = 4Z^2\alpha^2 \frac{E'^2}{|\mathbf{q}|^4} \, |\, F(\mathbf{q}) \,|^2. \tag{6.18}$$

In words: the intensity scattered by an immobile target is proportional to the square of the Fourier transform of its charge distribution.

Rutherford cross section   Let us consider a point-like target with charge Zq_e at the origin. The charge density function is $Zq_e\delta(0)$. Its transform is a constant. If zq_e is the beam charge, Eq. (6.18) becomes

$$\frac{d\sigma}{d\Omega} = \left(\frac{d\sigma}{d\Omega}\right)_{\text{Rutherford}} = 4z^2 Z^2 \alpha^2 \frac{E'^2}{|\mathbf{q}|^4}. \tag{6.19}$$

This is the well-known Rutherford cross section.

Rutherford found this expression to interpret the Geiger and Marsden experiment. The probe was a beam of alpha particles, with kinetic energy E_k of a few MeV. Therefore, we can write $E_k = p^2/(2m)$. Let us now find the cross section as a function of E_k and the scattering angle.

Looking at Fig. 6.9, we see that from $p' = p$ and $E' = E$ it follows that $q = 2p \sin \theta/2$. We can also set $E = m$, obtaining the well-known expression

$$\frac{d\sigma}{d\Omega} = \frac{z^2 Z^2 \alpha^2}{16 E_k^2} \frac{1}{\sin^4 \dfrac{\theta}{2}}. \tag{6.20}$$

The cross section is independent of the azimuth ϕ. We integrate on ϕ recalling that $d\Omega = d\phi\, d\cos\theta$, obtaining

$$\frac{d\sigma}{d\cos\theta} = \frac{\pi\, z^2 Z^2 \alpha^2}{8} \frac{1}{E_k^2} \frac{1}{\sin^4 \dfrac{\theta}{2}} = \frac{\pi\, z^2 Z^2 \alpha^2}{2} \frac{1}{E_k^2} \frac{1}{(1-\cos\theta)^2}. \tag{6.21}$$

Notice the divergence for $\theta \to 0$. This is a consequence of the divergence of the assumed potential for $r \to 0$, a situation that is never found in practice.

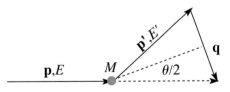

Fig. 6.9.   Geometric relation between three-momentum transfer and scattering angle in the Rutherford scattering.

Mott cross section To neglect the electron spin effects is a good approximation at low energies. However, if the speed of the electron is high, the spin effects become important. The expression of the cross section for electron scattering in the Coulomb potential of an infinite mass target, which implies $E' = E$, is due to Mott. It is given by Eq. (6.19) multiplied by $\cos^2(\theta/2)$

$$\left(\frac{d\sigma}{d\Omega}\right)_{\text{Mott}} = \left(\frac{d\sigma}{d\Omega}\right)_{\text{Rutherford}} \cos^2\frac{\theta}{2} = 4z^2Z^2\alpha^2\frac{E'^2}{|\mathbf{q}|^4}\cos^2\frac{\theta}{2}. \qquad (6.22)$$

The Mott cross section decreases with increasing angle faster than the Rutherford cross section and becomes zero at 180°. We shall give the reason for this in Section 7.4.

We finally consider the ultrarelativistic case, in which the projectile mass is negligible compared to its energy. The recoil energy can no longer be neglected, and $E' < E$. The expression valid for a point target is

$$\left(\frac{d\sigma}{d\Omega}\right)_{\text{point}} = \frac{E'}{E}\left(\frac{d\sigma}{d\Omega}\right)_{\text{Mott}}. \qquad (6.23)$$

6.3 Nucleon structure

In the 1960s a two-mile long linear electron accelerator (LINAC) was built at Stanford in California. Its maximum energy was 20 GeV. The laboratory, after that, was called the Stanford Linear Accelerator Center (SLAC).

J. Friedman, H. Kendall and collaborators at MIT and R. Taylor and collaborators at SLAC designed and built two electron spectrometers, up to 8 GeV and 20 GeV energy respectively. These instruments were set up to study the internal structure of the proton and the neutron. The layout and a picture are shown in Fig. 6.10.

The electron beam extracted from the LINAC is brought onto a target, which is of liquid hydrogen when the proton is being studied or liquid deuterium when the neutron is being studied. The beam is collimated and monochromatic with known energy E. The spectrometers measure the energy E' of the scattered electron and the scattering angle θ. The rest of the event, namely what happens to the nucleon, is not observed. We indicate it by X. This type of measurement is called an 'inclusive' experiment. The reaction is

$$e^- + p \rightarrow e^- + X. \qquad (6.24)$$

The 8 GeV spectrometer decouples the measurement of the angle from that of the momentum by using bending magnets that deflect in the vertical plane. Scaling up this technique to the 20 GeV spectrometer would have required a very large

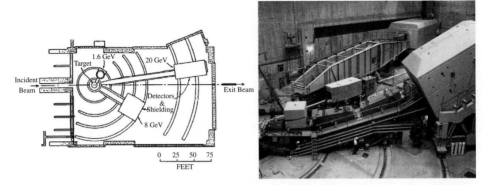

Fig. 6.10.   The spectrometers ride on rails and can be rotated about the target to change the angle of the detected electrons. The detectors are inside the heavy shielding structures visible at the ends of the spectrometers. (Taylor 1991. Courtesy of SLAC and © Nobel Foundation 1990)

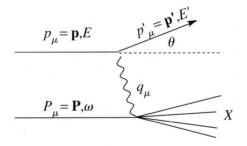

Fig. 6.11.   Sketch of the deep inelastic scattering.

vertical displacement. A brilliant solution was found by K. Brown and B. Richter who proposed a novel optics arrangement made of two vertical bending stages, the first upwards, the second downwards, but both contributing in the same direction to the dispersion. In such a way the vertical dimension was kept within bounds. The first experimental results for a 17 GeV energy beam were published in 1969.

Figure 6.11 shows the kinematics and defines the relevant variables. To detect small structures inside the nucleon we must hit them violently, breaking the nucleon. The process is called 'deep inelastic scattering' (DIS).

We must now define the kinematic variables that we shall use. The first is the four-momentum transfer to the nucleon, t, which is negative. To follow convention, we also define its opposite Q^2. With reference to Fig. 6.11 we have

$$-Q^2 \equiv t = q^\mu q_\mu \equiv (E' - E)^2 - (\mathbf{p}' - \mathbf{p})^2. \tag{6.25}$$

Being at high enough energy, we can neglect the electron mass, obtaining

$$q^\mu q_\mu = 2m_e^2 - 2(EE' - pp' \cos \theta) \approx -2EE'(1 - \cos \theta)$$
$$= -4EE' \sin^2 \frac{\theta}{2} = -Q^2. \qquad (6.26)$$

We see that to know Q^2 we must measure the scattered electron energy E' and its direction θ. Another invariant quantity that can be measured is the square of the mass of the hadronic system W^2

$$W^2 = \left(P_\mu + q_\mu\right)\left(P^\mu + q^\mu\right) = m_p^2 + 2P_\mu q^\mu - Q^2 = m_p^2 + 2m_p \nu - Q^2. \quad (6.27)$$

In the last member we have introduced a further Lorentz-invariant quantity

$$\nu \equiv P_\mu q^\mu / m_p. \qquad (6.28)$$

To see its physical meaning we look at its expression in the laboratory frame, where $P_\mu = (m_p, \mathbf{0})$ and $q_\mu = (E - E', \mathbf{q})$. Therefore

$$\nu = E - E'. \qquad (6.29)$$

We see that ν is the energy transferred to the target in the laboratory frame. We determine it by measuring E' and knowing the incident energy. We then use the two variables ν and Q^2 that are measured as just specified.

In the previous section we have given the Mott cross section, which is valid in conditions similar to those we are considering now for point-like targets. One can show that the scattering cross section from a target with a certain structure can be expressed in terms of two 'structure functions' $W_1(Q^2, \nu)$ and $W_2(Q^2, \nu)$. The former describes the interaction between the electron and nucleon magnetic moments, and as such is sensitive to the current density distribution in the nucleon; the latter describes the interaction between the charges and is sensitive to the charge distribution. In the kinematic conditions of the experiments that we shall consider the contribution of W_1 is negligible and we have

$$\frac{d\sigma}{d\Omega \, dE'} = \left(\frac{d\sigma}{d\Omega}\right)_{\text{point}} W_2\left(Q^2, \nu\right). \qquad (6.30)$$

To determine the function $W_2(Q^2, \nu)$ experimentally one measures the deep inelastic differential cross section at several values of Q^2 and ν, or, in practice, for different beam energies and scattering angles.

The main results of the measurements made at SLAC with the above-described spectrometers are shown in Fig. 6.12. Three sets of data points are shown, each for a different fixed value of the hadronic mass W. The points are the measured cross

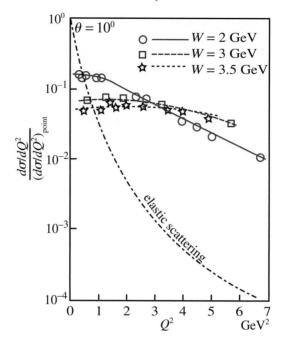

Fig. 6.12.   Deep inelastic cross sections measured at SLAC. (Breidenbach *et al.* 1969 © Nobel Foundation 1990)

section values divided by the computed point-like cross section, namely $W_2(Q^2, v)$, as functions of Q^2. Surprisingly, we see that in the deep inelastic region, namely for large enough values of W, the function W_2 does not vary, or varies just a little, with Q^2 and, moreover, is independent of W. It is difficult to avoid the conclusion that the nucleon contains point-like objects, as for the nucleus in the Geiger and Marsden experiment. Notice, for comparison, the steep decrease of the elastic cross section.

The deep physical implications of the experimental data were identified by R. Feynman (Feynman 1969). Initially he used the name 'partons' for the hard objects inside the nucleons, which were later identified as the quarks. We follow his argument and consider the scattering process in a frame in which the proton moves with a very large four-momentum, P_μ. In this frame we can neglect the transverse momenta of the partons and consider them to be moving all in the same direction with very large, but not necessarily equal, momenta, as schematically shown in Fig. 6.13.

Let us indicate by x the fraction of four-momentum of a given parton. Therefore its four-momentum is xP_μ.

Feynman put forward the hypothesis that the electron–parton collision can be considered as taking place on a free parton. We shall justify this 'impulse approximation' in Section 6.6.

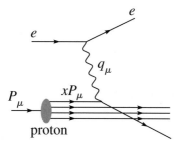

Fig. 6.13. Proton structure and kinematics of deep inelastic scattering in the infinite-momentum frame.

Let q_μ be the four-momentum transferred from the electron to the parton. Let us assume the mass m of the parton to be negligible and write

$$m^2 = (xP_\mu + q_\mu)(xP^\mu + q^\mu) \approx 0$$

i.e.

$$x^2 m_p^2 - Q^2 + 2xP_\mu q^\mu = 0$$

and, if $Q^2 \gg x^2 m_p^2$

$$x = \frac{Q^2}{2P_\mu q^\mu} = \frac{Q^2}{2\nu m_p}. \tag{6.31}$$

If this model is correct, the dependence of the structure function on Q^2 for a fixed x is the Fourier transform of the charge distribution in the parton that is found at x. If the parton is point-like then that transform is a constant, independent of Q^2. Moreover, the structure function should depend on x only. In other words, the function should not vary when ν and Q^2 vary, provided their ratio is kept constant. This property is known as the Bjorken 'scaling law' from the name of its discoverer (Bjorken 1969).

Let us now move on to another pair of kinematic variables, x and Q^2, and let us also define the dimensionless structure function F_2 (while W_2 is dimensionally an inverse energy)

$$F_2(x, Q^2) \equiv \nu W_2(Q^2, \nu). \tag{6.32}$$

The scaling law foresees that the values of F_2 measured for different values of Q^2 must be equal if x is the same. This is just what is shown by the data, as we shall see immediately, confirming that the scattering centres inside the nucleon are point-like and hard. Therefore, the quarks, which had been introduced to explain hadron spectroscopy, are physical, not purely mathematical, objects.

Moreover, the nucleons, and in general the hadrons, contain much more than what is shown by spectroscopy. In summary, high-resolution probes, principally electron and neutrino beams, have shown that the nucleons contain the following components:

- the three quarks that determine the spectroscopy, called 'valence' quarks
- the gluons that are the quanta of the colour field
- the quark and antiquark of the 'sea'. In fact, the following processes continually happen in the intense colour field: a gluon materialises in a quark–antiquark pair, which soon annihilates, two gluons fuse into one, etc. The sea contains quark–antiquark pairs of all flavours, with decreasing probability for increasing quark masses. Therefore, there are many $u\bar{u}$ and $d\bar{d}$ pairs, fewer $s\bar{s}$ and even fewer $c\bar{c}$. Obviously, there are as many sea quarks and antiquarks for each flavour.

We define $f(x)$ as the distribution of momentum fraction for the quark of f flavour; consequently, $f(x)\,dx$ is the probability that this quark carries a momentum fraction between x and $x+dx$ and $xf(x)\,dx$ is the corresponding amount of momentum fraction.

We also call $\bar{f}(x)$ the analogous function for the antiquark of f flavour and $g(x)$ that of the gluons. Having no electric and no weak charges, the gluons are not seen either by electrons or by neutrinos. These functions are called parton distribution functions (PDF).

Since the charm contribution is small, we shall neglect it for simplicity. We have 12 functions of x to determine experimentally, the distribution functions of the up, down and strange quarks and of their antiquarks in the proton and in the neutron. However, not all these functions are independent.

The isospin invariance gives the following relationships between proton and neutron distribution functions

$$u_p(x) = d_n(x) \qquad d_p(x) = u_n(x) \qquad \bar{d}_p(x) = \bar{u}_n(x) \qquad \bar{u}_p(x) = \bar{d}_n(x) \quad (6.33)$$

and

$$s_p(x) = s_n(x) \qquad \bar{s}_p(x) = \bar{s}_n(x). \tag{6.34}$$

Finally, the sea quarks have the same distributions of antiquarks of the same flavour and letting $s(x) \equiv s_p(x) = s_n(x)$, we have

$$s(x) = \bar{s}(x). \tag{6.35}$$

We are left with five independent functions. We call $u(x)$ the distribution of the u quark in the proton and of the d quark in the neutron, $u(x) \equiv u_p(x) = d_n(x)$, and similarly $d(x) \equiv d_p(x) = u_n(x)$, etc.

Notice that the u and d distribution functions contain the contributions both of the valence (u_v and d_v) and of the sea quarks (u_s and d_s). The sea contribution is equal to the distribution function of its antiquark

$$\bar{u}(x) = u_s(x) \quad \bar{d}(x) = d_s(x). \tag{6.36}$$

Experimentally, the distribution functions are obtained from the measurement of the deep inelastic differential cross sections of electrons, neutrinos and antineutrinos.

The electrons 'see' the quark charge, which, in units of the elementary charge, we call z_f for the quarks and $-z_f$ for the antiquarks. The measured structure function is the sum of the contributions of all the q and $\bar{q}$, weighted with the square of the charge z_f^2. Therefore the electrons do not distinguish quarks from antiquarks. We have

$$F_2(x) = x \sum_f z_f^2 [f(x) + \bar{f}(x)]. \tag{6.37}$$

With proton and neutron targets we have for the electron–proton scattering

$$\frac{F_2^{ep}(x)}{x} = \frac{4}{9}[u(x) + \bar{u}(x)] + \frac{1}{9}[d(x) + \bar{d}(x) + s(x) + \bar{s}(x)] \tag{6.38a}$$

and for the electron–neutron scattering

$$\frac{F_2^{en}(x)}{x} = \frac{4}{9}[d(x) + \bar{d}(x)] + \frac{1}{9}[u(x) + \bar{u}(x) + s(x) + \bar{s}(x)]. \tag{6.38b}$$

Both muon neutrino and antineutrino beams can be built at a proton accelerator as we have seen in Section 2.4. These are very powerful probes because they see different quarks. The reactions

$$\nu_\mu + d \to \mu^- + u \quad \nu_\mu + \bar{u} \to \mu^- + \bar{d} \quad \bar{\nu}_\mu + u \to \mu^+ + d \quad \bar{\nu}_\mu + \bar{d} \to \mu^+ + \bar{u} \tag{6.39}$$

are allowed, while

$$\nu_\mu + u \to \mu^+ + d \quad \nu_\mu + \bar{d} \to \mu^+ + \bar{u} \quad \bar{\nu}_\mu + d \to \mu^- + u \quad \bar{\nu}_\mu + \bar{u} \to \mu^- + \bar{d} \tag{6.40}$$

violate the lepton number and are forbidden. One might expect to have four independent processes, neutrino and antineutrino beams on proton and neutron targets, but only two of them are such, as can be easily seen. We then consider the proton targets only. By measuring the cross sections of the deep inelastic scatterings

$$\nu_\mu + p \to \mu^+ + X \quad \bar{\nu}_\mu + p \to \mu^- + X \tag{6.41}$$

one extracts the structure functions

$$\nu_\mu p \qquad \frac{F_2^{\nu_\mu p}(x)}{x} = 2[d(x) + \bar{u}(x)] \qquad (6.42a)$$

$$\bar{\nu}_\mu p \qquad \frac{F_2^{\bar{\nu}_\mu p}(x)}{x} = 2[u(x) + \bar{d}(x)] \qquad (6.42b)$$

where the factor 2 comes from the $V - A$ structure of the weak interaction that we shall study in Chapter 7. Actually, there are other pieces of experimental information that we shall not discuss. Without entering into details we give the results in Fig. 6.14. The bands are the uncertainties on the corresponding function.

We make the following observations: the valence quark distributions have a broad maximum in the range $x = 0.15$–0.3, and go to zero both for $x \to 0$ and for $x \to 1$. The probability of a valence quark having more than, say, 70% of the momentum is rather small. The sea quarks, on the contrary, have high probabilities at very low momentum fractions, less than $x \approx 0.3$.

Question 6.2   Show that the cross sections $\nu_\mu n$ and $\bar{\nu}_\mu n$ give the same relationships as (6.42).

One might think that the sum of the momenta carried by all the quarks and antiquarks is the nucleon momentum, but this is not so. Indeed, integrating the measured distribution functions one obtains

$$\int_0^1 x[u(x) + d(x) + \bar{u}(x) + \bar{d}(x) + s(x) + \bar{s}(x)]dx \approx 0.50. \qquad (6.43)$$

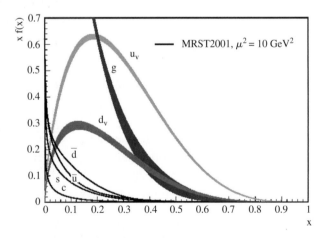

Fig. 6.14.   The parton distribution functions. (Yao *et al.* 2006 by permission of Particle Data Group and the Institute of Physics)

Half of the momentum is missing! We conclude that 50% of the nucleon momentum is carried by partons that have neither electric nor weak charges. These are the gluons.

Figure 6.14 includes the gluon distribution function, as obtained by difference from the total. It can be seen that the gluon contribution is important at low-momentum fractions, say below 0.2, becoming dominant below 0.1.

The HERA electron–proton collider was built at the DESY laboratory with the main aim of studying proton structure functions in a wide range of kinematic variables with a high resolving power. The colliding beams were electrons at 30 GeV and protons at 800 GeV. The two experiments ZEUS and H1 measured the structure function F_2 with high accuracy in the momentum transfer range $2.7 < Q^2 < 30\ 000\ \mathrm{GeV}^2$. Figure 6.15 shows the results at different x values for $6 \times 10^{-5} < x < 0.65$. Notice that data from fixed target experiments are also included.

We see that for the main range of x values, say for $x > 0.1$, the structure function F_2 is substantially independent of Q^2. The scaling law, as we anticipated, is experimentally verified.

However, the data show that at small x values the scaling law is falsified: the structure function increases with increasing Q^2, namely when we look into the proton with increasing resolving power. The scaling law violations bring us beyond the naïve parton model, which is only a first approximation. Indeed, they had been theoretically predicted by Yu. L. Dokshitzer (Dokshitzer 1977), V. N. Gribov and L. N. Lipatov (Gribov & Lipatov 1972), G. Altarelli and G. Parisi (Altarelli & Parisi 1977) between 1972 and 1977 (DGLAP). The theoretical predictions are the curves in Fig. 6.15 and, as we can see, they are in perfect agreement with the data, proof of the validity of the theory.

We try to understand the phenomenon with the help of Fig. 6.16. The quarks in the nucleon emit and absorb gluons, with higher probability at lower x.

Consider a quark with momentum fraction x emitting a gluon, which takes the momentum fraction $x - x'$. Therefore the quark momentum fraction becomes x', which is less than before the emission. If the resolving power is not sufficient (Q^2 not large) one sees the quark and the gluon as a single object and measures x. If Q^2 is large enough one resolves the two objects and measures the quark momentum fraction to be x'. Therefore, at small x the distribution functions increase with increasing resolving power Q^2.

The scaling law violations depend, as one can understand, on the coupling a_s (and on its Q^2 dependence). The curves in Fig. 6.15 have been calculated with a_s left as a free parameter, determined by the best fit of the curves to the data. This is one of the ways in which a_s is determined.

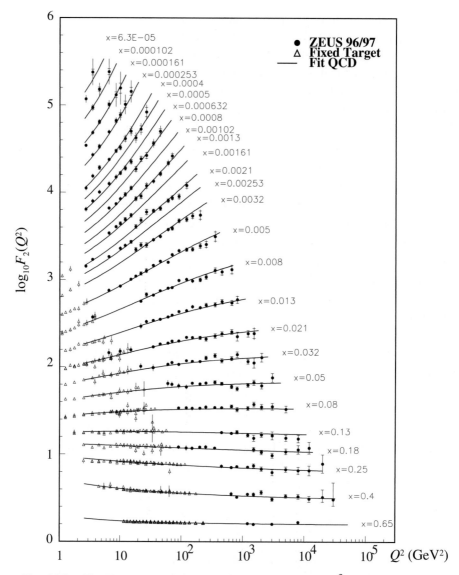

Fig. 6.15.   The F_2 proton structure function as a function of Q^2 at different values of x as measured by the ZEUS experiment at HERA and by several fixed target experiments. Lines are the DGLAP theoretical predictions. (Adapted from Chekanov *et al.* 2001)

Fig. 6.16.   Looking at partons with smaller and larger resolving power.

6.4 The colour charges

The gauge symmetry of the strong interaction is $SU(3)$. It is an exact symmetry or, in other words, the colour charges are absolutely conserved. Since this group is more complex than the $U(1)$ group of QED, the colour charge structure is more complex than that of the electric charge.

In both cases the charges are in a fundamental representation of the group. For $U(1)$ this is simply a singlet. Actually $SU(3)$ has two fundamental representations, **3** and **$\bar{3}$**. Correspondingly, there are three different charges, called red, green and blue (R, G, B). Each of them can have two values, say $+$ and $-$. The former are in the **3**, the latter in the **$\bar{3}$**. The quarks have colour charges $+$, the antiquarks $-$. By convention, instead of speaking of positive and negative colour, one speaks of colour and anticolour; for example a negative red charge is called antired. We shall use this convention but one can easily think of charges of both signs for every colour, as the reader prefers. The strong force depends only on the colour, it is independent of the flavour and the electric charge. However, the colour charge cannot be measured. Consequently, the probability of finding a quark in one of the colours is 1/3 in every instance.

The gluons belong to the octet that is obtained by 'combining' a colour and an anticolour

$$\mathbf{3} \otimes \bar{\mathbf{3}} = \mathbf{8} \oplus \mathbf{1}. \tag{6.44}$$

We see that the situation is similar to that which we met in the quark model. Indeed, we are dealing with the same symmetry group, i.e. $SU(3)$. We can then profit by the analogy, but keeping in mind that it is only formal. In this analogy the colour triplet **3** corresponds to the flavour quark triplet d, u, s and the anticolour antitriplet **$\bar{3}$** to the antiquark antitriplet. Note however that there is no analogue of the isospin.

Recalling Eqs. (4.46) the singlet is

$$g_0 = \frac{1}{\sqrt{3}}(R\bar{R} + B\bar{B} + G\bar{G}) \tag{6.45}$$

which is completely symmetric. In the singlet the colour charges neutralise each other. As a result it does not interact with the quarks. Consequently there is no singlet gluon.

By analogy with the meson octet, the eight gluons are

$$g_1 = R\bar{G} \quad g_2 = R\bar{B} \quad g_3 = G\bar{R} \quad g_4 = G\bar{B} \quad g_5 = B\bar{R} \quad g_6 = B\bar{G}$$

$$g_7 = \frac{1}{\sqrt{2}}(R\bar{R} - G\bar{G}) \quad g_8 = \frac{1}{\sqrt{6}}(R\bar{R} + G\bar{G} - 2B\bar{B}). \tag{6.46}$$

The meson octet contains three meson–antimeson pairs, i.e. π^+ and π^-, K^+ and K^- and K^0 and $\bar{K}^0$. Similarly, six gluons have a colour and a different anticolour and make up three particle–antiparticle pairs: g_1 and g_3, g_2 and g_5, g_4 and g_6. The other two are antiparticles of themselves (completely neutral). Notice that g_7, analogous to π^0, has two colours and two anticolours and that g_8, analogous to η_8, has three colours and three anticolours. There is no octet–singlet mixing because the $SU(3)$ symmetry is unbroken.

Figure 6.17(a) shows, for comparison, the vertex of the electromagnetic interaction. The ingoing and outgoing particles are equal, their charge is z_1 (elementary charges). The overall interaction amplitude between two charges z_1 and z_2 is proportional to $z_1 z_2 a$.

The chromodynamic vertex is more complex, as shown in Fig. 6.18. First of all, the incoming and outgoing fermions may be different, for example two quarks of the same flavour and different colours; the gluon has the colour of one of them and the opposite of the colour of the other. Secondly, the vertex contains not only the coupling $\sqrt{a_s}$ but also a 'colour factor' $\kappa_\lambda^{c_i \bar{c}_j}$, where c_i and c_j are the colours of the two quarks and λ is the gluon type. Finally, by convention, there is a factor $1/\sqrt{2}$.

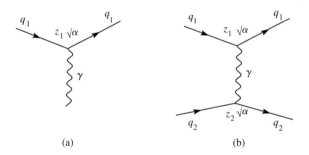

Fig. 6.17.   The electromagnetic vertex and scattering.

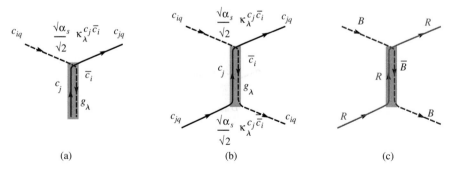

Fig. 6.18.   (a) General quark–gluon vertex, showing the colour lines; (b) quark–quark scattering with gluon exchange; (c) a blue-quark–red-quark scattering.

The gluon is usually represented by a helix, without a direction. However, it should have a direction for every colour it carries. For example g_1 moving backward in time is g_3 moving forward. In this section we shall highlight this by drawing the gluon as a grey band in which the colour flows are represented by arrows, as in Fig. 6.18.

Figure 6.18(c) shows an example of colour interaction, a diagram contributing to the scattering of a B (blue) quark and an R (red) quark. A blue quark changes to red by emitting a blue-antired gluon that is absorbed by a red quark changing to blue. The same process can be seen also as a red quark changing to blue by emitting a red-antiblue gluon that is absorbed by a blue quark changing to red. The general rule is that the colour lines are continuous through the diagram.

However, a quantitative evaluation requires not only following the colour but also including the appropriate colour factors, which are simply the numerical factors appearing in Eq. (6.46), namely

$$\kappa_1^{R\bar{G}} = 1 \quad \kappa_2^{R\bar{B}} = 1 \quad \kappa_3^{G\bar{R}} = 1 \quad \kappa_4^{G\bar{B}} = 1 \quad \kappa_5^{B\bar{R}} = 1 \quad \kappa_6^{B\bar{G}} = 1 \quad \kappa_7^{R\bar{R}} = \frac{1}{\sqrt{2}}$$

$$\kappa_7^{G\bar{G}} = -\frac{1}{\sqrt{2}} \quad \kappa_8^{R\bar{R}} = \frac{1}{\sqrt{6}} \quad \kappa_8^{G\bar{G}} = \frac{1}{\sqrt{6}} \quad \kappa_8^{B\bar{B}} = -\frac{2}{\sqrt{6}}. \tag{6.47}$$

The colour factors of antiquarks are the opposite of those of quarks.

We now look at two examples. Let us start with the interaction between two quarks of the same colour, BB for example

$$^{B}q + {}^{B}q \rightarrow {}^{B}q + {}^{B}q \tag{6.48}$$

shown in Fig. 6.19. From Eqs. (6.47) we see that only one gluon can mediate this interaction, g_8. We have

$$\frac{1}{\sqrt{2}}\kappa_8^{B\bar{B}} \frac{1}{\sqrt{2}}\kappa_8^{B\bar{B}} = \frac{1}{2}\left(\frac{-2}{\sqrt{6}}\right)\left(\frac{-2}{\sqrt{6}}\right) = \frac{1}{3}. \tag{6.49}$$

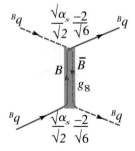

Fig. 6.19.   Interaction between two blue quarks.

Now consider *RR* (Fig. 6.20)

$$^{R}q + {}^{R}q \to {}^{R}q + {}^{R}q. \tag{6.50}$$

We have two contributions, g_7 and g_8. We sum them

$$\frac{1}{\sqrt{2}}\kappa_7^{R\bar{R}} \frac{1}{\sqrt{2}}\kappa_7^{R\bar{R}} + \frac{1}{\sqrt{2}}\kappa_8^{R\bar{R}} \frac{1}{\sqrt{2}}\kappa_8^{R\bar{R}} = \frac{1}{2}\left(\frac{1}{\sqrt{2}}\right)\left(\frac{1}{\sqrt{2}}\right) + \frac{1}{2}\left(\frac{1}{\sqrt{6}}\right)\left(\frac{1}{\sqrt{6}}\right) = \frac{1}{3}. \tag{6.51}$$

As expected from the symmetry, the force between *R* and *R* is the same as between *B* and *B*. The positive sign means that the force is repulsive. As in electrostatics, same-sign colour charges repel each other.

Question 6.3   Verify the intensity of the force between *R* and *G*.

Since gluons are coloured, they can interact coupled by continuous colour lines, as shown in Fig. 6.21(a). In this example a red-antiblue gluon 'splits' into a green-antiblue gluon and a red-antigreen gluon. Gluon–gluon scattering can

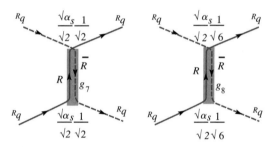

Fig. 6.20.   Interaction between two red quarks.

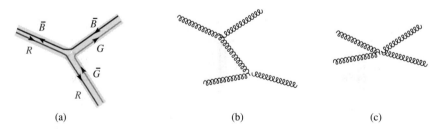

Fig. 6.21.   (a) Three-gluon vertex: a red-antiblue gluon splits into a red-antigreen and a green-antiblue gluons; (b) gluon–gluon scattering with gluon exchange; (c) direct four-gluon scattering.

happen by exchanging another gluon as shown in Fig. 6.21(b), which includes two vertices of the type in Fig. 6.21(a). A further contribution to the gluon–gluon scattering is the four-gluon coupling shown in Fig. 6.21(c).

6.5 Colour bound states

The hadrons do not have any colour charge, but are made up of coloured quarks. It follows that the colour charges of these quarks must form a 'neutral' combination. An electromagnetic analogue is the atom, which is neutral because it contains as many positive charges as negative ones. In QCD the neutrality is the colour singlet state. Let us see how this happens for mesons and baryons.

We start with the mesons, the simpler case. They are bound quark–antiquark states. The colour of the quark is in the **3** representation, the colour of the antiquark in **3̄**. They bind because their product contains the singlet

$$\mathbf{3} \otimes \mathbf{\bar{3}} = \mathbf{8} \oplus \mathbf{1}. \tag{6.52}$$

The singlet state is

$$(q\bar{q})_{\text{singlet}} = \frac{1}{\sqrt{3}} \left({}^B q\, {}^{\bar{B}}\bar{q} + {}^R q\, {}^{\bar{R}}\bar{q} + {}^G q\, {}^{\bar{G}}\bar{q} \right). \tag{6.53}$$

We notice that, by symmetry, the interactions between the three pairs in this expression are equal. So it is enough to compute one of them, say ${}^B q\, {}^{\bar{B}}\bar{q}$, and multiply by 3. In the calculation we must take all the possibilities into account; the initial state is ${}^B q\, {}^{\bar{B}}\bar{q}$, but the final state can be any quark–antiquark pair. Consequently, we have the diagrams of Fig. 6.22.

Recalling that the antiquark colour factors are opposite to those of the quarks, and including the normalisation factor $(1\sqrt{3})^2$, we find the total colour factor for

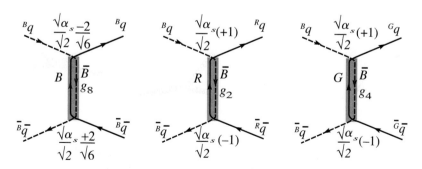

Fig. 6.22. Diagrams for blue-quark–antiquark interaction.

the (6.53) interaction

$$3\left(\frac{1}{\sqrt{3}}\right)^2 \frac{1}{2}\left[\kappa_8^{B\bar{B}}\kappa_8^{\bar{B}B} + \kappa_2^{R\bar{B}}\kappa_2^{\bar{R}B} + \kappa_4^{G\bar{B}}\kappa_4^{\bar{G}B}\right]a_s = \frac{1}{2}\left(-\frac{4}{6} - 1 - 1\right)a_s = -\frac{4}{3}a_s.$$

$$(6.54)$$

Notice in particular the negative sign. As in electrostatics, two opposite charges attract each other.

Now let us consider the baryons, which contain three quarks. Their colours are in the **3** representation. As the product $\mathbf{3} \otimes \mathbf{3} \otimes \mathbf{3} = \mathbf{10} \oplus \mathbf{8} \oplus \mathbf{8} \oplus \mathbf{1}$ contains a singlet (the neutral colour combination) three quarks can bind together. Let us see the structure in detail, starting from the first product

$$\mathbf{3} \otimes \mathbf{3} = \mathbf{6} \oplus \bar{\mathbf{3}}. \tag{6.55}$$

The **6** is symmetric, the $\bar{\mathbf{3}}$ is antisymmetric. Taking the second product we have

$$(\mathbf{3} \otimes \mathbf{3}) \otimes \mathbf{3} = \mathbf{6} \otimes \mathbf{3} \oplus \bar{\mathbf{3}} \otimes \mathbf{3}. \tag{6.56}$$

There is no singlet in the product $\mathbf{6} \otimes \mathbf{3}$, the only one is in $\bar{\mathbf{3}} \otimes \mathbf{3}$ as shown by Eq. (6.52). In conclusion, every quark pair inside a baryon is in the *antisymmetric* colour $\bar{\mathbf{3}}$ and couples with the third quark to form the singlet. Recalling the discussion in Section 4.8 we have

$$(qqq)_{\text{singlet}} = \frac{1}{\sqrt{6}}\left[\left({}^R q^B q - {}^B q^R q\right){}^G q + \left({}^G q^R q - {}^R q^G q\right){}^B q + \left({}^B q^G q - {}^G q^B q\right){}^R q\right].$$

$$(6.57)$$

It is easy to show that the colour factors of the three addenda are equal. We then calculate one of them, say the first, and multiply by 6. The two contributing diagrams are shown in Fig. 6.23, corresponding to ${}^R q + {}^B q \to {}^R q + {}^B q$ and ${}^R q + {}^B q \to {}^B q + {}^R q$.

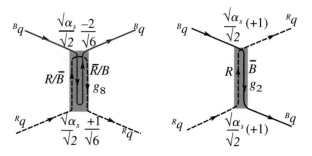

Fig. 6.23. Diagrams for blue-quark–red-quark interaction.

We must be careful with the signs. The contribution of the second process must be taken with a minus sign because the final quarks are inverted and because the wave function is antisymmetric. We obtain

$$6 \left(\frac{1}{\sqrt{6}}\right)^2 \frac{1}{2} \left[\kappa_8^{R\bar{R}} \kappa_8^{B\bar{B}} - \kappa_2^{R\bar{B}} \kappa_2^{R\bar{B}} \right] a_s = \frac{1}{2} \left[-\frac{2}{6} - 1 \right] a_s = -\frac{2}{3} a_s. \tag{6.58}$$

The negative result implies a very important difference from the electric charges. Two different colour charges in an antisymmetric combination attract the charge of the third colour of the same sign. For example the combination $(RB - BR)$ attracts G roughly as $\bar{G}$ does.

Question 6.4 Calculate the contribution of the third addendum in (6.57).

Question 6.5 Given three objects R, G and B, how many symmetric combinations of the three, equal or different, can be made? How many antisymmetric combinations?

The characteristics of the colour charges that we have seen are demonstrated by the 'hyperfine structure' of the meson and baryon spectra. We start by recalling the hyperfine structure of the hydrogen atom, which is made of two opposite charge spin 1/2 particles, bound by the electromagnetic interaction. Since the photon is a vector particle, the hyperfine structure term appears as the interaction between the magnetic moments of the proton and the electron

$$\Delta E \propto -\boldsymbol{\mu}_e \cdot \boldsymbol{\mu}_p \propto -q_e q_p \mathbf{s}_1 \cdot \mathbf{s}_2 \tag{6.59}$$

where q_p and q_e are their equal and opposite electric charges. Consider in particular the S states. The two spins can be parallel ($^3S_1$, $J = 1$) or antiparallel ($^1S_0$, $J = 0$). The energy difference $E(^3S_1) - E(^1S_0)$ is very small and positive.

The mesons are also bound states of two spin 1/2 opposite colour particles. The $^3S_1$ states are the vector mesons, the $^1S_0$ states the pseudoscalar mesons. The differences are now large, and positive; for example $m(K^*) - m(K) = 395$ MeV. Now consider a baryon and take two of its quarks. If their total spin is 1 the baryon is in the decimet, if it is 0, it is in the octet. The separation between the levels is again large and again positive; for example $m(\Delta) - m(p) = 293$ MeV.

The interaction responsible for the separation between the levels is mediated for QCD, as for QED, by massless vector bosons and appears as an interaction between 'colour magnetic moments'. These have the direction of the spins, but the charges to be considered are the colour charges. Therefore, we have

$$\Delta E \propto -\boldsymbol{\mu}_e \cdot \boldsymbol{\mu}_p \propto -\kappa_1 \kappa_2 \mathbf{s}_1 \cdot \mathbf{s}_2 \tag{6.60}$$

where κ_1 and κ_2 are the colour factors. There are two important differences with respect to electrodynamics. Firstly, the separation is much larger, because the colour coupling is bigger. The second difference requires more discussion.

Let us start with the mesons. Recall that if $\mathbf{J}$ is the sum of $\mathbf{s}_1$ and $\mathbf{s}_2$, from

$$\langle \mathbf{J}^2 \rangle = \langle (\mathbf{s}_1 + \mathbf{s}_2)^2 \rangle = \langle \mathbf{s}_1^2 \rangle + \langle \mathbf{s}_1^2 \rangle + 2 \langle \mathbf{s}_1 \cdot \mathbf{s}_2 \rangle$$

we have

$$\langle 2\mathbf{s}_1 \cdot \mathbf{s}_2 \rangle = J(J+1) - s_1(s_1+1) - s_2(s_2+1) = J(J+1) - \frac{3}{2}$$

and

$$\langle 2\mathbf{s}_1 \cdot \mathbf{s}_2 \rangle = -\frac{3}{2} \quad \text{for } J = 0; \qquad \langle 2\mathbf{s}_1 \cdot \mathbf{s}_2 \rangle = +\frac{1}{2} \quad \text{for } J = 1. \qquad (6.61)$$

We shall now see that if the colour charges were to behave like the electric charges the resulting hyperfine structure would be wrong. In this hypothesis the product of the charges would be $\kappa_1 \kappa_2 = -1$ and

$$\Delta E \propto - \kappa_1 \kappa_2 \mathbf{s}_1 \cdot \mathbf{s}_2 = + \mathbf{s}_1 \cdot \mathbf{s}_2 \propto -\frac{3}{2} \quad \text{for } J = 0 \qquad \Delta E \propto +\frac{1}{2} \quad \text{for } J = 1. \qquad (6.62)$$

Calling K a positive proportionality constant, we have

$$m(^3S_1) - m(^1S_0) = + 2K. \qquad (6.63)$$

In the case of the baryons we must consider the contributions of all the quark pairs and sum them. Let us start with the sum of the internal products

$$\Sigma \equiv \langle 2(\mathbf{s}_1 \cdot \mathbf{s}_2 + \mathbf{s}_2 \cdot \mathbf{s}_3 + \mathbf{s}_3 \cdot \mathbf{s}_1) \rangle$$
$$= \langle (\mathbf{s}_1 + \mathbf{s}_2 + \mathbf{s}_3) - s_1(s_1+1) - s_2(s_2+1) - s_3(s_3+1) \rangle = J(J+1) - \frac{9}{4}$$

and

$$\Sigma = -\frac{3}{2} \quad \text{for } J = \frac{1}{2} \qquad \Sigma = +\frac{3}{2} \quad \text{for } J = \frac{3}{2}. \qquad (6.64)$$

If it were as for the electric charges we would have $\kappa_1 \kappa_2 = +1$ and consequently

$$\Delta E \propto -\kappa_1 \kappa_2 \, \mathbf{s}_1 \cdot \mathbf{s}_2 \propto +\frac{3}{2} \quad \text{for } J = \frac{1}{2} \qquad \Delta E \propto -\frac{3}{2} \quad \text{for } J = \frac{3}{2}$$

and

$$m(\mathbf{10}) - m(\mathbf{8}) = -3K. \tag{6.65}$$

In conclusion, if the colour charges were to behave like the electric charges, the vector meson masses should be larger than the pseudoscalar meson masses, which is correct, and the masses of the decimet should be smaller than those of the octet, which is wrong. Moreover, in absolute value, the hyperfine structure of the baryons would be one and a half times larger than that of the mesons; instead it is somewhat smaller.

However, the colour force structure is given by $SU(3)$ not by $U(1)$ and the colour factors given by (6.54) and (6.58) must be considered. We have

$$\begin{aligned} m({}^3S_1) - m({}^1S_0) &= -2K \times (-4/3)a_s = +a_s K \times 8/3 \\ m(\mathbf{10}) - m(\mathbf{8}) &= -3K \times (-2/3)a_s = +a_s K \times 2. \end{aligned} \tag{6.66}$$

The predicted mass splittings have the same sign and the splitting for the mesons is larger than for the baryons, as experimentally observed.

6.6 The evolution of a_s

The strong interaction coupling constant a_s, which is dimensionless as is a, is renormalized in a similar manner, but with a fundamental difference. As shown in Fig. 6.24, we must include in the vertex expansion not only fermion loops, but also gluon loops, due to the fact that the gluons carry colour charges. The theory shows that bosonic and fermionic loop contributions have opposite signs.

The effect of vacuum polarisation due to the quarks is similar to that which we have seen in electrodynamics, with the colour charges in place of the electric charge. The quark–antiquark pairs coming out of vacuum shield the colour charge, reducing its value for increasing distance, or for increasing momentum transfer in the measuring process.

However, the action of gluons is a smearing of the colour charge, which results in an effect of the opposite sign from that of quarks, called 'antiscreening'. The net

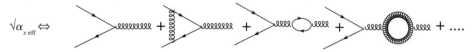

Fig. 6.24. The lowest-order diagrams contributing to the QCD vertex, illustrating the relationship between the 'naked' coupling constant and the 'effective' (measured) one.

result is that the colour charges *decrease* with decreasing distance. D. Politzer (Politzer 1973), D. Gross and F. Wilczek (Gross & Wilczek 1973) discovered this property theoretically in 1973. The expression they found for the evolution of a_s is

$$a_s\left(|Q|^2\right) = \frac{a_s(\mu^2)}{1 + \dfrac{a_s(\mu^2)}{12\pi}(33 - 2n_f)\ln\left(|Q|^2/\mu^2\right)}. \tag{6.67}$$

As in QED, the theory does not specify the constant μ, a parameter that must be determined experimentally. The quantity n_f is the number of quark flavours effectively contributing to the loops, namely those with mass about $m_f < |Q|$. We see that the coupling constant decreases when $|Q|^2$ increases because $(33 - 2n_f)$ is always positive since n_f is never larger than 6. As we saw for a, the dependence of the reciprocal of a_s on $\ln(|Q|^2/\mu^2)$ is linear, but for the important effect of the opening of thresholds, it is given by

$$a_s^{-1}\left(|Q|^2\right) = a_s^{-1}(\mu^2) + \frac{33 - 2n_f}{12\pi}\ln\left(|Q|^2/\mu^2\right). \tag{6.68}$$

Equation (6.67) can be usefully written in an equivalent form, defining a scale λ_{QCD}, with the dimension of a mass, as the free parameter in lieu of μ

$$\lambda_{\mathrm{QCD}}\left(n_f\right) \equiv \mu^2 \exp\left[-\frac{12\pi}{(33 - 2n_f)a_s(\mu^2)}\right]. \tag{6.69}$$

With this definition we have

$$a_s\left(Q^2\right) = \frac{12\pi}{(33 - 2n_f)\ln\left(|Q|^2/\lambda_{\mathrm{QCD}}^2\right)}. \tag{6.70}$$

We showed explicitly in (6.68) the dependence of λ_{QCD} on the number of excited flavours, which, in turn, depends on $|Q|^2$. The fundamental λ_{QCD} parameter, called 'lambda-QCD', is obtained from the experimentally measured dependence of a_s on $|Q|^2$. Its values for three and four excited quarks are

$$\lambda_{\mathrm{QCD}}(3) \approx 400\,\mathrm{MeV} \qquad \lambda_{\mathrm{QCD}}(4) \approx 200\,\mathrm{MeV}. \tag{6.71}$$

Lambda-QCD is important because it separates two energy regimes. For energies less than λ_{QCD} the coupling constant is large and a perturbative development of the physical quantities is impossible. When two quarks are very close, namely when the momentum transfer is large, their interaction is feeble, a property called 'asymptotic freedom'.

Like a, α_s cannot be measured directly, but must be extracted from measured quantities by a theoretical calculation. An important difference from the case of QED is that in QCD one cannot use a perturbative expansion. Therefore the determination of α_s is intrinsically less accurate than that of a. However, α_s has been extracted in a coherent way from observables measured in a wealth of processes. We quote, for example: (1) the probability of observing a third jet in e^+e^- hadronic processes, which is proportional to α_s; (2) the excess of hadronic production noticed with reference to Fig. 6.3; and (3) the scaling law violations in deep inelastic scattering. Without entering into any detail, we show only the principal measurements in Fig. 6.25. The figure also shows the QCD theoretical prediction. The width of the band is the theoretical uncertainty. Notice the rapid decrease of the coupling. The value of α_s at a few hundred MeV (not shown in the figure) is around 10, but already at 1 GeV it is less than 1. Figure 6.25 shows also, for comparison, the evolution of a, which is slower and increasing.

The antiscreening action of the coloured gluons deserves further discussion, which we shall do by following the arguments of Wilczek. Consider a free quark, with its colour charge. In its neighbourhood the quantum vacuum pulsates; quark–antiquark pairs form and immediately disappear, gluons appear from nothing and fade away. This cloud of virtual particles antiscreens the central quark making the colour charge grow indefinitely with increasing distance from the quark. However, this would

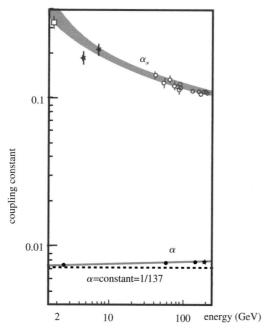

Fig. 6.25.   The evolution of a and α_s. (Courtesy of Mele 2005, CERN)

require an infinite energy, which is impossible. This catastrophic growth can be avoided if near the quark its antiquark is present, because their clouds neutralise each other where they overlap. Therefore, a quark and its antiquark can exist in a finite energy system. The same result as with the antiquark is obtained with a pair of quarks of the two complementary colours, in an antisymmetric state. However, neither a quark, nor an antiquark, nor a quark pair can exist alone for an appreciable time.

The mechanism that keeps quarks and antiquarks permanently inside the hadrons is called confinement. Let us consider the mesons, which are simpler. In a first approximation a meson is made up of a quark–antiquark pair and the colour field, with all its virtual particles, between them. The distance between quark and antiquark oscillates continuously with a maximum elongation of the order of one fermi. Indeed, the attractive force increases when the distance increases, because the cancellation of the two antiscreening clouds decreases. Suppose now that we try to break the meson by sending into it a high-energy particle, an electron for example. If the electron hits, for example, the quark, this will start moving further apart from the antiquark. What happens then?

We try to give a simplified description of a very complex phenomenon. We start with the analogy of the electrostatic force. Figure 6.26 shows the electrostatic field between two equal and opposite charges. When the distance increases, the energy density of the field decreases.

The behaviour of the colour field is different, for reasons we cannot explain here. Figure 6.27(a) shows the colour field-lines between a quark and an antiquark. At distances of about one fermi the colour field is concentrated in a narrow 'tube'. When the separation between quark and antiquark increases, the length of

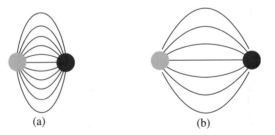

(a)   (b)

Fig. 6.26.   The electrostatic field-lines between two equal and opposite charges. The lines going to infinity are not drawn for simplicity.

(a)   (b)   (c)

Fig. 6.27.   Sketch of the colour field-lines between a quark and an antiquark.

the tube increases, but its diameter remains approximately constant. Therefore the field energy density remains constant and the total energy in the tube increases proportionally to its length. When the energy in the tube is large enough it becomes energetically convenient to break the tube producing a new quark–antiquark pair at the two new ends, as in Fig. 6.27(c). We now have a second meson, which is colour neutral. The process continues and more hadrons are created out of the colour field energy. It is the hadronisation process.

The situation is similar for the quark confinement in a baryon, in which there are three colour tubes.

We can now answer another question. In the SLAC deep inelastic experiments of Section 6.3 a quark is hit by an electron and is suddenly accelerated. Why does it not radiate? Why, in other words, is the impulse approximation a good one? The explanation is again the antiscreen. At small distances from the quark its charge is small and therefore the virtual particle cloud is only feebly attached to the quark. The hit quark darts away leaving its cloud behind, almost as if it had no charge. Later on, when the virtual particles respond to the change, a new cloud forms around the quark and moves with it. However this last process does not imply significant momentum and energy radiation. This is why in the inclusive experiments, which measure only energy and momentum fluxes, the quarks behave as free, even if they are confined in a nucleon.

We also understand now the claim we made in Section 6.1, when we said that soft radiation from a quark is frequent, whilst hard radiation is rare. Indeed, at small momentum transfer the interaction constant is large, but it is small at large momentum transfer.

We saw in Chapter 4 that the decays of the hidden-flavour particles, the ϕ ($s\bar{s}$), the ψs ($c\bar{c}$) and the Υs ($b\bar{b}$), into final states not containing the 'hidden' quark are suppressed. This property was noticed by several authors and became known after the names of three of them (Okubo, Zweig and Iizuka) as the OZI rule. The rule remained purely heuristic until QCD gave the reason for it.

Figure 6.28 shows, as an example, the case of the J/ψ. In the process shown in (a) a soft gluon radiated by one of the quarks materialises in a quark–antiquark

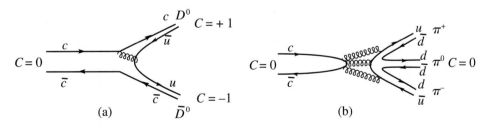

Fig. 6.28. Two diagrams for the J/ψ decay.

pair. This process is favoured by QCD, but forbidden by energy conservation. Therefore, the charm–anticharm pair must annihilate into gluons. How many? The original pair, being colourless, cannot annihilate into one gluon, which is coloured; it cannot annihilate into two gluons, because the process violates charge conjugation. The minimum number of gluons is three, as in Fig. 6.28(b). The norm of their four-momentum is the square of the mass of the decaying meson, m_V. Since this is rather large, the gluons are hard. The decay probability is proportional to $a_s^3(m_V^2)$, which is small. We have

$$a_s^3\left(m_\phi^2\right) \approx 0.5^3 = 0.13; \quad a_s^3\left(m_{J/\psi}^2\right) \approx 0.3^3 = 0.03; \quad a_s^3\left(m_\Upsilon^2\right) \approx 0.2^3 = 0.008.$$

6.7 The origin of hadron mass

In Table 4.5 we gave a summary of the quark quantum numbers, including the values, or ranges of values, of their masses. We have already emphasised that quark masses are not measurable (with the exception of the top), because the quarks are never free. Quark 'masses' can be determined only indirectly through their influence on the properties of the hadrons. This implies that the definition of quark mass has a quantitative meaning only within a specific theoretical framework. Note that historically the first definitions of quark masses were given within the framework of particular quark models. This gave rise to the so-called 'constituent masses'. We have never used this concept; it cannot be rigorously related to the mass parameters of QCD, which are those that we have used. More specifically, the mass of a quark is defined as the parameter that appears in the Lagrangian or, equally, as the mass parameter that appears in the quark propagator.

We may now observe that three of the quarks have masses that are large compared to λ_{QCD}, hence in an energy region in which a_s is substantially smaller than 1, while the opposite is true for the three small-mass quarks. A consequence is that the masses of the large-mass quarks are much less sensitive to the details of the theoretical scheme than those of the small-mass quarks. This explains the differences in the uncertainties in the table.

A fundamental theoretical problem is the calculation of the hadron spectrum and, more generally, of the basic hadronic properties, *ab initio* from QCD. The problem is difficult because the energy scale of the masses of the hadrons made of u, d and s quarks is in the region where the coupling constant a_s is large. As a consequence the contributions of the diagrams of increasing order do not decrease and a perturbative development is not possible. The problem is solved by numerical methods: powerful theoretical techniques have been developed. These are suitable for application on parallel supercomputers, which are, in some cases, designed and built by the theorists themselves. Computing power has reached tens of Teraflops (one Teraflop is 10^{12} floating point operations per second), a level that allows the calculation of the mass spectrum with a small percentage accuracy. This is happening at the time of writing (2007).

Here we can only show the physical essence of the problem on the basis of a very simple model. Consider the proton that has a mass, about 1 GeV, much larger than the sum of the masses of the component u and d quarks, namely, about 10 MeV. What then is the origin of the proton mass?

Let us start with a well-known problem: the mass of the hydrogen atom. Its size, the distance a between electron and proton, is dictated by the uncertainty principle. In fact, the electron potential energy is negative and decreases with decreasing distance between electron and proton. But the localisation of the wave function has an energy cost. The smaller the uncertainty of the electron position, the greater the uncertainty of its momentum, implying that the average value of the momentum itself is larger and, finally, that the average kinetic energy is larger. The atomic radius is the distance at which the sum of potential and kinetic energy is at a minimum. This fact, known from atomic physics, is recalled in Problem 5.2.

If $E(a)$ is the total (kinetic plus potential) electron energy at the distance a, the atom mass is

$$m_{\mathrm{H}} = m_p + m_e + E(a) = m_p + m_e - 13.6\,\mathrm{eV}. \tag{6.72}$$

In words: the mass of the hydrogen atom is the sum of the masses of its constituents and of the work that must be done on the system to move the constituents into a configuration in which their interaction is zero. This configuration, for the atom, is when the constituents are far apart. The work is negative and small in comparison with the masses of the atom's constituents.

Having recalled a familiar case, let us go back to the proton. The QCD interaction amongst the three valence constituent quarks is strong at distances of the order of the proton radius (a little less than a femtometre). On the other hand, if the three quarks were located at the same point they would not interact because the three antiscreening clouds would cancel each other out exactly (in the $SU(3)$ singlet configuration in which they are). This cannot happen precisely because of the energy cost of the localisation of the wave functions. The three quarks adjust their positions at the average distances that minimize the energy, as in the case of the atom. We can take this distance as the proton radius r_p.

We start with the evaluation of the proton mass. Again this is the sum of the masses of the constituent quarks (a small fraction of the total) and of the work that must be done on the system to bring the constituents into a configuration in which they do not interact; this is now where the quarks are *very close* to each other. The work is positive because it corresponds to the extraction of energy from the system (the 'spring' is contracting) and is by far the largest contribution to the proton mass.

The scale of the energy difference between an intense and negligible inter-action is, of course, lambda-QCD. In order of magnitude, the work to bring one quark into a non-interacting configuration is $\lambda_{QCD} \approx 300–400$ MeV. In total (three quarks) we have $m_p \approx 3\,\lambda_{QCD} \approx 1$ GeV.

Having obtained a reasonable value for the proton mass, let us now check if we find a reasonable value for the proton radius. Following the arguments of Section 6.6, we assume that the energy of the colour field increases proportionally to the average distance r between the quarks, say as kr. The quark velocities are close to the speed of light and we can assume their kinetic energy to be equal to their momentum p. In conclusion the energy of the three quarks is $E = 3p + kr$. The uncertainty principle now gives $pr \approx 1$ and we have

$$E = \frac{3}{r} + kr. \tag{6.73}$$

The proton radius is the distance r_p that makes the energy minimum

$$\left(\frac{dE}{dr}\right)_{r_p} = 0 = -\frac{3}{r_p^2} + k \tag{6.74}$$

a relationship between k and r_p

$$k = \frac{3}{r_p^2}. \tag{6.75}$$

We obtain k by stating that the minimum energy must be equal to the proton mass. From (6.73)

$$m_p = E(r_p) = \frac{6}{r_p}. \tag{6.76}$$

For $m_p \approx 1$ GeV we obtain $r_p \approx 1.2$ fm, one and a half times the correct value, which is a good agreement, considering the approximation of our calculation.

In conclusion, the proton (the nucleon) mass would be very small if the three constituent quarks were in exactly the same position, because the antiscreen clouds would cancel each other out. The distance between the quarks is imposed by the energetic cost of the localisation and, in turn, determines, due to the incomplete overlap of the clouds, the proton (nucleon) mass. The largest fraction, 97%, of the proton mass, the largest fraction of the mass of the matter we know, is the energy of the colour field.

The situation is similar for all hadrons that contain only u and d as valence quarks; in the other cases the mass of the quarks makes an appreciable contribution in the case of s, a large contribution in the case of c, and a dominant contribution in the case of b.

Question 6.6   Protons contain two charged quarks, neutrons only one. Evaluate the difference in electrostatic energy between two u quarks when their separation is 0.8 fm and when it is 0.3 fm (where they can be considered asymptotically free).

Question 6.7   What would the Universe have been if the values of the masses of the u and d quarks were inverted?

6.8 The quantum vacuum

We have already discussed in Chapter 5, even if only qualitatively, the 'vacuum polarisation' phenomenon in the vicinity of a particle with electric charge and in this chapter of a particle with colour charge. However, this phenomenon occurs even if no particle is present, as we shall now see.

We say that a region of space is empty, on large scales, if it does not contain particles or fields. Macroscopically, the electromagnetic field is zero and the colour field too. The latter condition is obvious because this field exists only inside hadrons that are absent in vacuum. Quantum mechanics teaches us that the vacuum is not empty at all, but, on small scales, contains virtual particles, their antiparticles, and the quanta of their interactions.

The positron–electron pairs we have met in our discussion of the evolution of α and drawn in Fig. 5.29 are also present in the absence of the central particle of that figure, even in vacuum. Figure 6.29(a) shows a positron–electron pair popping out of the vacuum. It recombines after a time Δt short enough to allow energy conservation to be compatible with the uncertainty of the measurement process, namely

$$\Delta t \leq \frac{1}{2m}. \tag{6.77}$$

Similar processes happen for every fermion–antifermion pair ($\mu^+\mu^-$, $\tau^+\tau^-$ and quark–antiquark), as exemplified in Fig. 6.29(b). In general, the mass m in Eq. (6.77) is the mass of the fermion.

(a)   (b)

Fig. 6.29.   Diagrams of the vacuum polarisation by (a) a positron–electron pair, (b) a quark–antiquark pair.

To fix the scale, a positron–electron pair with $2m \approx 1\,\text{MeV}$ typically lives $\Delta t \approx 6.6 \times 10^{-22}\,\text{s}$, hence in a region smaller than $c\Delta t \approx 200$ fm, while $u\bar{u}$ or $d\bar{d}$ pairs, with masses of an order of magnitude larger, have lifetimes ten times shorter in volumes within a ten femtometre radius.

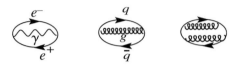

Fig. 6.30. Higher-order diagrams of the vacuum polarisation by positron–electron pairs and quark–antiquark pairs.

Fig. 6.31. Diagrams of the vacuum polarisation by gluons.

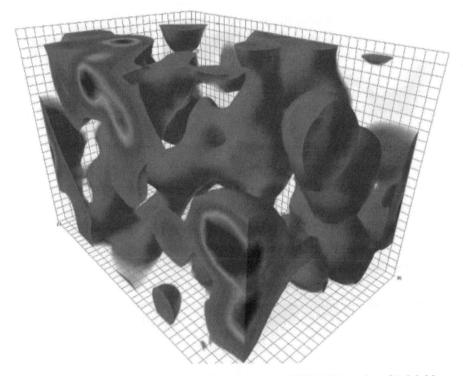

Fig. 6.32. The quantum vacuum. (D. Leinweber, CSSM, University of Adelaide http://www.physics.adelaide.edu.au/theory/staff/leinweber/VisualQCD/)

During the short life of the couple, one or more photons or one or more gluons may be present, as in Fig. 6.30. There is more to it than that: as gluons carry colour charge themselves, only-gluon processes can occur, as in those of Fig. 6.31. Since the gluon mass is zero, these processes take place at all energy scales.

In conclusion, the vacuum, when seen at the scale of a femtometre or less, is alive. It contains mass and energy fluctuations that grow larger at decreasing time and space scales. The fluctuations can be calculated using powerful parallel computers, smoothing over the very small scales. Figure 6.32 shows an example of such calculations made by D. Leinweber. It is a snapshot of the energy fluctuations in a volume of about 2 fm per side. In the time-dependent simulation the energy 'lumps' evolve, changing shape, merging, disappearing and reappearing, but keeping the same general appearance.

Quantum vacuum is an extremely dynamic medium; its properties determine, to a large extent, the properties of matter itself.

The vacuum contains energy, in the same way a hadron does. But the presence of the quarks, real not virtual, in the hadron fosters the materialisation of energy as mass. As we have seen this is 97% of the mass of the matter we know.

Problems

6.1. How many gluons exist? Give the electric charges of each of them. Give the values of their strangeness, charm and beauty. What is the gluon spin? How many different quarks exist for every flavour? What are their charges? Does QCD define the number of families?

6.2. Evaluate $R \equiv \sigma(e^+e^- \to \text{hadrons})/\sigma(e^+e^- \to \mu^+\mu^-)$ at $\sqrt{s} = 2.5$ GeV and at $\sqrt{s} = 4$ GeV.

6.3. Consider the reaction $e^+ + e^- \to q + \bar{q}$ at a collider with CM energy $\sqrt{s} = 20$ GeV. Give a typical value of the hadronic jet opening angle in a two-jet event. If θ is the angle of the common jet direction with the beams, what is the ratio between the counting rates at $\theta = 90°$ and $\theta = 30°$?

6.4. Consider an electron beam of energy $E = 2$ GeV hitting an iron target (assume it is made of pure ^{56}Fe). How large is the maximum four-momentum transfer?

6.5. Geiger and Marsden observed that alpha particles, after hitting a thin metal foil, not too infrequently bounced back. Calculate the ratio between the scattering probabilities for $\theta > 90°$ and for $\theta > 10°$.

6.6. An alpha particle beam of kinetic energy $E = 6$ MeV and intensity $R_i = 10^3$/s goes through a gold foil ($Z = 79$, $A = 197$, $\rho = 1.93 \times 10^4 \, \text{kg/m}^3$) of thickness $t = 1$ μm. Calculate the number of particles per unit time scattered at angles larger than 0.1 rad.

6.7. Electrons with 10 GeV energy are scattered by protons initially at rest at 30°. Find the maximum energy of the scattered electrons.

6.8. $E = 20$ GeV electrons scatter elastically, emerging with energy $E' = 8$ GeV. Find the scattering angle.

6.9. Find the ratio between the Mott and Rutherford cross sections for the scattering of the same particles at the same energy at 90°.

6.10. In a deep inelastic scattering experiment aimed at studying the proton structure, an $E = 100$ GeV electron beam hits a liquid hydrogen target. The energy E' and the direction of the scattered electrons are measured. If x and Q^2 are, respectively, the momentum fraction and the four-momentum transfer, find E' for $Q^2 = 25$ GeV^2 and for $x = 0.2$.

6.11. What is the value of the x variable in elastic scattering?

6.12. Find the expression (6.11) $E' = E/[1 + \frac{E}{M}(1 - \cos\theta)]$, taking the elastic cross section as the limit of the inelastic cross section.

6.13. Consider the scattering of ν_μ and $\bar{\nu}_\mu$ by nucleons in the quark model, in terms of scattering by quarks. Consider the d, u and s quarks and antiquarks. Write the contributing weak processes with a muon in the final state.

6.14. As in the previous problem but considering the quarks c and $\bar{c}$.

6.15. In a deep inelastic scattering experiment aimed at studying the proton structure, an $E = 100$ GeV electron beam hits a liquid hydrogen target. Find the expression for the momentum transfer Q^2 as a function of the scattering angle θ in the L frame and of the momentum fraction x. What is the maximum momentum transfer for $x = 0.2$?

6.16. In the HERA collider an electron beam of energy $E_e = 30$ GeV hits a proton beam with energy $E_p = 820$ GeV. The energy and the direction of the scattered electron are measured in order to study the proton structure. Calculate the CM energy $\sqrt{s}$ and the energy E_{ef} an electron beam must have to reach the same $\sqrt{s}$ at a fixed target. Calculate the maximum four-momentum transfer of the electron Q^2_{max} for $x = 0.4$, 0.01 and 0.0001. Compare with Fig. 6.15.

6.17. Evaluate the ratio a/a_s at $Q^2 = (10 \text{ GeV})^2$ and at $Q^2 = (100 \text{ GeV})^2$. Take $\lambda_{QCD} = 200$ MeV, $a^{-1}(m_Z^2) = 129$ and $M_Z = 91$ GeV.

6.18. Calculate a_s at 1 TeV ($\lambda_{QCD} = 200$ MeV).

6.19. Why can the quark and gluon jets be clearly observed only at energies much higher than the hadron masses?

6.20. A non-charmed baryon has strangeness $S = -2$ and electric charge $Q = 0$. What are the possible values of its isospin I and of its third component I_z? What is it usually called if $I = 1/2$?

6.21. The proton has uud as valence quarks. Write down the triplet wave function in its spin, isospin and colour factors, taking into account that all the orbital momenta are zero.

6.22. As in the previous problem but for the Λ hyperon.

6.23. Consider the processes (1) $e^+ + e^- \rightarrow \mu^+ + \mu^-$ and (2) $e^+ + e^- \rightarrow$ hadrons at the two CM energies $\sqrt{s} = 2$ GeV and $\sqrt{s} = 20$ GeV. Calculate the ratio of the cross section of process 1 at the two energies. Calculate (approximately) the ratios of the cross sections of the two processes at each of the two energies. What is the ratio of the cross section of process 2 at the two energies?

6.24. We observe the elastic scattering of $E = 5$ GeV electrons from protons at the angle $\theta = 8°$ and we measure their energy. What is its expected value (neglecting the electron mass)? What is the scattered electron energy in the CM frame?

6.25. A beam of alpha particles of kinetic energy $E = 10$ MeV and intensity $I = 1\,\mu\mathrm{A}$ hits a lead target ($A = 207$, $Z = 82$, $\rho = 1.14 \times 10^4$ kg/m^3) of thickness $t = 0.2$ mm. We locate a counter of area $S = 1$ cm^2 at a distance of $l = 0.5$ m beyond the target at the angle $\theta = 40°$. Neglecting, when necessary, the variation of the angle on the detector, find: (a) the number of incident particles per second R_i; (b) the solid angle $\Delta\Omega$ under which the target sees the detector; (c) the differential cross section at the detector; (d) how many hits the detector counts per second.

Further reading

Friedman, J. I. (1990); Nobel Lecture, *Deep Inelastic Scattering: Comparison with the Quark Model* http://nobelprize.org/nobel_prizes/physics/laureates/1990/friedman-lecture.pdf

Taylor, R. E. (1990); Nobel Lecture, *Deep Inelastic Scattering: The Early Years* http://nobelprize.org/nobel_prizes/physics/laureates/1990/taylor-lecture.pdf

Wilczek, F. A. (2004); Nobel Lecture, *Asymptotic Freedom: From Paradox to Paradigm* http://nobelprize.org/nobel_prizes/physics/laureates/2004/wilczek-lecture.pdf

7

Weak interactions

7.1 Classification of weak interactions

The weak interaction is the only one, including gravitation, that does not produce bound states. This is a consequence of its weakness compared to strong and electromagnetic interactions, and to its very small range compared to gravitation. Weak interaction effects can be observed in decays and in collisions only when they are not hidden by the presence of strong or electromagnetic forces. Purely weak probes are neutrinos. There are two main artificial sources of neutrinos, proton accelerators, which produce beams containing mainly ν_μ or $\bar{\nu}_\mu$ and fission reactors that produce $\bar{\nu}_e$.

Three vector bosons mediate the weak interactions. Two are electrically charged, the W^+ and the W^-, each the antiparticle of the other, and one is neutral, the Z^0. They do not have colour charge. In the weak interactions vertex, two fermions join a vector boson. If this is a W, the charges of the initial and final fermions differ by one unit and we talk of 'charged-current' (CC) interaction, if it is a Z the two electric charges are equal and we talk of 'neutral current' (NC).

We know three types of processes:

1. **Leptonic processes**. Only leptons are present, both in the initial and final state. Examples are the μ decay, which proceeds via CC

$$\text{CC} \quad \mu^- \rightarrow e^- \nu_\mu \bar{\nu}_e \tag{7.1}$$

and the neutrino–electron scattering, to which both CC and NC contribute

$$\text{CC} \quad \nu_\mu e^- \rightarrow \nu_e \mu^- \qquad \text{NC} \quad \nu_\mu e^- \rightarrow \nu_\mu e^-. \tag{7.2}$$

The corresponding tree-level diagrams are shown in Fig. 7.1

2. **Semileptonic processes**. Both hadrons and leptons are present. An important example is beta decay. In particular, the beta decays of the nucleons and the corresponding decays at the quark level are given in (7.3). The diagrams are

234

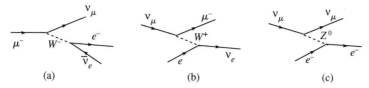

Fig. 7.1. Three leptonic processes.

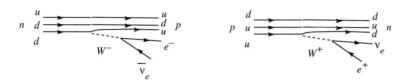

Fig. 7.2. Beta decay of the nucleons.

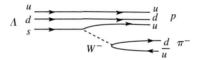

Fig. 7.3. A non-leptonic decay.

shown in Fig. 7.2.

$$\begin{aligned} \text{CC} \quad & n \to p + e^- + \bar{v}_e \qquad u \to d + e^- + \bar{v}_e \\ \text{CC} \quad & p \to n + e^+ + v_e \qquad d \to u + e^+ + v_e. \end{aligned} \tag{7.3}$$

Another important example of a semileptonic process is neutrino scattering from nucleons, or, correspondingly, from quarks. There are two cases, as shown in (7.4), via CC and via NC. The final state of the former contains the charged lepton of the same family as the initial neutrino, the final state of the latter contains a neutrino equal to the initial one.

$$\begin{aligned} \text{CC} \quad & v_\mu + n \to \mu^- + p \qquad v_\mu + d \to \mu^- + u \\ \text{NC} \quad & v_\mu + p \to v_\mu + p \qquad v_\mu + u \to v_\mu + u. \end{aligned} \tag{7.4}$$

3. **Non-leptonic processes**. Only hadrons are present both in the initial and in the final state. Still the process is weak. This class contains only decays, as, for example

$$\Lambda^0 \to p + \pi^- \qquad s \to u + \bar{u} + d. \tag{7.5}$$

Figure 7.3 shows that the process is mediated by a *W*.

As we have already discussed, the weak nature of the process is easily recognised from the long decay times and also from the flavour violation.

7.2 Low-energy lepton processes and the Fermi constant

The leptonic processes are the only purely weak processes, 'uncontaminated', at the tree-level, by the strong interaction. Their probabilities, cross sections or decay rates can therefore be calculated with high accuracy. Let us see a few important cases.

The probability amplitudes of the weak processes at energies much lower than the W mass are proportional to a fundamental quantity, the Fermi constant G_F. This is the case for the decays of all the fermions, except the top, because the masses are much smaller than the mass of the W, which is about 80 GeV. It is also the case for the low-energy scattering processes.

Like all fundamental quantities, the Fermi constant must be measured with high accuracy. This is done using the μ^+ beta decay, a purely weak process

$$\mu^+ \to e^+ \bar{\nu}_\mu \nu_e. \tag{7.6}$$

Figure 7.4(a) shows the lowest-order diagram. The constant g in the vertices is the 'weak charge'. It is dimensionless and its magnitude is comparable to the electromagnetic coupling $\sqrt{a}$.

The matrix element is, apart from numerical factors, the product of the two couplings and of the propagator

$$M \propto \frac{gg}{M_W^2 - t}. \tag{7.7}$$

We now use the fact that all the values of the momentum transfer are very small, $-t \ll M_W^2$ and write, with an extremely good approximation,

$$M \propto \frac{g^2}{M_W^2}. \tag{7.8}$$

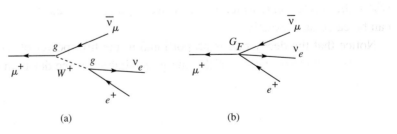

(a) (b)

Fig. 7.4. Muon beta decay.

We see that the matrix element is small, and the interaction is feeble, because M_W is large. For example, with a value of g^2 of the same order of magnitude of a, M is of the order of $10^{-6}\,\mathrm{GeV}^{-2}$.

In these conditions the momentum transfer is far too small to resolve the two vertices and the interaction behaves like a point-like four-fermion interaction. This is the type of interaction originally introduced by Fermi (Fermi 1934) and is shown as a diagram in Fig. 7.4(b).

As is evident from (7.8), the Fermi constant has physical dimensions. It is convenient to start the discussion in SI units. The Fermi constant is defined in such a way that the quantity $G_F/(\hbar c)^3$ has the dimensions of 1/[energy^2]. Specifically, its relationship to the weak charge is, by definition

$$\frac{G_F}{(\hbar c)^3} = \frac{\sqrt{2}}{8} \frac{g^2}{(M_W c^2)^2} \qquad \text{(SI)} \qquad (7.9a)$$

becoming in NU

$$G_F = \frac{\sqrt{2}}{8} \frac{g^2}{M_W^2} \qquad \text{(NU)}. \qquad (7.9b)$$

In the above expressions the numerical factors are due to historical reasons. In NU the dimensions are

$$[G_F] = [E^{-2}] = [L^2]. \qquad (7.10)$$

The calculation of the μ lifetime gives

$$\frac{\hbar}{\tau_\mu} = \Gamma(\mu^+ \to e^+ \bar{\nu}_\mu \nu_e) = \frac{1}{192\pi^3} \frac{G_F^2}{(\hbar c)^6} (m_\mu c^2)^5 (1 + \varepsilon) \qquad (7.11)$$

or, in NU

$$\Gamma(\mu^+ \to e^+ \bar{\nu}_\mu \nu_e) = \frac{1}{192\pi^3} G_F^2 m_\mu^5 (1 + \varepsilon) \qquad (7.12)$$

where the correction ε, which is zero if we neglect the electron mass, is small and can be calculated exactly.

Notice that the decay rate is proportional to the fifth power of the mass of the decaying particle or, rather, of the energy available for the decay in the centre of mass frame. This property is general and is due to dimensional reasons. Indeed, the decay rate has the dimensions of energy and is the product of G_F^2 by a constant. Neglecting the electron mass, the only available constant is the muon mass, which must consequently appear at the fifth power to make the dimensions right.

Looking at (7.12) we see that the determination of the Fermi constant requires an accurate measurement of the muon lifetime and an extremely precise measurement of its mass. The present value is

$$G_F/(\hbar c)^3 = 1.166\,37 \pm 0.000\,01 \times 10^{-5}\,\text{GeV}^{-2}\,\text{[9 ppm]}. \qquad (7.13)$$

Question 7.1 Evaluate the distance between the vertices in Fig. 7.4.

Lepton universality

The charged weak interaction is universal, and is equal for all fermions. This property is evident for leptons, but not at all for quarks. Let us see a few examples.
    The e-μ universality can be checked on the two leptonic decays of the τ

$$\tau^+ \to e^+ \bar{\nu}_\tau \nu_e \qquad \tau^+ \to \mu^+ \bar{\nu}_\tau \nu_\mu. \qquad (7.14)$$

Let us neglect, for simplicity, the electron and muon masses. As we are searching for possible differences, let us indicate the weak charges by different symbols, g_e, g_μ and g_τ (Fig. 7.5).
    The two partial widths are, not mentioning constants that are the same for both

$$\Gamma\left(\tau^- \to \mu^- \bar{\nu}_\mu \nu_\tau\right) \propto \frac{g_\tau^2}{M_W^2}\frac{g_\mu^2}{M_W^2}m_\tau^5 \qquad \Gamma\left(\tau^- \to e^- \bar{\nu}_e \nu_\tau\right) \propto \frac{g_\tau^2}{M_W^2}\frac{g_e^2}{M_W^2}m_\tau^5. \qquad (7.15)$$

We measure their ratio by measuring the ratio between the corresponding branching ratios BR

$$\frac{\Gamma\left(\tau^- \to \mu^- \bar{\nu}_\mu \nu_\tau\right)}{\Gamma\left(\tau^- \to e^- \bar{\nu}_e \nu_\tau\right)} = \frac{\text{BR}\left(\tau^- \to \mu^- \bar{\nu}_\mu \nu_\tau\right)}{\text{BR}\left(\tau^- \to e^- \bar{\nu}_e \nu_\tau\right)} = \frac{g_\mu^2 \rho_\mu}{g_e^2 \rho_e} \qquad (7.16)$$

where the last factor is the ratio of the phase space volumes, which can be precisely calculated. Using the measured quantities (Yao *et al.* 2006) we have

$$\frac{\text{BR}\left(\tau^- \to \mu^- \bar{\nu}_\mu \nu_\tau\right)}{\text{BR}\left(\tau^- \to e^- \bar{\nu}_e \nu_\tau\right)} = \frac{(17.36 \pm 0.05)\%}{(17.84 \pm 0.05)\%} = 0.974 \pm 0.004 \qquad (7.17)$$

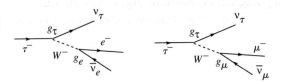

Fig. 7.5.   Leptonic decays of the τ.

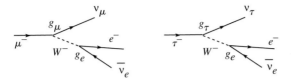

Fig. 7.6. Beta decay of the μ and the τ.

which gives

$$g_\mu/g_e = 1.001 \pm 0.002. \tag{7.18}$$

The μ-τ universality can be checked from the muon and tau beta decay rates (Fig. 7.6). Taking into account that the μ decays 100% of the time in this channel, we have

$$\frac{\Gamma\left(\mu^- \to e^- \bar{\nu}_e \nu_\mu\right)}{\Gamma\left(\tau^- \to e^- \bar{\nu}_e \nu_\tau\right)} = \frac{1}{\tau_\mu} \frac{\tau_\tau}{\mathrm{BR}\left(\tau^- \to e^- \bar{\nu}_e \nu_\tau\right)}. \tag{7.19}$$

On the other hand the theoretical ratio is

$$\frac{\Gamma\left(\mu^- \to e^- \bar{\nu}_e \nu_\mu\right)}{\Gamma\left(\tau^- \to e^- \bar{\nu}_e \nu_\tau\right)} = \frac{g_e^2 g_\mu^2 m_\mu^5 \rho_\mu}{g_e^2 g_\tau^2 m_\tau^5 \rho_\tau} = \frac{g_\mu^2 m_\mu^5 \rho_\mu}{g_\tau^2 m_\tau^5 \rho_\tau} \tag{7.20}$$

and we have

$$\frac{g_\mu^2}{g_\tau^2} = \frac{1}{\tau_\mu} \frac{\tau_\tau}{\mathrm{BR}\left(\tau^- \to e^- \bar{\nu}_e \nu_\tau\right)} \frac{m_\tau^5 \rho_\tau}{m_\mu^5 \rho_\mu}. \tag{7.21}$$

In conclusion we need to measure the two lifetimes, the two masses and the branching ratio $\mathrm{BR}(\tau^- \to e^- \bar{\nu}_e \nu_\tau)$. The measurements give

$$g_\mu/g_\tau = 1.001 \pm 0.003. \tag{7.22}$$

Consider now another important, purely leptonic process, sometimes called 'quasi-elastic' scattering namely

$$\nu_\mu + e^- \to \mu^- + \nu_e. \tag{7.23}$$

The corresponding diagram is shown in Fig. 7.7(a). At values of $-t$ much smaller than M_W (t is the four-momentum transfer), the diagram is well approximated by the four-fermion point interaction in Fig. 7.7(b).

The centre of mass energy squared is $s = m_e^2 + 2m_e E_\nu \approx 2m_e E_\nu$, where E_ν is the neutrino energy in the laboratory frame. A consequence of the smallness of the electron mass is that $\sqrt{s} \ll M_W$ at all the available neutrino beam energies.

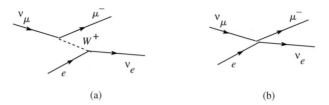

Fig. 7.7.   Muon-neutrino–electron 'quasi-elastic' scattering.

In these conditions, the calculation gives

$$\sigma\left(\nu_\mu e^- \to \nu_e \mu^-\right) = \frac{G_F^2}{\pi}s = \frac{G_F^2}{\pi}2m_e E_\nu = 1.7 \times 10^{-45} E_\nu(\text{GeV})\,\text{m}^2. \qquad (7.24)$$

The cross section grows linearly with the neutrino energy in the laboratory frame. This behaviour is radically different from what we saw in Section 5.7. We can understand the difference, again with a dimensional argument. Indeed, the cross section has the dimension of $[1/E]^2$ and is proportional to the square of the Fermi constant G_F^2, which has the dimension of $[1/E^2]^2$. Consequently G_F^2 must be multiplied by an energy squared. The only available such quantity is s.

However, no cross section can increase indefinitely with energy, because the scattering probability cannot be larger than 100%. Actually (7.24) is not valid if $\sqrt{s}$ is comparable to or larger than M_W because under these circumstances we must use the complete expression (7.7) of the propagator. This appears squared in the differential cross section, i.e. as

$$\left(\frac{g^2}{M_W^2 - t}\right)^2.$$

Considering that the maximum momentum transfer $-t$ increases linearly with s, we understand how the propagator can stop the increase of the cross section and ultimately, when $s \gg M_W^2$, make it decrease proportionally to $1/s$.

7.3 Parity violation

We begin by giving a few historical hints, going back to 1953. We saw in Section 4.5 that the G-stack cosmic ray exposure and the first experiments at accelerators had shown the existence of two apparently identical particles, which were different only in their decay mode, namely the θ^+ decaying into $\pi^+\pi^0$, and the τ^+ decaying into $\pi^+\pi^+\pi^-$. The spin-parity of the former final state, a two-pion system, belongs to the sequence $J^P = 0^+$, 1^-, 2^+, ... , while, as we have seen in Section 4.5, the analysis made by Dalitz of the three-pion final state of the τ decay gave $J^P = 0^-$.

The problem became known as the $\theta\tau$ puzzle. The puzzle could be solved if parity were not conserved in the decay. This hypothesis was fairly acceptable to the experimentalists but sounded almost like blasphemy to theoreticians. Indeed, parity is a symmetry of space-time itself, just as the rotations are; 'it had to be absolutely conserved'. At the general conference on particle physics, the 'Rochester Conference' of 1956, R. Feynman asked C. N. Yang, after his speech, a question that he had been asked by M. Bloch, the codiscoverer of the η meson: '*is it possible to think that parity is not conserved?*' Yang answered that T. D. Lee and himself had had a look at the issue, but without reaching any conclusions. The conclusion came a few months later, when Lee and Yang showed that no experimental proof existed of parity conservation in weak interactions (Lee & Yang 1956).

Following their reasoning, let us consider the beta decay of a nucleus $N \to N' + e + \nu$ in the centre of mass frame. The kinematic quantities are the three momenta $\mathbf{p}_{N'}$, $\mathbf{p}_e$, $\mathbf{p}_\nu$. With them we can build:

- scalar products such as $\mathbf{p}_{N'} \cdot \mathbf{p}_e$; being scalar they do not violate P
- the mixed product $\mathbf{p}_{N'} \cdot \mathbf{p}_e \times \mathbf{p}_\nu$; it is a pseudoscalar and, added to a scalar, would violate P, but it is zero because the three vectors are coplanar.

Lee and Yang concluded that parity conservation could be tested only using an axial vector. Such an axial vector is provided by polarisation. One must polarise a sample of nuclei, inducing a non-zero expectation value of the intrinsic angular momentum $\langle \mathbf{J} \rangle$, and measure an observable proportional to the pseudoscalar $\langle \mathbf{J} \rangle \cdot \mathbf{p}_e$.

The experiment was done by C. S. Wu and collaborators at the National Bureau of Standards (Wu *et al.* 1957) with ^{60}Co. Polarisation can be achieved by orienting the nuclear magnetic moments $\boldsymbol{\mu}$, which are parallel to the spins, in a magnetic field $\mathbf{B}$. The probability of a certain direction of the magnetic moment relative to the field is given by the Boltzmann factor

$$\exp\left(-\frac{\boldsymbol{\mu} \cdot \mathbf{B}}{kT}\right). \tag{7.25}$$

The problem of the experiment follows from the inverse proportionality of the magnetic moment to the mass of the particle. Since the nuclear mass is large (compared to the mass of the electron), nuclei are difficult to polarise. We see from (7.25) that a very low temperature, in practice a few millikelvin, and a strong magnetic field are needed. The latter was obtained by imbedding the cobalt in a paramagnetic crystal. If the crystal is in a magnetic field, even a weak one, the electronic magnetic moments, which are large, become oriented in the field and generate inside the crystal local fields of dozens of tesla.

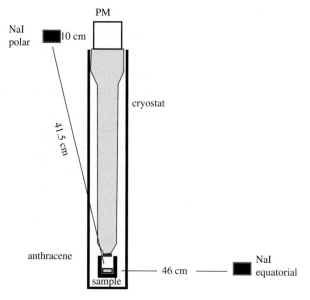

Fig. 7.8.   Sketch of the experiment. (Simplified from Wu *et al.* 1957)

Figure 7.8 shows a sketch of the experiment. The spin-parity of the ^{60}Co nucleus is $J^P = 5^+$. The polarised nuclei beta decay into an excited state of ^{60}Ni with $J^P = 4^+$. The daughter nucleus keeps the polarisation of the parent nucleus

$$^{60}\text{Co}\left(J^P = 5^+\right) \Uparrow \to {}^{60}\text{Ni}^{**}\left(J^P = 4^+\right) \Uparrow + e^- + \bar{\nu}_e. \qquad (7.26)$$

Two gamma decays of Ni follow in cascade to the fundamental level, maintaining polarisation

$$^{60}\text{Ni}^{**}\left(J^P = 4^+\right) \Uparrow \to {}^{60}\text{Ni}^*\left(J^P = 2^+\right) \Uparrow + \gamma\left(1.173\,\text{MeV}\right) \qquad (7.27)$$

$$^{60}\text{Ni}^*\left(J^P = 2^+\right) \Uparrow \to {}^{60}\text{Ni}\left(J^P = 0^+\right) \Uparrow + \gamma\left(1.332\,\text{MeV}\right). \qquad (7.28)$$

The two electromagnetic decays are not isotropic, the gamma emission probability is a function of the angle θ with the field. Therefore, we can monitor the polarisation of the sample by measuring this anisotropy.

The polarising magnetic field is oriented along the vertical axis of Fig. 7.8; its direction can be chosen to be upward or downward. The photons are detected using two counters made of NaI crystals, which scintillate when absorbing a photon. One counter (equatorial) is at 90° to the field, the other ('polar') at about 0°. The electrons must be detected inside the cryostat. To this aim a scintillating anthracene crystal is located at the tip of a plastic bar that guides the scintillation light to a photomultiplier (PM). In this way, the experiment counts the electrons emitted in the direction of the polarisation or opposite to it, depending on the orientation of the polarising field.

The operations start by switching the magnetic field on to polarise the nuclei. Once the polarisation is obtained, in a few seconds, the field is switched off (at time zero) and counting of the photons and the electrons starts. The polarisation slowly decays, to disappear in a few minutes. The photon flux depends on the direction to the polarisation axis but does not change when it is reversed, because parity is conserved by electromagnetic interactions. As anticipated, the degree of polarisation is measured by the gamma anisotropy, which, if W_γ are the counting rates, is defined as

$$\varepsilon_\gamma \equiv \frac{W_\gamma(90°) - W_\gamma(0°)}{W_\gamma(90°) + W_\gamma(0°)}. \tag{7.29}$$

Figure 7.9(a) shows the measurements of $W_\gamma(0°)$ and $W_\gamma(90°)$, divided by their values at zero field, as functions of time. Both show the decay of the polarisation giving us the shape of the decay curve.

If the beta decay violates parity, the angular distribution of the emitted electrons is asymmetric under $\theta \Leftrightarrow \pi - \theta$. Therefore, the counting rate is expected to depend on the angle as

$$W_e(\theta) \propto 1 + P\beta_e a \cos \theta \tag{7.30}$$

where the constant a is zero if parity is conserved and ± 1 if it is maximally violated. The latter situation corresponds to a $V \pm A$ structure of the interaction. The initial polarisation P of the Wu experiment was about 0.6. We shall explain in

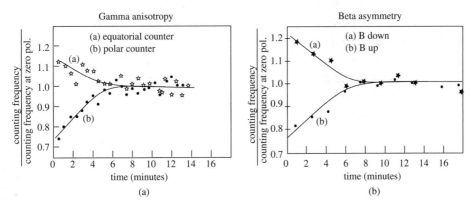

Fig. 7.9.   (a) The measurements of $W_\gamma(0°)$ and $W_\gamma(90°)$, divided by their values at zero field, as functions of time. (b) The electron counting rates with the field direction upward and downward, divided by the counting rate without polarisation. (Adapted from Wu *et al.* 1957)

the next section the presence of the factor β_e, which is the speed of the electrons divided by the speed of light.

Figure 7.9(b) shows the electron counting rates with the field direction upward and downward, divided by the counting rate without polarisation. In the former configuration the detectors count the electrons emitted at about 0°, in the latter at about 180°. Both ratios decay following the curve of the polarisation. The fundamental observation is that the two rates are different: the electrons are emitted in directions (almost) opposite to the field much more frequently than (almost) along it. This was the experimental proof of parity violation. Moreover, the measurement of a gave

$$a \approx -1 \qquad (7.31)$$

which shows that the parity violation is, within the errors, maximal. If we assume the interaction to be $V + xA$, the result is compatible with $x = -1$. Taking the uncertainties of the measurement into account, the experiment gave

$$-1 < x < -0.7. \qquad (7.32)$$

The conclusion of the analysis of the pion decay in Section 3.5 was that the space-time structure of the charged-current weak interaction is V or A or any combination of them. The Wu experiment chooses the combination $V - A$. We try to illustrate the point in Fig. 7.10.

The thick arrows indicate the 'directions' of the spins; their lengths are such as to satisfy the conservation of the third component of the angular momentum. The thin arrows are the preferential directions of the motions. The result of the experiment is that the preferential motion of the electron is opposite to the field, and consequently of its spin. As the nuclei decay at rest, the preferential direction of the antineutrino is opposite to that of the electron. Therefore, the antineutrino spin is in the preferential direction of its velocity. The $V - A$ structure of the charged-current weak interaction implies that this behaviour is common to all fermions and to all antifermions respectively. We have used imprecise language here; we shall make it accurate in the next section.

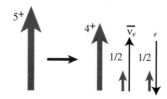

Fig. 7.10. Schematic of the spin 'directions' in the decay

$$^{60}\text{Co}\left(J^P = 5^+\right) \, \Uparrow \rightarrow \, ^{60}\text{Ni}^{**}\left(J^P = 4^+\right) \, \Uparrow + e^- + \bar{\nu}_e.$$

7.4 Helicity and chirality

The four-component Dirac bi-spinor describes a fermion and its antiparticle and can be written in terms of two corresponding two-component spinors ϕ and χ

$$\psi(x) = \begin{pmatrix} \psi_1 \\ \psi_2 \\ \psi_3 \\ \psi_4 \end{pmatrix} = \begin{pmatrix} \phi \\ \chi \end{pmatrix} \qquad \phi = \begin{pmatrix} \phi_1 \\ \phi_2 \end{pmatrix} \qquad \chi = \begin{pmatrix} \chi_1 \\ \chi_2 \end{pmatrix}. \tag{7.33}$$

There are three possibilities in order to give a physical meaning to the two components of ϕ and χ, depending on the quantity we wish to be defined.

Polarisation The two states have a defined third component of the spin on an axis. This axis must be physically defined, typically by the direction of a magnetic or electric field. Taking for example ϕ, the two eigenstates are $\phi^+ = \begin{pmatrix} 1 \\ 0 \end{pmatrix}$ and $\phi^- = \begin{pmatrix} 1 \\ 0 \end{pmatrix}$.

$$\begin{aligned} \frac{1}{2}\sigma_z \begin{pmatrix} 1 \\ 0 \end{pmatrix} &= \frac{1}{2}\begin{pmatrix} 1 & 0 \\ 0 & -1 \end{pmatrix}\begin{pmatrix} 1 \\ 0 \end{pmatrix} = +\frac{1}{2}\begin{pmatrix} 1 \\ 0 \end{pmatrix} \\ \frac{1}{2}\sigma_z \begin{pmatrix} 0 \\ 1 \end{pmatrix} &= \frac{1}{2}\begin{pmatrix} 1 & 0 \\ 0 & -1 \end{pmatrix}\begin{pmatrix} 0 \\ 1 \end{pmatrix} = -\frac{1}{2}\begin{pmatrix} 0 \\ 1 \end{pmatrix}. \end{aligned} \tag{7.34}$$

Helicity Even in the absence of an external field, the velocity of the particle defines a direction in any reference frame different from the rest frame. The states of definite helicity are the eigenstates of the third component of the spin in that direction. If **p** is the momentum of the particle, the helicity operator is

$$\frac{1}{2}\frac{\mathbf{p} \cdot \boldsymbol{\sigma}}{p}. \tag{7.35}$$

The two helicity eigenvalues are $+1/2$, if the spin is in the direction of the motion, and $-1/2$ if in the opposite direction. We draw the reader's attention to the fact that the helicity eigenstates are two-component spinors, describing a fermion or an antifermion, not both.

As evident from its definition, the helicity is not, in general, a Lorentz-invariant observable. Consider for example a particle with negative helicity in a certain reference frame. If the particle is massive, namely if its speed is less than c, we can always find another frame in which the particle travels in the opposite direction. In this reference frame the helicity is positive. Only if the fermion is rigorously massless is its velocity frame independent and its helicity Lorentz invariant.

However, rigorously massless fermions do not exist in nature, to the best of our knowledge.

Chirality Whilst polarisation and helicity are properties of a fermion, formally of a two-component spinor, chirality is a property of the four-component spinors. The two 'chiral' states are the eigenstates of γ_5, with two possible eigenvalues, $+1$ and -1; the chiral states are called right (R) and left (L) respectively. If the wave function ψ is a solution of the Dirac equation, the projectors of the positive (R) and negative (L) chirality states are respectively $\frac{1}{2}(1 + \gamma_5)$ and $\frac{1}{2}(1 - \gamma_5)$

$$\psi_L = \frac{1}{2}(1 - \gamma_5)\psi \qquad \psi_R = \frac{1}{2}(1 + \gamma_5)\psi. \tag{7.36}$$

The conjugated states are

$$\bar{\psi}_L = \bar{\psi}\frac{1}{2}(1 + \gamma_5) \qquad \bar{\psi}_R = \bar{\psi}\frac{1}{2}(1 - \gamma_5). \tag{7.37}$$

Question 7.2 Verify that these operators are projectors, namely that applying one of them twice gives the same result as applying it once and that applying both of them results in the null state.

Chirality is important because, as we shall see in the following sections, only the left bi-spinor, namely the first one in (7.36) and (7.37), is both source and receptor of the charged-current weak interaction. Notice that the chiral states are not stationary states, because γ_5 does not commute even with the free Hamiltonian, with its mass term to be precise. Therefore, chirality is not conserved even in the free particle motion. However, γ_5 commutes with the Hamiltonian and chirality is a good quantum number for massless particles. Even if these do not exist the mass of the fermion is negligible at high enough energy, as is very often the case.

We must now examine the fermion and antifermion components of the chiral states. We start with the terminology. Consider first the **left chiral** four-component spinor, namely the eigenstate of γ_5 with negative eigenvalue. Its **fermion** two-component spinor is called **left**, whilst its **antifermion** two-component spinor is called **right**. This terminology is unfortunate, but is the accepted one. Similarly, the fermion two-component spinor of the right four-component bi-spinor is called right, the antifermion is called left.

We shall now study the helicity content of the chiral states. We start by recalling that the two two-component spinors ϕ and χ of a solution ψ of the Dirac equation are completely correlated. Actually

$$\left(\gamma_\mu p^\mu - m\right)\psi = (E\gamma_0 - \mathbf{p} \cdot \boldsymbol{\gamma} - m)\psi = 0 \tag{7.38}$$

or explicitly

$$\begin{pmatrix} E-m & -\mathbf{p}\cdot\boldsymbol{\sigma} \\ \mathbf{p}\cdot\boldsymbol{\sigma} & -(E+m) \end{pmatrix} \begin{pmatrix} \phi \\ \chi \end{pmatrix} = \begin{pmatrix} 0 \\ 0 \end{pmatrix} \tag{7.39}$$

which gives the two relationships

$$\phi = \frac{\mathbf{p}\cdot\boldsymbol{\sigma}}{E-m}\chi \qquad \chi = \frac{\mathbf{p}\cdot\boldsymbol{\sigma}}{E+m}\phi. \tag{7.40}$$

Consider now the chiral states, for example the left one

$$\psi_L = \frac{1}{2}(1-\gamma_5)\psi = \frac{1}{2}\begin{pmatrix} 1 & -1 \\ -1 & 1 \end{pmatrix}\begin{pmatrix} \phi \\ \chi \end{pmatrix} = \frac{1}{2}\begin{pmatrix} \phi-\chi \\ \chi-\phi \end{pmatrix}. \tag{7.41}$$

We see that it contains the combination $\phi - \chi$ and not $\phi + \chi$. Consider the upper component (i.e. the fermion, as opposed to the antifermion). Let us take the z-axis in the direction of motion and write (7.41) in terms of the helicity eigenstates $\phi^{+1/2}$ and $\phi^{-1/2}$. We have

$$\frac{1}{2}(\phi-\chi) = \frac{1}{2}\left(1-\frac{\mathbf{p}\cdot\boldsymbol{\sigma}}{E+m}\right)\phi = \frac{1}{2}\left(1-\frac{p_z}{E+m}\right)\phi^{+1/2} + \frac{1}{2}\left(1+\frac{p_z}{E+m}\right)\phi^{-1/2}. \tag{7.42}$$

We see that the upper component of the left bi-spinor is not an eigenstate of helicity and that the amplitudes of its helicity components depend on the reference frame. However, if the particle is massless, we can write $p_z = E$ and

$$\frac{1}{2}(\phi-\chi) = \phi^{-1/2}. \tag{7.43}$$

The upper component of the left bi-spinor of a zero-mass particle is the negative helicity eigenstate. If the particle is massive, the 'wrong' helicity component, the positive one $\phi^{+1/2}$, vanishes when $E \gg m$, but is appreciable at low energy.

Let us now consider the antifermion in ψ_L. We have

$$\frac{1}{2}(\chi-\phi) = \frac{1}{2}\left(1-\frac{\mathbf{p}\cdot\boldsymbol{\sigma}}{E-m}\right)\chi = \frac{1}{2}\left(1-\frac{p_z}{E-m}\right)\chi^{+1/2} + \frac{1}{2}\left(1+\frac{p_z}{E-m}\right)\chi^{-1/2}. \tag{7.44}$$

The antiparticle can be considered as a negative energy solution of the Dirac equation. If the mass is zero $E = -p_z$, and we have

$$\frac{1}{2}(\chi-\phi) = \chi^{+1/2}. \tag{7.45}$$

We see that the massless antiparticle of a negative chirality bi-spinor is the helicity eigenstate with positive eigenvalue. Again, if the particle is massive, the 'wrong' helicity component vanishes when $E \gg m$.

The following observation is in order. The reader might have had the impression from the above discussion of an asymmetry between matter and antimatter. However this is not true, the definition of which is the particle and which is the antiparticle component of the Dirac bi-spinor is completely arbitrary.

Chirality cannot be directly determined, rather the helicity of the particle or of the antiparticle is measured, as in the Wu experiment and in those we shall discuss in the next section. To be precise, we measure the expectation value of the helicity. Considering, for example, the particle spinor of the left bi-spinor, let Π_+ and Π_- be the probabilities of finding it in each of the two helicity states (spin in the direction of motion or opposite to it). These probabilities are the squares of the two amplitudes in (7.42). The helicity expectation value is then

$$h = \frac{\Pi_+ - \Pi_-}{\Pi_+ + \Pi_-} = \frac{(E+m-p)^2 - (E+m+p)^2}{(E+m-p)^2 + (E+m+p)^2} = \frac{-p}{E} = -\beta. \tag{7.46}$$

In conclusion, the expectation value of the helicity, or simply the helicity, of a fermion of negative chirality is the opposite of the ratio between its speed and the speed of light. A similar calculation shows that the helicity of a positive chirality fermion is $+\beta$.

Question 7.3 Demonstrate the last statement, after having found the equation analogous to (7.42) for a right fermion.

Going back to the Wu experiment, we see that it implies that the electrons produced in the beta decay are left, while the antineutrinos are right. In both cases they are part of a left bi-spinor. We shall see how these conclusions are confirmed by helicity measurements in the next section.

We now recall the space-time structure of the electromagnetic interaction that is given by

$$\sqrt{a} A_\mu \bar{f} \gamma^\mu f. \tag{7.47}$$

It graphically corresponds to the electromagnetic vertex shown in Fig. 7.11, which represents the coupling of initial and final equal fermions and of a fermion–antifermion pair to a photon.

Let us now go back to the process $e^+ e^- \to \mu^+ \mu^-$. In Section 5.7 we anticipated that the vertex in Fig. 7.11(b) couples an electron and a positron if their helicities are opposite, not if they are equal, when the energy is much higher than the masses of the particles. We now prove this statement. Consider for example the

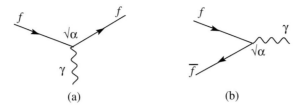

Fig. 7.11.   The electromagnetic vertex.

case in which both helicities are positive. Since the masses can be neglected, the electron is a positive chirality fermion, the positron is a negative chirality anti-fermion, namely

$$f_R = \frac{1 + \gamma_5}{2} f \qquad \bar{f}_L = \bar{f} \frac{1 + \gamma_5}{2}.$$

We then write the electromagnetic interaction and, taking into account that $\gamma_5^2 = 1$, we have

$$\bar{f}_L \gamma_\mu f_R = \bar{f} \frac{1 + \gamma_5}{2} \gamma_\mu \frac{1 + \gamma_5}{2} f = \bar{f} \gamma_\mu \frac{1 + \gamma_5}{2} \frac{1 - \gamma_5}{2} f = 0.$$

In a similar way one proves that the initial and final fermions in Fig. 7.11(a) are coupled only if their helicities are equal, provided they can be considered as massless. We can now understand why the probability of the elastic scattering of electrons by a massive target becomes zero at 180°, as foreseen by the Mott formula (6.22). Indeed, in such conditions the incoming and outgoing electrons would have opposite velocities, but spins in the same direction, hence opposite helicities.

7.5 Measurement of the helicity of leptons

The conclusion we reached at the end of the previous section implies that the space-time structure of the charged-current weak interaction is $V - A$.

The charged-current weak interaction is mediated by two charged bosons, the W^+ and the W^-. Therefore, the initial and final fermions are different. For example, an initial electron disappears while a W^- and a ν_e appear, and similarly for the other families as shown in Fig. 7.12.

The expressions of the $V - A$ interaction corresponding to the three vertices are

$$g\bar{\nu}_e \gamma^\mu (1 - \gamma_5) e \qquad g\bar{\nu}_\mu \gamma^\mu (1 - \gamma_5)\mu \qquad g\bar{\nu}_\tau \gamma^\mu (1 - \gamma_5)\tau. \tag{7.48}$$

Here, in the first case for example, e is the annihilation operator of the initial electron and $\bar{\nu}_e$ the creation operator of the final antineutrino. The coupling constant g is, as we know, universal. The $V - A$ structure implies the following. From the properties

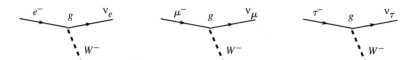

Fig. 7.12.   The charged-current weak interaction lepton vertices.

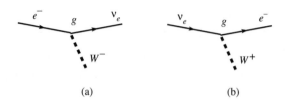

(a)   (b)

Fig. 7.13.   Weak charged-current electron–electron-neutrino vertex.

of the γ matrices, we have

$$\gamma^{\mu}(1-\gamma_5) = \frac{1}{2}(1+\gamma_5)\gamma^{\mu}(1-\gamma_5). \tag{7.49}$$

We see that only the left four-component spinor operators appear in the expressions (7.48). In general, let us call f and i the final and initial ones respectively. Their left projections are

$$\bar{f}_L \equiv \bar{f}\left(\frac{1+\gamma_5}{2}\right) \qquad i_L \equiv \left(\frac{1-\gamma_5}{2}\right)i. \tag{7.50}$$

We can then write Eq. (7.48) as

$$g\bar{f}\gamma^{\mu}(1-\gamma_5)i = 2g\bar{f}\left(\frac{1+\gamma_5}{2}\right)\gamma^{\mu}\left(\frac{1-\gamma_5}{2}\right)i = 2g\bar{f}_L\gamma^{\mu}i_L. \tag{7.51}$$

In the latter form the charged-current weak interaction is very similar to the electromagnetic one, but with a fundamental difference: the states that couple to the W are the left fermions and the right antifermions.

   Let us look more closely at the vertex, at the electron vertex of Fig. 7.13(a), for example. In the vertex an e^- enters and v_e exits, hence the charge of the final lepton is larger than that of the initial lepton. Therefore, the W must be negative, as in Fig. 7.13(a), corresponding to the current

$$j_{\mu}^- = 2g\bar{v}_{eL}\gamma_{\mu}e_L^-. \tag{7.52}$$

In Fig. 7.13(b) a v_e enters and an e^- exits; the W must be positive and the corresponding current is

$$j_{\mu}^+ = 2g\bar{e}_L^-\gamma_{\mu}v_{eL}. \tag{7.53}$$

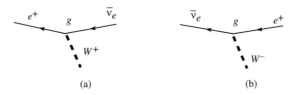

Fig. 7.14.   Same as Fig. 7.13 but with antiparticles.

In both cases the incoming and outgoing particles are left. The same two currents describe the vertices with antiparticles, as shown in Fig. 7.14. The antiparticles are right.

The experimental verification of these fundamental properties is based on the measurement of the helicity of the neutrinos and the electrons produced in beta decays. Neutrinos are not massless, but their mass is so small as to be observable only in the phenomena that we shall discuss in Chapter 10. Until that chapter we shall consider neutrinos as massless.

We shall now describe the experiment of M. Goldhaber, L. Grodznis and A. Sunyar (Goldhaber *et al.* 1958) on the measurement of the helicity of the neutrino. The experiment was carried out at the Brookhaven National Laboratory in 1957. Let us see its logical steps.

The first element is the gamma resonant emission and absorption by nuclei.

Consider a medium and let N be its nuclei. A nucleus can be excited to the N^* level, and subsequently decays to the fundamental level by emitting a photon

$$N^* \rightarrow \gamma + N. \tag{7.54}$$

The 'resonance' process of interest is this emission followed by the absorption of the photon by another nucleus, which becomes excited in the N^* level.

$$\gamma + N \rightarrow N^*. \tag{7.55}$$

To be in resonance the photon must have the right energy to give the transition energy E to the nucleus N. This is E augmented by the recoil energy of the final state. However, the energy of the photon from reaction (7.54) is E diminished by the recoil kinetic energy of the emitting nucleus. Therefore, the resonance process cannot take place if the excited nucleus N^* is at rest. This condition is necessary for the experiment to succeed.

The resonant conditions can be satisfied if the initial N^* moves relative to the medium when it decays. The energy of the photon in the reference frame of the medium depends on its direction relative to that of N^*. As the photons have larger energies in the forward directions (Doppler effect) they can induce the resonance only in these directions. This is the second necessary condition for the experiment. We shall see in Questions 4 and 5 how to satisfy the two conditions.

The second element of the experiment is the transfer of the neutrino helicity (h_ν) to a photon and the measurement of the helicity of the latter (h_γ). To do this we first need a nuclide, which we call A, producing by K-capture the excited state N^* and the neutrino, the helicity of which we shall determine. Remember that K-capture is the capture by the nucleus of an atomic electron in the S wave. The process is

$$A + e^- \rightarrow N^* + \nu_e. \tag{7.56}$$

Obviously the condition that the energy of the N^* so produced is in resonance must be satisfied. This is not yet enough, because, as we shall immediately see, the angular momentum of A must be $J=0$ and that of N^*, $J=1$. At this point, one might conclude that there are so many conditions that it is hopeless to seek two nuclides satisfying all of them. However, Goldhaber, Grodznis and Sunyar found that ^{152}Eu and ^{152}Sm have all the required characteristics. Fortune favours the bold!

Let us see how helicity is transferred from the neutrino to the photon. There are three steps.
1. The ^{152}Eu decays by K-capture of an S wave electron

$$^{152}\text{Eu}(J=0) + e^- \rightarrow \, ^{152}\text{Sm}^*(J=1) + \nu_e. \tag{7.57}$$

Let the neutrino direction be the quantisation axis z. The Sm* direction is $-z$.
2. Select the cases in which the Sm* decays emitting a photon in the forward direction, namely $-z$, by use of the resonant emission–absorption process. The emission process is

$$^{152}\text{Sm}^*(J=1) \rightarrow \, ^{152}\text{Sm}(J=0) + \gamma. \tag{7.58}$$

The first three columns of Table 7.1 give all the combinations of the third components of the spins that satisfy the angular momentum conservation in the reaction (7.57). The fourth column gives the corresponding neutrino helicity. Taking into account that the projection of the photon angular momentum on its velocity cannot be zero, only two cases remain. We observe that in both the photon and neutrino have the same helicity.
3. Measurement of the photon circular polarisation, namely of its helicity. Figure 7.15 shows a sketch of the experiment.

The europium source is located above an iron slab in a vertical magnetic field **B**, used to analyse the polarisation state of the photon. The flight direction of the samarium nuclei is approximately the downward vertical. Neglecting the small difference, we take the vertical as the z-axis. The direction of the magnetic field can be chosen as z or $-z$. Remember now that the spins of the electrons responsible for the ferromagnetism are oriented opposite to **B**. These electrons can easily absorb the photons, by flipping their spin, if the photon spin has the direction of **B**.

Table 7.1

$s_z(e)$	$s_z(\mathrm{Sm}^*)$	$s_z(v)$	h_v	$s_z(\gamma)$	h_γ
$+1/2$	1	$-1/2$	$-$	1	$-$
$+1/2$	0	$+1/2$	$+$	0	$\times$
$-1/2$	-1	$+1/2$	$+$	-1	$+$
$-1/2$	0	$-1/2$	$-$	0	$\times$

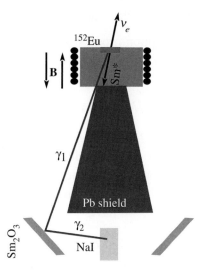

Fig. 7.15.   Sketch of the neutrino helicity experiment.

However, they cannot do so for photons with spin in the direction opposite to **B**. Therefore, the iron slab absorbs the former substantially more than the latter.

The photon detector is a NaI crystal. This cannot be reached directly by the photon because it is shielded by a lead block. An adequately shaped samarium ring surrounds the detector.

If the resonance process takes place, a photon, call it γ_1, emitted by Sm* is absorbed by a Sm nucleus in the ring. The latter immediately de-excites emitting a photon, γ_2, which reaches the detector (in a fraction of cases). The process can happen only if the Sm* was travelling in the right forward direction.

The measured quantity is the asymmetry R between the counting rates with the field oriented in one direction and the other, I_+ and I_- , i.e.

$$R = \frac{I_+ - I_-}{I_+ + I_-}. \qquad (7.59)$$

From this measurement the longitudinal polarisation of the photon is easily extracted. The final result is that the helicity of the neutrino is negative and compatible with -1. This proves the $V - A$ structure of the CC weak interaction.

Question 7.4   Prove that the resonance condition is not satisfied by Sm* decay at rest.

The energy difference between the two Sm levels is $E_{Sm} = 963$ keV. The recoil energy, $E_{K,Sm}$, is small and to find it we can use non-relativistic expressions. The recoil momentum p_{Sm} is equal and opposite to the photon momentum p_γ, which is also the photon energy. Since we are calculating a correction, we can approximate the photon energy with E_{Sm}. The Sm recoil kinetic energy is

$$E_{K,Sm} = \frac{p_{Sm}^2}{2M_{Sm}} = \frac{p_\gamma^2}{2M_{Sm}} \approx \frac{E_{Sm}^2}{2M_{Sm}} = \frac{0.963^2 \times 10^{12}\,\text{eV}^2}{2 \times 1.52 \times 10^{11}\,\text{eV}} = 3\,\text{eV}. \qquad (7.60)$$

The recoil energy in the absorption process is substantially equal to this and, in conclusion, the photon energy is below the resonance energy by twice $E_{K,Sm}$, namely

$$\delta E = E_{Sm}^2 / M_{Sm} \quad \Rightarrow \quad \delta E = 6\,\text{eV}. \qquad (7.61)$$

Is this difference small or large? To answer this question we must compare δE with the resonance width. The natural width is very small, as for all nuclear electromagnetic transitions, about 20 meV, much less than δE. However, we must consider the Doppler broadening of the resonance due to thermal motion. At room temperature, $kT = 26$ meV, we have

$$\frac{\Delta E_{Sm}(\text{thermic})}{E_{Sm}} = 2\sqrt{\frac{2\ln 2 \times kT}{M_{Sm}}} = 2\sqrt{\frac{1.4 \times 26 \times 10^{-3}}{1.52 \times 10^{11}}} \approx 10^{-6}$$

and

$$\Delta E_{Sm}(\text{thermic}) = 10^{-6} \times 963\,\text{keV} \approx 1\,\text{eV}. \qquad (7.62)$$

In conclusion, the energy of the photons emitted by Sm* at rest is smaller than the resonance energy by six times its width. The process does not take place. The first necessary condition is therefore satisfied.

Question 7.5   Prove that the resonance condition is satisfied for forward γ emission by Sm* in flight.

We compute the kinetic energy of the Sm* produced by the Eu K-capture. The reasoning is very similar to that above. The energy released in the transition is $E_{Eu} = 911$ keV. The recoil momentum and the neutrino momentum are equal and opposite. The neutrino momentum is equal to its energy, which we approximate by E_{Eu}. Consequently, the Sm* recoil kinetic energy is

$$E_{K,Sm^*} \approx \frac{E_{Eu}^2}{2M_{Sm}} = \frac{0.911^2 \times 10^{12}\,\text{eV}^2}{2 \times 1.52 \times 10^{11}\,\text{eV}} = 2.7\,\text{eV}. \qquad (7.63)$$

The recoil speed in the laboratory reference frame is

$$\beta_{Sm^*} = \sqrt{\frac{E_{K,Sm^*}}{M_{Sm}}} = \sqrt{\frac{2\times2.7}{1.52\times10^{11}}} = 5.8\times10^{-6}.$$

In this frame the photon energy E_L is maximum when the photon is emitted in the flight direction of the nucleus. Recalling Example 1.1 and setting $\gamma = 1$, we have

$$E_L - E_{Sm^*} = \beta_{Sm^*}\times E_{Sm^*} = 5.8\times10^{-6}\times911\,\text{keV} = 5.3\,\text{eV}. \tag{7.64}$$

This is within about 1 eV from resonance; hence, taking into account the Doppler broadening, the resonance condition is satisfied. The second necessary condition is satisfied.

In the design stages of the experiment it was not at all guaranteed that these miraculous conditions would indeed be satisfied. To check this crucial point, L. Grodznis (Grodznis 1958) performed a preliminary experiment to measure the 'resonant' cross section, using the same apparatus as shown in Fig. 7.15 without the magnet. In this way he used the photons coming from Sm* produced in reaction (7.58), exactly as in the final experiment. In conclusion, the success of the experiment was due to the exceptional kindness offered by Nature and to the equally exceptional boldness of the experimenters.

The experimental verification of the relationship (7.46) for the electrons of the β^- decay of a nucleus is much easier. We need to measure the helicity of the electrons, namely their longitudinal polarisation. In practice, it is much easier to measure the transverse polarisation using a thin high-Z metal plate immersed in a magnetic field. The analysing power is due to the fact that, in the above conditions, the electron scattering cross section in a metal depends on its transverse

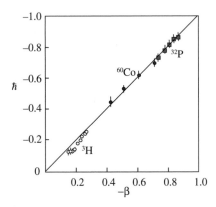

Fig. 7.16.   The helicity of the electron as a function of its speed. (Adapted from Koks & van Klinken 1976)

polarisation. To change the electron polarisation from longitudinal to transverse we let the electrons go through a quarter of a circumference in a magnetic field. The direction of the momentum vector changes by $90°$ whilst the spin direction remains unaltered.

Figure 7.16 shows the measurements of the helicity of electrons of different speeds coming from the beta decays of three nuclei. These have been chosen to cover three different speed intervals: tritium at small velocities, cobalt at intermediate ones and phosphorus at high velocities. We see that the agreement with the theoretical prediction is good.

7.6 Violation of the particle–antiparticle conjugation

Weak interactions violate not only parity but also the particle–antiparticle conjugation.

In the discussion of parity violation we have used the fact that the Lagrangian contains the correlation term $\boldsymbol{\sigma} \cdot \mathbf{p}$ between the spin and the momentum, which is not invariant under inversion of the axes. Observing that $\boldsymbol{\sigma} \cdot \mathbf{p}$ does not vary under charge conjugation, the same reasoning would lead to the naïve conclusion that charge conjugation is conserved, but the conclusion is wrong.

Indeed, the operator $\boldsymbol{\sigma} \cdot \mathbf{p}$ is, within a factor $1/p$, the helicity operator. The presence of this operator in the Lagrangian selects the left spinors and therefore violates $\mathcal{C}$.

The Wu experiment gives only indirect evidence for $\mathcal{C}$ violation, assuming $\mathcal{CPT}$ invariance. We observe that under time reversal $\mathcal{T}$: $\mathbf{p} \rightarrow -\mathbf{p}$, $\boldsymbol{\sigma} \rightarrow -\boldsymbol{\sigma}$, and under $\mathcal{P}$: $\mathbf{p} \rightarrow -\mathbf{p}$, $\boldsymbol{\sigma} \rightarrow \boldsymbol{\sigma}$. Therefore, under $\mathcal{PT}$ we have $\boldsymbol{\sigma} \cdot \mathbf{p} \rightarrow -\boldsymbol{\sigma} \cdot \mathbf{p}$. In conclusion, if $\mathcal{CPT}$ is conserved, $\boldsymbol{\sigma} \cdot \mathbf{p}$ violates $\mathcal{C}$.

To have direct evidence of $\mathcal{C}$ violation in a Wu-type experiment one would need to study the beta decay of an antinucleus, which is obviously impossible. However, we can use the decays of the mesons, in particular the pions. The decay chains $\pi^- \rightarrow \mu^- + \bar{\nu}_\mu$ followed by $\mu^- \rightarrow e^- + \nu_\mu + \bar{\nu}_e$ and $\pi^+ \rightarrow \mu^+ + \nu_\mu$ followed by $\mu^+ \rightarrow e^+ + \bar{\nu}_\mu + \nu_e$ are reciprocally charge conjugated. The experiments show that the helicities of the electron in the former and of the positron in the latter have opposite expectation values. This is a direct proof of $\mathcal{C}$ violation.

More generally, the $\mathcal{C}$ operator transforms a process with emission of a left neutrino into a process with emission of an antineutrino, which is also left. Indeed, $\mathcal{C}$ only changes particles into antiparticles leaving all the rest unvaried. However, experiments show that the antineutrinos are right. Namely, $\mathcal{C}$ is violated exactly as required by the $V - A$ structure of the interaction. In conclusion, the charged-current weak interactions violate $\mathcal{P}$ and $\mathcal{C}$, both maximally.

L. Landau (Landau 1957), in order to recover the particle–antiparticle symmetry, observed in 1957 that all interactions were invariant under the 'combined parity' CP, namely under the space inversion and the simultaneous particle–antiparticle substitution. The Alice image in the CP mirror is Antialice with her right and left hands interchanged. The counterpart of, say, a right particle is not its right antiparticle (P), but its left antiparticle (CP). For example, the β^- decay of a nucleon produces an antineutrino that is right; in the mirror the β^+ decay of the antinucleus would produce a neutrino that is left.

The matter–antimatter symmetry that the discoveries of P and C non-conservation had broken was thus re-established. However, seven years later J. Christenson, J. Cronin, V. Fitch and R. Turlay (Christenson *et al.* 1964) observed that the 'long-lifetime' neutral K meson, which is called K_L and is the CP eigenstate with $CP = -1$, decays, even if rarely, into two pions, namely into a state with $CP = +1$. Not even CP is a perfect symmetry. Matter and antimatter are not exactly equal. We shall see this in Chapter 8.

7.7 Cabibbo mixing

We saw in Section 7.2 that the CC weak interaction is universal in the lepton sector. This means that the couplings of the W mesons to the neutrino–lepton pairs of all the families are identical. In the quark sector the W meson couples to up-type and down-type quark pairs. However, the interaction is universal only if these quark states are not the states of definite flavour, but appropriate quantum superpositions of the latter. This property, now known as 'quark mixing', was discovered by N. Cabibbo in 1963 (Cabibbo 1963). Notice that at that time only the non-strange and strange hadrons were known. The other flavours were still to be discovered.

The problem was the following. There are two types of beta decays of the strange hadrons, those that conserve strangeness and those that violate it by $|\Delta S| = 1$. Whilst universality requires the corresponding matrix elements to be equal, the latter are substantially smaller than the former. For example the $\Delta S = 0$ decay

$$n \to pe^- \bar{\nu}_e \tag{7.65}$$

has a much higher probability than the similar $|\Delta S| = 1$ case

$$\Lambda \to pe^- \bar{\nu}_e. \tag{7.66}$$

Using the knowledge we have today, Fig. 7.17 shows the diagrams at the quark level of the two decays. The two quarks present in both cases before and after the decay are, in a first approximation, simple 'spectators'. The final quarks are the

Fig. 7.17.   Strangeness-changing and non-changing beta decays.

Fig. 7.18.   Strangeness-changing and non-changing meson decays.

same in the two cases; the only difference is that in one case an s quark decays, in the other a d quark.

Universality would require their matrix elements to be

$$M \propto G_F \cdot \bar{v}_{eL}\gamma_a e_{eL} \cdot \bar{d}_L\gamma^a u_L \qquad M \propto G_F \cdot \bar{v}_{eL}\gamma_a e_{eL} \cdot \bar{s}_L\gamma^a u_L \tag{7.67}$$

with the same coupling constant.

As a second example consider the $\Delta S = 0$ decay of the pion

$$\pi^- \rightarrow \mu^- \bar{v}_\mu \tag{7.68}$$

and the similar $|\Delta S| = 1$ decay of the K meson

$$K^- \rightarrow \mu^- \bar{v}_\mu. \tag{7.69}$$

Figure 7.18 shows their quark diagrams. Again the only difference is the decaying quark: s or d.

The 'universal' matrix elements would be

$$M \propto G_F \cdot \bar{\mu}_L\gamma_a v_{\mu L} \cdot \bar{d}_L\gamma^a u_L \qquad M \propto G_F \cdot \bar{\mu}_L\gamma_a v_{\mu L} \cdot \bar{s}_L\gamma^a u_L. \tag{7.70}$$

Let us focus on the meson case, which is simpler. The measured partial decay rates are

$$\begin{aligned}
\Gamma(\pi \rightarrow \mu v) &= \mathrm{BR}(\pi \rightarrow \mu v)/\tau_{\pi^+} = 1/(2.6 \times 10^{-8})\,\mathrm{s}^{-1} \\
\Gamma(K \rightarrow \mu v) &= \mathrm{BR}(K \rightarrow \mu v)/\tau_{K^+} = 0.64/(1.24 \times 10^{-8})\,\mathrm{s}^{-1}
\end{aligned} \tag{7.71}$$

giving the ratio

$$\Gamma(K \rightarrow \mu v)/\Gamma(\pi \rightarrow \mu v) = 1.34. \tag{7.72}$$

However, if the coupling constants of the $\bar{u}s$ pair and of the $\bar{u}d$ pair to the W are the same as in (7.70), the ratio of the decay rates is equal to the ratio of the phase space volumes, namely

$$\frac{\Gamma(K \to \mu v)}{\Gamma(\pi \to \mu v)} = \frac{m_K \left[1 - \left(m_\mu / m_K \right)^2 \right]^2}{m_\pi \left[1 - \left(m_\mu / m_\pi \right)^2 \right]^2} = 8.06. \tag{7.73}$$

Actually, the situation is not so simple, because the quarks decay inside the hadrons. We discussed in Section 3.6, for the pion decay, how the effects of the strong interaction can be factorised into the pion decay constant f_π. The same can be done for the K meson decay with another decay constant f_K. These factors cannot be measured directly and are very difficult to calculate, but we can say something about their ratio, which is what we need. Actually if the $SU(3)_f$ symmetry were exact, we would have $f_K/f_\pi = 1$. It can be shown that the observed symmetry breaking implies $f_K/f_\pi > 1$. Therefore, the effect of the strong interactions is to worsen the disagreement between the experiment and the universality. The ratio between the semileptonic decay rates of the K and the pion is an order of magnitude smaller than expected.

The analysis of the semileptonic decays of the nucleons and the hyperons, with and without change of strangeness, must also take into account the hadronic structure and its approximate $SU(3)_f$ symmetry. We only say here that the conclusion is that, again, the $|\Delta S| = 1$ decays are suppressed by about an order of magnitude compared to the $\Delta S = 0$ ones. Notice that it is the change in strangeness that matters, not the strangeness itself. For example the decay $\Sigma^\pm \to \Lambda e^\pm v$ is not suppressed.

Another problem is that the value of the coupling constant in the beta decay of the neutron is somewhat smaller than that of the muon decay.

All of this is explained if we assume, like Cabibbo, that the down-type quarks entering the CC weak interactions are not d and s, but, say, d' and s'. Each (d, s) and (d', s') pair is an ortho-normal base. The latter is obtained from the former by the rotation of a certain angle, called the 'Cabibbo angle' θ_C. This is shown schematically in Fig. 7.19. In a formula, the down-type quark that couples to the W is a quantum superposition of d and s, namely the state

$$d' = d \cos \theta_C + s \sin \theta_C. \tag{7.74}$$

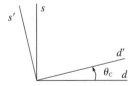

Fig. 7.19.   The Cabibbo rotation.

Indeed, the coefficients of d and s must satisfy the normalisation condition, namely the sum of their square must be one. Therefore, they can be thought of as the sine and the cosine of an angle.

In the Cabibbo theory there is only one matrix element for (7.65) and (7.66) in which d' appears, namely

$$M \propto G_F \cdot \bar{e}_L \gamma_a \nu_{eL} \cdot \bar{d}'_L \gamma^a u_L. \tag{7.75}$$

Using (7.74) we obtain for the two decays

$$
\begin{aligned}
M &\propto G_F \cos\theta_C \cdot \bar{e}_L \gamma_a \nu_{eL} \cdot \bar{d}_L \gamma^a u_L \quad \text{for } \Delta S = 0 \\
M &\propto G_F \sin\theta_C \cdot \bar{e}_L \gamma_a \nu_{eL} \cdot \bar{s}_L \gamma^a u_L \quad \text{for } \Delta S = 1.
\end{aligned}
\tag{7.76}
$$

Since the angle θ_C is small, the $|\Delta S| = 1$ transition probabilities, which are proportional to $\sin^2\theta_C$ are smaller than the $\Delta S = 0$ ones that have the factor $\cos^2\theta_C$ by about an order of magnitude. Moreover, the constant of the neutron decay is $G_F^2 \cos^2\theta_C$, which is somewhat smaller than the pure G_F^2 of the muon decay.

If the theory is correct, a single value of the Cabibbo angle must agree with the rates of all the semileptonic decays, of the nuclei, of the neutron, of the hyperons and of the strange and non-strange mesons. Both experimental and theoretical work is needed for this verification. Experiments must measure decay rates and other relevant kinematic quantities with high accuracy. Theoretical calculations must consider the fact that the elementary processes at the quark level, such as those shown in Figs. 7.17 and 7.18, take place inside hadrons. Consequently, the transition probabilities are not given simply by the matrix elements in (7.76). The evaluation of the interfering strong interaction effects is not easy because the QCD coupling constant α_s is large in the relevant momentum transfer region.

We shall discuss the measurement of $\sin\theta_C$ and $\cos\theta_C$ in Section 7.9 on two examples. We mention here that all the measurements give consistent results. The values are

$$\theta_C = 12.9° \qquad \cos\theta_C = 0.974 \qquad \sin\theta_C = 0.221. \tag{7.77}$$

In conclusion, the CC weak interactions are also universal in the quark sector, provided that the 'quark mixing' phenomenon is taken into account.

7.8 The Glashow, Iliopoulos and Maiani mechanism

An immediate consequence of the Cabibbo theory is the presence, in the Lagrangian, of the term

$$\bar{d}'_L \gamma_a d'_L = \cos^2\theta_C \bar{d}_L \gamma_a d_L + \sin^2\theta_C \bar{s}_L \gamma_a s_L + \cos\theta_C \sin\theta_C [\bar{d}_L \gamma_a s_L + \bar{s}_L \gamma_a d_L] \tag{7.78}$$

which describes neutral-current transitions. In particular, the last term implies neutral currents that change strangeness (SCNC, strangeness-changing neutral currents) because they connect s and d quarks. However, the corresponding physical processes are strongly suppressed. For example, the two NC and CC decays

$$K^+ \to \pi^+ + v_e + \bar{v}_e \qquad K^+ \to \pi^0 + v_e + e^+ \qquad (7.79)$$

should proceed with similar probabilities, as understood from the diagrams shown in Fig. 7.20.

On the contrary, the former decay is strongly suppressed, the measured values of the branching ratios (Yao *et al.* 2006) being

$$\mathrm{BR}(K^+ \to \pi^+ v\bar{v}) = \left(1.5^{+1.3}_{-0.9}\right) \times 10^{-10}$$
$$\mathrm{BR}\left(K^+ \to \pi^0 e^+ v_e\right) = (4.98 \pm 0.07) \times 10^{-2}. \qquad (7.80)$$

S. Glashow, I. Iliopoulos and L. Maiani observed in 1970 (Glashow *et al.* 1970) that the d' and u states can be thought of as the members of the doublet $\binom{u}{d'}$. Now, they thought, a fourth quark might exist, the 'charm' c as the missing partner of s', to form a second similar doublet $\binom{c}{s'}$.

Since s' is orthogonal to d' we have

$$s' = -d \sin \theta_C + s \cos \theta_C. \qquad (7.81)$$

We anticipated this situation in Fig. 7.19. Clearly, the relationship between the two bases is the rotation

$$\begin{pmatrix} d' \\ s' \end{pmatrix} = \begin{pmatrix} \cos \theta_C & \sin \theta_C \\ -\sin \theta_C & \cos \theta_C \end{pmatrix} \begin{pmatrix} d \\ s \end{pmatrix}. \qquad (7.82)$$

From the historical point of view this was the prediction of a new flavour. We saw in Section 4.9 how it was discovered.

Let us now see how the 'GIM' mechanism succeeds in suppressing the strangeness-changing neutral currents. In addition to the terms (7.78) we now

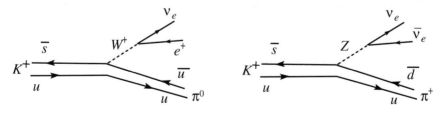

Fig. 7.20.   Strangeness-changing charged- and neutral-current decays.

have

$$\bar{s}'_L\gamma_a s'_L = \sin^2\theta_C \bar{d}_L\gamma_a d_L + \cos^2\theta_C \bar{s}_L\gamma_a s_L - \cos\theta_C \sin\theta_C[\bar{d}_L\gamma_a s_L + \bar{s}_L\gamma_a d_L]. \quad (7.83)$$

Summing the two, we obtain

$$\bar{s}'_L\gamma_a s'_L + \bar{d}'_L\gamma_a d'_L = \bar{d}_L\gamma_a d_L + \bar{s}_L\gamma_a s_L. \quad (7.84)$$

The SCNC cancel out. However, a NC term remains in the Lagrangian, namely the NC between equal quarks or, in other words, the strangeness-conserving neutral current. As we shall see in Section 7.10 the corresponding physical processes were indeed discovered, in 1973. We observe here that the Cabibbo rotation is irrelevant for the NC term. In other words this term is the same in the two bases.

7.9 The quark mixing matrix

The GIM mechanism explains the suppression of the SCNC in the presence of two families. Later on, the third family with its two additional quark flavours was discovered, as we have seen. It was also found that the flavour-changing neutral currents (FCNC) for all the flavours, not only for strangeness, are suppressed. Therefore, we need to generalise the concepts of the preceding sections.

Equation (7.82) is a transformation between two orthogonal bases. The doublet $\binom{d}{s}$ is the base of the down-type quarks with definite mass. These are the states, let us say, that would be stationary if they could be free. The doublet $\binom{d'}{s'}$ is the base of down-type quarks that are the weak interaction eigenstates, namely the states produced by such interaction. The two bases are connected by a unitary transformation that we now call V, to develop a formalism suitable for generalisation to three families. The elements of V are real in the two-family case, as we shall soon show. We rewrite (7.82) as

$$\begin{pmatrix} d' \\ s' \end{pmatrix} = \begin{pmatrix} V_{ud} & V_{us} \\ V_{cd} & V_{cs} \end{pmatrix} \begin{pmatrix} d \\ s \end{pmatrix} = \begin{pmatrix} \cos\theta_C & \sin\theta_C \\ -\sin\theta_C & \cos\theta_C \end{pmatrix} \begin{pmatrix} d \\ s \end{pmatrix}. \quad (7.85)$$

The generalisation to three families was done by M. Kobaiashi and K. Maskawa in 1973 (Kobaiashi & Maskawa 1973). The quark mixing transformation is

$$\begin{pmatrix} d' \\ s' \\ b' \end{pmatrix} = \begin{pmatrix} V_{ud} & V_{us} & V_{ub} \\ V_{cd} & V_{cs} & V_{cb} \\ V_{td} & V_{ts} & V_{tb} \end{pmatrix} \begin{pmatrix} d \\ s \\ b \end{pmatrix}. \quad (7.86)$$

The matrix is called the Cabibbo–Kobaiashi–Maskawa (CKM) matrix. It is unitary, namely

$$VV^+ = 1. \tag{7.87}$$

The three-family expression of the charged-current interaction is

$$\sum_{i=1}^{3} \bar{u}^i \gamma_\mu (1 - \gamma_5) V_{ik} d^k = \sum_{i=1}^{3} \bar{u}_L^i \gamma_\mu V_{ik} d_L^k \tag{7.88}$$

where we have set $u^1 = u$, $u^2 = c$, $u^3 = t$, $d^1 = d$, $d^2 = s$, $d^3 = b$. Focussing on the flavour indices, the structure is

$$(\bar{u} \quad \bar{c} \quad \bar{t}) \begin{pmatrix} V_{ud} & V_{us} & V_{ub} \\ V_{cd} & V_{cs} & V_{cb} \\ V_{td} & V_{ts} & V_{tb} \end{pmatrix} \begin{pmatrix} d \\ s \\ b \end{pmatrix}. \tag{7.89}$$

This justifies the names of the indices of the matrix elements.

We shall now determine the number of independent elements of the matrix. A complex 3×3 matrix has in general 18 real independent elements, 9 if it is unitary. If it were real, it would be orthogonal, with 3 independent elements, corresponding to the three rotations, namely the Euler angles. The 6 remaining elements of the complex matrix are therefore phase factors of the $\exp(i\delta)$ type. Not all of them are physically meaningful.

Indeed, the particle fields, the quarks in this case, are defined modulo an arbitrary phase factor. Moreover, (7.89) is invariant for the substitutions

$$d^k \rightarrow e^{i\theta_k} d^k \qquad V_{ik} \rightarrow e^{-i\theta_k} V_{ik}. \tag{7.90}$$

With three such substitutions we can absorb a global phase for each row in the d type quarks, eliminating three phases. Similarly, we can absorb a global phase factor for each column in a u type quark. It seems, at first, that the other three phase factors can be eliminated, but only two of them are independent. Indeed V does not change when all the d and all the u change by the same phase. Consequently, the six phases we used to redefine the fields must satisfy a constraint; only five of them are independent. In conclusion, the number of phases physically meaningful is $6 - 5 = 1$. Summing up, the three-family mixing matrix has four free parameters, which can be taken to be three rotation angles and one phase factor $\exp(i\delta)$.

Going back to two families, the 2×2 unitary matrix has four independent real parameters. One of them is the Cabibbo rotation. The other three are phase factors. Two of them can be absorbed in the d type quarks and two in the u type ones. This makes four and, subtracting one constraint makes three. As anticipated, the matrix is real.

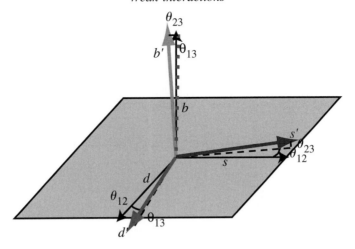

Fig. 7.21.   The quark rotations.

Coming back to three families, we define the rotations as follows. We take three orthogonal axes (x, y, z) and we let each of them correspond to a down-type quark as (d, s, b), as in Fig. 7.21. We rotate in the following order: the first rotation is by θ_{12} around z, the second by θ_{13} around the new y, the third by θ_{23} around the last x. The product of three rotation matrices, which are orthogonal, describes the sequence. Writing, to be brief, $c_{ij} = \cos \theta_{ij}$ and $s_{ij} = \sin \theta_{ij}$, we have

$$V = \begin{pmatrix} 1 & 0 & 0 \\ 0 & c_{23} & s_{23} \\ 0 & -s_{23} & c_{23} \end{pmatrix} \begin{pmatrix} c_{13} & 0 & s_{13} \\ 0 & 1 & 0 \\ -s_{13} & 0 & c_{13} \end{pmatrix} \begin{pmatrix} c_{12} & -s_{12} & 0 \\ s_{12} & c_{12} & 0 \\ 0 & 0 & 1 \end{pmatrix}.$$

We must still introduce the phase. This cannot be simply a factor, which would be absorbed by a field, becoming non-observable. Actually, there are several equivalent procedures. We shall use the following expression

$$V = \begin{pmatrix} 1 & 0 & 0 \\ 0 & c_{23} & s_{23} \\ 0 & -s_{23} & c_{23} \end{pmatrix} \begin{pmatrix} c_{13} & 0 & s_{13} e^{-i\delta_{13}} \\ 0 & 1 & 0 \\ -s_{13} e^{+i\delta_{13}} & 0 & c_{13} \end{pmatrix} \begin{pmatrix} c_{12} & -s_{12} & 0 \\ s_{12} & c_{12} & 0 \\ 0 & 0 & 1 \end{pmatrix}.$$

$$(7.91)$$

Notice that the last expression is valid only if the mixing matrix is unitary. This must be experimentally verified. 'New physics', not included in the theory, may induce violations of unitarity and consequently invalidate the expression of the mixing matrix in terms of three rotation angles and a phase factor. The most obvious example is the presence of a fourth family.

Therefore, the theory must be tested by measuring all the elements of the mixing matrix (7.86), nine amplitudes and a phase, and by checking if the unitary

conditions between them are satisfied or not. These are amongst the strictest tests passed by the Standard Model.

The absolute values of six matrix elements, $|V_{ud}|$, $|V_{us}|$, $|V_{cd}|$, $|V_{cs}|$, $|V_{ub}|$ and $|V_{cb}|$ have been determined by measuring the semileptonic decay rates of the hadrons of different flavour, strangeness, charm and beauty, as we shall see in two examples. The values are (Ceccucci *et al.* 2006)

$$|V_{ud}|^2 = (973.77 \pm 0.27) \times 10^{-3} \qquad |V_{us}|^2 = (225.7 \pm 2.1) \times 10^{-3}$$
$$|V_{ub}|^2 = (4.31 \pm 0.30) \times 10^{-3} \qquad |V_{cd}|^2 = (230 \pm 11) \times 10^{-3} \qquad (7.92)$$
$$|V_{cs}|^2 = (957 \pm 95) \times 10^{-3} \qquad |V_{cb}|^2 = (41.6 \pm 0.6) \times 10^{-3}.$$

Moreover, the following three products are extracted from the mass differences of three neutral-meson systems $K^0 \bar{K}^0$, $B^0 \bar{B}^0$ and $B_s^0 \bar{B}_s^0$, i.e. from the measured oscillation periods, as we shall discuss in the next chapter: $|V_{ud}||V_{us}|$ from Δm_{K^0}, $|V_{td}||V_{tb}|$ from Δm_{B^0} and $|V_{tb}||V_{ts}|$ from $\Delta m_{B_s^0}$. The ninth element, $|V_{tb}|$, has not yet been measured, but is known to be very close to 1. The imaginary parts, when present, are measured in $\mathcal{CP}$ violation phenomena, as we shall discuss in Section 8.5. The determination of the mixing matrix elements needs not only measurements but also theoretical input, often difficult due to the always-present QCD effects.

Using the independently measured absolute values of the elements, three unitarity checks can be done. Summing up the measured values we obtain

$$|V_{ud}|^2 + |V_{us}|^2 + |V_{ub}|^2 = 0.9992 \pm 0.0011 \qquad \text{1st row}$$
$$|V_{cd}|^2 + |V_{cs}|^2 + |V_{cb}|^2 = 0.968 \pm 0.181 \qquad \text{2nd row} \qquad (7.93)$$
$$|V_{ud}|^2 + |V_{cd}|^2 + |V_{td}|^2 = 1.001 \pm 0.005 \qquad \text{1st column.}$$

We see that the conditions are satisfied.

The most accurate values of the matrix elements are obtained by a global fit that uses all the available measurements and assumes unitarity. The procedure also gives a (very accurate) value of $|V_{tb}|$. The result (Ceccucci *et al.* 2006) is

$$\begin{pmatrix} |V_{ud}| & |V_{us}| & |V_{ub}| \\ |V_{cd}| & |V_{cs}| & |V_{cb}| \\ |V_{td}| & |V_{ts}| & |V_{tb}| \end{pmatrix} = \begin{pmatrix} 973.83^{+0.24}_{-0.23} & 227.2 \pm 1.0 & 3.96 \pm 0.09 \\ 227.1 \pm 1.0 & 972.96 \pm 0.24 & 42.21^{+0.10}_{-0.80} \\ 8.14^{+0.32}_{-0.64} & 41.61^{+0.12}_{-0.78} & 999.100^{+0.034}_{-0.004} \end{pmatrix} \times 10^{-3}.$$

$$(7.94)$$

It is also useful to consider the angles and their sines, namely

$$\sin\theta_{12} = 0.2229 \pm 0.0022 \qquad \theta_{12} = 12.9°$$
$$\sin\theta_{23} = 0.0412 \pm 0.0002 \qquad \theta_{23} = 2.4° \qquad (7.95)$$
$$\sin\theta_{13} = 0.0036 \pm 0.0007 \qquad \theta_{13} = 0.2°.$$

We see that the angles are small or very small (the rotations in Fig. 7.21 have been exaggerated to make them visible). Moreover, there is a hierarchy in the angles, namely $s_{12} \gg s_{23} \gg s_{13}$. We do not know why.

Therefore, the diagonal elements of the matrix are very close to one; the mixing between the third and second families is smaller than that between the first two, and even smaller is the mixing between the first and third families. In practice, we might say, the hadrons prefer to decay semileptonically into the nearest family. This implies that the 2×2 submatrix of the first two families is very close to being unitary and therefore the Cabibbo angle is almost equal to θ_{12} and, finally, that $|V_{ud}| \approx |V_{cs}| \approx \cos \theta_C$ and $|V_{us}| \approx |V_{cd}| \approx \sin \theta_C$.

We now give two examples of measurement of the absolute values of the mixing matrix elements, namely of $|V_{ud}|$ and $|V_{us}|$.

$|V_{ud}| \approx \cos \theta_C$ is measured in three different types of processes:

- in the super-allowed beta transitions of several nuclei (i.e. $J^P = 0^+ \to J^P = 0^+$ transitions between two members of the same isospin multiplet, with $\Delta I_z = \pm 1$); this is currently the most accurate method;
- in the beta decay of the neutron;
- in the so-called $\pi e3$ decay of the pion.

The measurements give equal values within per mil uncertainties. We now give some hints on $\pi e3$, which is

$$\pi^+ \to \pi^0 + e^+ + \nu_e. \tag{7.96}$$

Even if this channel does not provide the most precise value of $|V_{ud}|$, it is theoretically very clean, being free from nuclear physics effects and having a simple matrix element. However, it is experimentally challenging because it is extremely rare, with a branching ratio of 10^{-8}. Consequently, in order to have a statistical uncertainty of, say, 10^{-3} one needs to collect a total of 10^{14} pion decays and to be able to discriminate with the necessary accuracy the $\pi e3$s. Clearly, the reason for the rareness of the decay is the smallness of the Q-value, considering that (Yao *et al.* 2006)

$$\Delta \equiv m_{\pi^+} - m_{\pi^0} = 4.5936 \pm 0.0005 \, \text{MeV}. \tag{7.97}$$

Notice that the uncertainty is only 10^{-4}.

Figure 7.22 shows the two Feynman weak-interaction diagrams at the quark level.

However, the non-decaying quark does not behave simply as a 'spectator' as Fig. 7.22 suggests. On the contrary, it strongly interacts with the companions, before and after the decay, in a non-perturbative QCD regime. The contribution of diagrams such as the one sketched in Fig. 7.23 must be calculated.

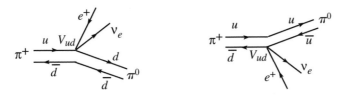

Fig. 7.22. Quark level diagrams for $\pi e3$ decay.

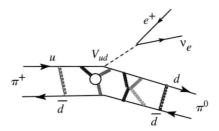

Fig. 7.23. Hadronic complications in $\pi e3$.

We set $p_\mu = p_\mu^{\pi^+} + p_\mu^{\pi^0}$ and $p_\mu^{\pi^+} - p_\mu^{\pi^0}$, where $p_\mu^{\pi^+}$ and $p_\mu^{\pi^0}$ are the four-momenta of the initial and final pions. It can be shown that the matrix element is

$$M \propto f_+\left(q^2\right) G_F V_{ud} p^\mu \bar{v}_e \gamma_\mu \left(1 - \gamma_5\right) e \approx f_+(0) G_F V_{ud} p^\mu \bar{v}_e \gamma_\mu \left(1 - \gamma_5\right) e \qquad (7.98)$$

where $f_+(q^2)$ is a function of the norm of the four-momentum transfer, which takes into account the effects of strong interactions. Since in the $\pi e3$ the Q-value is so small, we have approximated the form factor with its value at $q^2 = 0$ at the last member. Moreover, it turns out that $f_+(0)$ is determined by the $SU(2)$ symmetry, because π^0 and π^+ belong to the same isospin multiplet.

The partial width is given by

$$\Gamma\left(\pi^+ \to \pi^0 e^+ v_e\right) \equiv \frac{\mathrm{BR}\left(\pi_{e3}^+\right)}{\tau_\pi} = \frac{G_F^2 \Delta^5}{30\pi^3} |V_{ud}|^2 \left(1 - \frac{\Delta}{2m_{\pi^+}}\right)^3 f(\varepsilon)(1 + \delta_{\mathrm{EM}}) \quad (7.99)$$

where δ_{EM} is a 'radiative correction' at the loop level of a small percentage and $\varepsilon \equiv (m_e/\Delta)^2 \approx 10^{-2}$.

The function f is the 'Fermi function' and is known as a series of powers of ε. The overall theoretical uncertainty is small $\sim 10^{-3}$. As we have seen in the case of muon decay in Section 7.2, the Q-value of the decay, Δ in this case, appears at the fifth power. As we have seen Δ is known with good precision. The same is true for the pion lifetime. The most precise experiment on $\mathrm{BR}\left(\pi_{e3}^+\right)$ is PIBETA performed at the PSI Laboratory in Zurich (Pocanic *et al.* 2004). The positive pions are stopped in the middle of a sphere of CsI crystals, used to detect the two γs from the π^0 decay, measure their energies and normalise to the $\pi^+ \rightarrow e^+ \nu_e$ rate. The measured value is

$$\mathrm{BR}\left(\pi^+_{e3}\right) = (1.036 \pm 0.006) \times 10^{-8}. \qquad (7.100)$$

With this value and the above-mentioned inputs from theory we obtain

$$|V_{ud}| = 0.9728 \pm 0.0030. \qquad (7.101)$$

We now consider $|V_{us}| \approx \sin\theta_C$. It can be determined in different processes:

- semileptonic decays of hyperons with change of strangeness;
- semileptonic decays of the K mesons, which currently give the most precise values. The initial particle can be a charged or a neutral kaon, the daughters can be $\pi e \nu_e$ or $\pi e \nu_\mu$ in their different charge states. We anticipate here that the neutral K meson states of definite lifetime are not the states of definite strangeness, K^0 and $\bar{K}^0$, but two linear combinations of them, called K_S and K_L, having shorter and longer lifetimes respectively.

We now discuss a precise result obtained by the KLOE experiment (Ambrosino *et al.* 2006) by measuring the branching ratio of the decay

$$K_S \rightarrow \pi^\mp + e^\pm + \nu_e. \qquad (7.102)$$

The two diagrams shown in Fig. 7.24 contribute at the quark level.

Of course, strong interaction complications are present and they are more difficult to handle than for the $\pi e3$ decays. The calculation gives the result

$$\Gamma\left(K_S \rightarrow \pi^\mp e^\pm \nu_e\right) \equiv \frac{\mathrm{BR}(K_{Se3})}{\tau_S}$$
$$= \frac{G_F^2 m_K^5}{128\pi^3} |V_{ud}|^2 f_-(0) I_+ (1 + \delta_{\mathrm{EW}})(1 + \delta_{\mathrm{EM}}). \qquad (7.103)$$

As in the case of $\pi e3$ the form factor $f_+(q^2)$ is a function of the square of the momentum transfer $q_\mu = p_\mu^{K^0} - p_\mu^{\pi^\pm}$ that takes into account the strong interaction effects. However, $f_+(q^2)$ cannot now be considered as a constant because the range of momentum transfer is wide. Its q^2 dependence is measured up to the constant factor $f_+(0)$, which is theoretically calculated. This calculation is

Fig. 7.24.   Quark level diagrams for $Ke3$.

somewhat more uncertain compared to the calculation for $\pi e3$ because the initial and final mesons do not belong to an isospin multiplet but only to an $SU(3)_f$ multiplet. The current value is $f_+(0) = 0.961 \pm 0.008$. The function $f_+(q^2)$ does not appear in (7.103) because the integration on q^2 has already been done, giving the factor I_+. Finally, δ_{EW} and δ_{EM} are electroweak and electromagnetic 'radiative corrections'. They are about 2% and about 0.5% respectively.

The KLOE experiment collected a pure sample of about 13 000 semileptonic K_S decay events working at the DAΦNE ϕ-factory at Frascati. A ϕ-factory is a high luminosity e^+e^- collider that operates at the centre of mass energy $\sqrt{s} = m_\phi = 1.020\,\text{GeV}$. The pure ϕ-meson initial state decays $\sim 34\%$ of the time into neutral K mesons. A single wave function describes the time evolution of both particles. It is antisymmetric under their exchange because the orbital angular momentum is $L=1$. Consequently the two bosons cannot be equal, if one of them decays as K_S the other one must decay as K_L. Moreover, being in the centre of mass frame, the two decays are back-to-back. Consequently, if we detect a K_L we know that a K_S is present in the opposite direction. The mean decay paths of K_S and K_L are $\lambda_S \approx 0.6\,\text{cm}$ and $\lambda_L \approx 350\,\text{cm}$. The latter is small enough to allow an efficient detection of the K_Ls.

The KLOE detector consists mainly of a large cylindrical TPC in a magnetic field surrounded by a lead-scintillating-fibre sampling calorimeter. A sample of about 400 million $K_S K_L$ pairs was collected. The two rates $K_S \to \pi^+ e^- \bar{\nu}_e$ and $K_S \to \pi^- e^+ \nu_e$ were separately measured and normalised to the branching ratio into $K_S^0 \to \pi^+ \pi^-$, which is known with $\sim 0.1\%$ accuracy. The result is

$$\text{BR}(K_S \to \pi^- e^+ \nu_e) + \text{BR}(K_S \to \pi^+ e^- \bar{\nu}_e) = (7.046 \pm 0.091) \times 10^{-4}. \quad (7.104)$$

With this value and the necessary theoretical input one obtains

$$|V_{us}| = 0.2240 \pm 0.0024. \quad (7.105)$$

Example 7.1 Estimate the following ratios: $\Gamma(D^0 \to K^+ K^-)/\Gamma(D^0 \to \pi^+ K^-)$, $\Gamma(D^0 \to \pi^+ \pi^-)/\Gamma(D^0 \to \pi^+ K^-)$ and $\Gamma(D^0 \to K^+ \pi^-)/\Gamma(D^0 \to \pi^+ K^-)$.

We start by recalling the valence quark composition of the hadrons of the problem: $D^0 = c\bar{u}, K^+ = u\bar{s}, K^- = s\bar{u}, \pi^+ = u\bar{d}, \pi^- = d\bar{u}$. We draw the tree-level diagrams for the decaying quark for each process (Fig. 7.25).

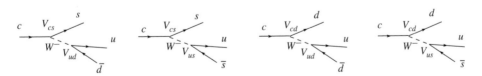

Fig. 7.25. Quark diagrams for c decays.

The first diagram is favoured because it has at the two vertices $|V_{cs}|$ and $|V_{ud}|$ that are both large, $\sim \cos \theta_C$. In the second and third diagrams, the coefficient at one vertex is large while the coefficient at the other is small: respectively $|V_{us}|$ and $|V_{cd}| \approx \sin \theta_C$ (they are said to be 'Cabibbo suppressed'). In the fourth diagram the coefficients at both vertices are small, its amplitude is proportional to $|V_{us}||V_{cd}| \approx \sin^2 \theta_C$ ('doubly Cabibbo suppressed').

Summing up, we have

$$\frac{\Gamma(D^0 \to K^+ K^-)}{\Gamma(D^0 \to \pi^+ K^-)} \propto \frac{|V_{cs}|^2 |V_{us}|^2}{|V_{cs}|^2 |V_{ud}|^2} \approx \tan^2 \theta_C \approx 0.05$$

$$\frac{\Gamma(D^0 \to \pi^+ \pi^-)}{\Gamma(D^0 \to \pi^+ K^-)} \propto \frac{|V_{cd}|^2 |V_{ud}|^2}{|V_{cs}|^2 |V_{ud}|^2} \approx \tan^2 \theta_C \approx 0.05$$

$$\frac{\Gamma(D^0 \to K^+ \pi^-)}{\Gamma(D^0 \to \pi^+ K^-)} \propto \frac{|V_{cd}|^2 |V_{us}|^2}{|V_{cs}|^2 |V_{ud}|^2} \approx \tan^4 \theta_C \approx 0.0025.$$

In a proper calculation of the decay rates one must take into account the phase space (easy) and the colour field effects (difficult). With this caveat, the experimental values confirm the hierarchy

$$\frac{\Gamma(D^0 \to K^+ K^-)}{\Gamma(D^0 \to \pi^+ K^-)} \approx 0.10 \qquad \frac{\Gamma(D^0 \to \pi^+ \pi^-)}{\Gamma(D^0 \to \pi^+ K^-)} \approx 0.04 \qquad \frac{\Gamma(D^0 \to K^+ \pi^-)}{\Gamma(D^0 \to \pi^+ K^-)} < 0.02.$$

Example 7.2 Estimate the ratio: $\Gamma(B^- \to D^0 K^{*-})/\Gamma(B^- \to D^0 \rho^-)$.

The valence quark compositions are: $B^- = b\bar{u}, D^0 = c\bar{u}, \rho^- = d\bar{u}, K^{*-} = s\bar{u}$. We draw the diagrams in Fig. 7.26.

Looking at the vertex coefficients we have

$$\frac{\Gamma(B^- \to D^0 K^{*-})}{\Gamma(B^- \to D^0 \rho^-)} \propto \frac{|V_{us}|^2 |V_{cb}|^2}{|V_{ud}|^2 |V_{cb}|^2} = \frac{|V_{us}|^2}{|V_{ud}|^2} \approx \tan^2 \theta_C \approx 0.05.$$

The experimental value is ~ 0.05.

M. Kobaiashi and K. Maskawa observed in 1972 (Kobaiashi & Maskawa 1973) that the phase factor present in the mixing matrix for three (and not for two)

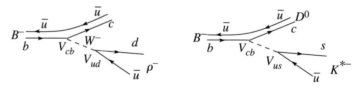

Fig. 7.26. Quark diagrams for two b decays.

families implies $\mathcal{CP}$ violation. This is due to the fact that the phase factor $\exp(i\delta)$ appears in the wave function that becomes $\exp[i(\omega t + \delta)]$. The latter expression is obviously not invariant under time reversal if $\delta \neq 0$ and $\delta \neq \pi$. Since $\mathcal{CPT}$ is conserved, $\mathcal{CP}$ must be violated. We shall see in Section 8.5 a measurement of the phase of one of the mixing matrix elements. We report here that $\mathcal{CP}$ violating phase in (7.91) is large and not known with high precision. Its value is

$$\delta_{13} \approx 60°. \tag{7.106}$$

7.10 Weak neutral currents

We have seen that flavour-changing neutral-current processes are strongly suppressed. However flavour-conserving neutral-current processes exist in Nature. The experimental search for such processes went on for a very long time. In the 1970s, groups engaged in neutrino physics at CERN built a neutrino beam from the CPS proton synchrotron, a large bubble chamber called Gargamelle, and the associated instrumentation. Gargamelle was filled with 15 t of CF_3Br, which is a freon, a heavy liquid that provides both the mass necessary for an appreciable neutrino interaction rate and a good photon detection probability. The experiments made with this instrument made many contributions to neutrino physics, in particular the discovery of neutral currents in 1973 (Hasert *et al.* 1973). Let us see how.

The incident beam contains mainly ν_μ (with a small ν_e contamination). All the CC events have a μ^- in the final state, which is identified by its straight non-interacting minimum ionising track.

If neutral currents exist, the following process can happen on a generic nucleus N

$$\nu_\mu + N \rightarrow \nu_\mu + \text{hadrons}. \tag{7.107}$$

This type of event is identified by the absence of the muon in the final state, which contains only hadrons (the neutrino cannot be seen). Figure 7.27 is an example.

Analysing the image, we identify all the tracks as hadrons and none as a following muon (Perkins 2004). Neutrinos enter from the left of the picture and one of them interacts. Around the vertex we see: a short dark track directed upward, which is recognised as a stopping proton; two e^+e^- pairs that are the materialisation of the two photons from a decay $\pi^0 \rightarrow \gamma\gamma$; and two charged tracks of opposite signs. The track moving upwards is negative (as inferred by the known direction of the magnetic field) and interacts (it passes below two eye-shaped images; the interaction is near the second one), therefore it is a hadron. The positive track is a π^+ that ends with a charge-exchange reaction producing a π^0,

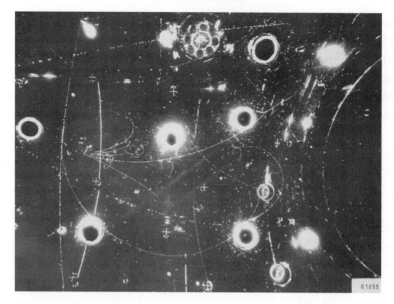

Fig. 7.27. A neutral-current event in Gargamelle. (Photo CERN)

as recognised from the electron originating from the Compton scattering of one of the photons from its decay. The electron is the small vertical track under the 'eye' pointing to the end point of the π^+ track.

The discovery of the NC weak interactions clearly suggested that the weak interactions might be very similar to the electromagnetic ones. Rapid theoretical development followed, leading to electroweak unification, as we shall see in Chapter 9.

Problems

7.1. Draw the Feynman quark diagrams of the following strong and weak decays:
$K^{*+} \rightarrow K^0 + \pi^+$; $n \rightarrow p + e^- + \bar{\nu}_e$; $\pi^+ \rightarrow \mu^+ + \nu_\mu$.

7.2. Draw the Feynman quark diagrams of the following strong and weak decays:
$\pi^+ \rightarrow \pi^0 + e^+ + \nu_e$; $\rho^+ \rightarrow \pi^0 + \pi^+$; $K^0 \rightarrow \pi^- + \pi^+$; $\Lambda \rightarrow p + e^- + \bar{\nu}_e$.

7.3. Find the value of the Fermi constant G_F in SI units, knowing that $G_F/(\hbar c)^3 = 1.17 \times 10^{-5} \, \text{GeV}^{-2}$.

7.4. The PEP was a collider in which the two beams of e^+ and e^- collided in the CM reference frame. Consider the beam energy $E_{cm} = 29 \, \text{GeV}$ and the reaction $e^+ + e^- \rightarrow \tau^+ + \tau^-$. Find the average distance the τ will fly before decaying.

7.5. Consider the decays $\mu^+ \rightarrow e^+ + \nu_e + \bar{\nu}_\mu$ and $\tau^+ \rightarrow e^+ + \nu_e + \bar{\nu}_\tau$. The branching ratios are 100% for the first, 16% for the second. The μ lifetime is $\tau_\mu = 2.2 \, \mu\text{s}$. Calculate the τ_τ lifetime.

7.6. Neglecting the masses, calculate the cross section of the process: $e^+e^- \rightarrow \tau^+\tau^-$ at $\sqrt{s} = 10\,\text{GeV}$ and at $\sqrt{s} = 100\,\text{GeV}$.

7.7. What are the differences between neutrinos and antineutrinos? What are the conserved quantities in neutrino scattering? Complete with the missing particle $\nu_\mu + e^- \rightarrow \mu^- + ?$ If neutrinos are massless, what is the direction of their spin? And for antineutrinos? The Universe is full of neutrinos at a temperature of about 2 K. What is the neutrino average speed if their mass is 50 meV?

7.8. Write the reaction (or the reactions if there is more than one) by which a ν_μ can produce a single pion hitting: (a) a proton; (b) a neutron. Do the decays $\mu^+ \rightarrow e^+ + \gamma$ and $\mu^+ \rightarrow e^+ + e^+ + e^-$ exist? Give the reason for your answer.

7.9. We send a π^- beam onto a target and we observe the inclusive production of Λ. We measure the momentum $\mathbf{p}_\Lambda$ and the polarisation σ_Λ of the hyperon. How can we check if parity is conserved in these reactions? What do you expect to happen?

7.10. How can you observe parity violation in the decay $\pi \rightarrow \mu\nu$?

7.11. The muons have the same interactions, electromagnetic and weak, as the electrons. Why does a muon with energy of a few GeV pass through an iron slab, while an electron of the same energy does not?

7.12. What is the minimum momentum of the electron from a muon at rest? What is the maximum momentum?

7.13. Cosmic rays are mainly protons. Their energy spectrum decreases with increasing energy. Their interactions with the atmospheric nuclei produce mesons, which give rise, by decaying, to ν_μ and ν_e. On a sample of $N_\nu = 10^6$ ν_μs with 1 GeV energy, how many interact in crossing the Earth along its diameter? ($\sigma \approx 7$ fb, $\rho \approx 5 \times 10^3$ kg/m^3, $R \approx 6000$ km.)

7.14. Consider the neutrino cross section on an electron $\sigma(\nu_\mu e^- \rightarrow \nu_\mu e^-) \approx \frac{G_F^2}{\pi}s$ and on an 'average nucleon' (namely the average between the cross sections on a proton and a neutron) $\sigma(\nu_\mu N \rightarrow \mu^- h) \approx 0.2 \times \frac{G_F^2}{\pi}s$ at energies $\sqrt{s} \gg m$, where m is the target mass and h any hadronic state (the factor 0.2 is due to the quark distribution inside the nucleus). Calculate their ratio at $E_\nu = 50\,\text{GeV}$. How does this ratio depend on energy? Calculate σ/E_ν for the two reactions.

7.15. Draw the Feynman diagrams at tree-level for the elastic scattering $\bar{\nu}_e e^-$. What is different in $\nu_e e^-$?

7.16. The GALLEX experiment at the Gran Sasso laboratory measured the ν_e flux from the Sun by counting the electrons produced in the reaction $\nu_e + {}^{71}\text{Ga} \rightarrow {}^{71}\text{Ge} + e^-$. Its energy threshold is $E_{\text{th}} = 233$ keV. From the solar luminosity one finds the expected neutrino flux $\phi = 6 \times 10^{14}$ m^{-2}s^{-1}. For a rough calculation, assume the whole flux to be above threshold and the average cross section $\sigma = 10^{-48}$ m^2. Assuming the detection efficiency $\varepsilon = 40\%$, how many ^{71}Ga nuclei are necessary to have one neutrino interaction per day? What is the corresponding ^{71}Ga mass? What is the natural gallium mass if the abundance of the ^{71}Ga isotope is $a = 40\%$? (The measured flux turned out to be about one-half of the expected value. This was a fundamental observation in the process of discovering neutrino oscillations.)

7.17. How many metres of iron must a ν_μ of 1 GeV penetrate to interact, on average, once? How long does this take? Compare that distance with the diameter of the Earth orbit ($\sigma = 0.017$ fb, $\rho = 7.7 \times 10^3$ kg m^{-3}, $Z = 26$, $A = 56$).

7.18. Write down a Cabibbo allowed and a Cabibbo suppressed semileptonic decay of the c quark. Write three allowed and three suppressed decays of D^+.

7.19. Draw the Feynman diagram for antibottom quark decay, favoured by the mixing. Write three favoured decay modes of the B^+.

7.20. Draw the principal Feynman diagrams for the top quark decay.

7.21. Draw the Feynman diagrams for bottom and charm decays. Estimate the ratio $\Gamma(b \to c + e + \nu_e)/\Gamma(b \to c)$.

7.22. Consider the measured decay rates $\Gamma(D^+ \to \bar{K}^0 e^+ \nu_e) = (7 \pm 1) \times 10^{10}$ s and $\Gamma(\mu^+ \to e^+ \nu_e \bar{\nu}_\mu) = 1/(2.2 \ \mu\text{s})$. Justify the ratio of the two quantities.

7.23. Consider the decays: (1) $D^+ \to \bar{K}^0 + \pi^+$; (2) $D^+ \to K^+ + \bar{K}^0$; (3) $D^+ \to K^+ + \pi^0$. Find the valence quark composition and establish whether it is favoured, suppressed or doubly suppressed for each of them.

7.24. Consider the measured values of the ratio $\Gamma(\Sigma^- \to ne^-\bar{\nu}_e)/\Gamma_{\text{tot}} \approx 10^{-3}$ and of the upper limit $\Gamma(\Sigma^+ \to ne^+\nu_e)/\Gamma_{\text{tot}} < 5 \times 10^{-6}$. Give the reason for such a difference.

7.25. Consider the decays: (1) $B^0 \to D^- + \pi^+$; (2) $B^0 \to D^- + K^+$; (3) $B^0 \to \pi^- + K^+$; (4) $B^0 \to \pi^- + \pi^+$. Find the valence quark composition of each of them, establish the dependence of the partial decay rates on the mixing matrix element and sort them in decreasing order of these rates.

7.26. A pion with momentum $p_\pi = 500$ MeV decays in the channel $\pi^+ \to \mu^+ + \nu$. Find the minimum and maximum values of the μ momentum. What are the flavour and the chirality of the neutrino?

7.27. Consider a large water Cherenkov detector for solar neutrinos. The electron neutrinos are detected by the reaction $\nu_e + e^- \to \nu_e + e^-$. Assume the cross section (at about 10 MeV) $\sigma = 10^{-47}$ m^2 and the incident flux in the energy range above threshold $\Phi = 10^{10}$ m^{-2} s^{-1}. What is the water mass in which the interaction rate is 10 events a day if the detection efficiency is $\varepsilon = 50\%$?

7.28. The iron-core stars end their life in a supernova explosion, if their mass is large enough. The atomic electrons are absorbed by nuclei by the process $e + Z \rightarrow (Z - 1) + \nu_e$. The star core implodes and its density grows enormously. Assume an iron core with density $\rho = 100\,000$ t/mm^3. Consider the neutrino energy $E_\nu = 10$ MeV and the cross section on iron $\sigma \approx 3 \times 10^{-46}$ m^2. Find the neutrino mean free path. ($A_{\mathrm{Fe}} = 56$.)

Further reading

Lee, T. D. (1957); Nobel Lecture, *Weak Interactions and Nonconservation of Parity*
http://nobelprize.org/nobel_prizes/physics/laureates/1957/lee-lecture.pdf

Okun, L. B. (1981); *Leptony i kwarki*. Nauka Moscow [English translation: *Leptons and Quarks*. North-Holland (1982)]

Pullia, A. (1984); *Structure of charged and neutral weak interactions at high energy. Rivista del Nuovo Cimento* **7** Series 3

Yang, C. N. (1957); Nobel Lecture, *The Law of Parity Conservation and other Symmetry Laws of Physics* http://nobelprize.org/nobel_prizes/physics/laureates/1957/yang-lecture.pdf

8

The neutral K and B mesons and $\mathcal{CP}$ violation

8.1 The states of the neutral K system

The physics of the flavoured, electrically neutral meson–antimeson pairs is an important chapter in weak interactions. These doublets are beautiful examples of quantum two-state systems.

Since the top quark does not bind inside hadrons, there are four such meson doublets, the K^0s, the D^0s, the B^0s and the B_s^0s. In each case, the states with definite flavour differ from the states with definite mass and lifetime, i.e. the stationary states. The quantum oscillation phenomenon takes place between the latter states. This 'flavour oscillation' has been observed in historical order in the K^0 system (strangeness oscillations), in the B^0 system (beauty oscillations), in the B_s^0 system (strangeness and beauty oscillations) and in the D^0 system (charm oscillations). The last was discovered by the BABAR (Aubert *et al.* 2007) and BELLE (Starich *et al.* 2007) experiments, only recently.

Whilst the phenomenon is basically the same for all the flavours, there are important quantitative differences because the oscillation pattern depends on the difference between the masses of the eigenstates (which is, in NU, the angular frequency of their oscillation) and on their widths. These quantities differ considerably in the four cases. We begin with an elementary discussion of the neutral K system. We defer to Section 8.5 the B^0 oscillation, which needs a somewhat more advanced formalism. In the same section we shall mention the recent discovery of the B_s^0 oscillation. We shall not discuss the D^0 oscillations.

The K meson system, which is the lightest one, was historically the first to be studied, at beam energies of a few GeV. K^0 and $\bar{K}^0$ are distinguished by only one quantum number, the strangeness flavour, which, while conserved in strong and electromagnetic interactions, is violated by weak interactions.

Specifically, strong interactions produce two different neutral K mesons, one with strangeness $S = +1$, the $K^0 = d\bar{s}$ and one with $S = -1$, the $\bar{K}^0 = s\bar{d}$, through, for example, the reactions

$$K^+ + n \rightarrow K^0 + p \text{ and } K^- + p \rightarrow \bar{K}^0 + n. \tag{8.1}$$

The two states can be distinguished, not only by the reaction that produced them, but also by the strong reactions they can induce. For example, a K^0 may have the 'charge exchange' reaction on protons $K^0 + p \rightarrow K^+ + n$, but not the hyperon production reaction $K^0 + p \rightarrow \pi^0 + \Sigma^+$, while a $\bar{K}^0$ produces a hyperon $\bar{K}^0 + p \rightarrow \pi^0 + \Sigma^+$ but does not have charge exchange with protons $\bar{K}^0 + p \rightarrow K^+ + n$.

Question 8.1 Does the K^0 charge exchange with neutrons? Does the $\bar{K}^0$ do so?

The two mesons are each the antiparticle of the other, namely

$$CP|K^0\rangle = |\bar{K}^0\rangle, \qquad CP|\bar{K}^0\rangle = |K^0\rangle. \tag{8.2}$$

K^0 and $\bar{K}^0$ can change one into the other via virtual common decay modes, mainly as $K^0 \leftrightarrow 2\pi \leftrightarrow \bar{K}^0$ and $K^0 \leftrightarrow 3\pi \leftrightarrow \bar{K}^0$.

The two CP eigenstates are the following linear superpositions of K^0 and $\bar{K}^0$

$$\begin{aligned}
|K_1^0\rangle &= \frac{1}{\sqrt{2}}\left(|K^0\rangle + |\bar{K}^0\rangle\right) & CP = +1 \\
|K_2^0\rangle &= \frac{1}{\sqrt{2}}\left(|K^0\rangle - |\bar{K}^0\rangle\right) & CP = -1.
\end{aligned} \tag{8.3}$$

Let us now consider the 2π and 3π neutral systems. As we know, the CP eigenvalue of a neutral 2π system is positive. Actually we recall that

$$\begin{aligned}
CP(\pi^0\pi^0) &= \left[CP(\pi^0)\right]^2 = (-1)^2 = +1 \\
CP(\pi^+\pi^-) &= C(\pi^+\pi^-)P(\pi^+\pi^-) = (-1)^l(-1)^l = +1.
\end{aligned} \tag{8.4}$$

As a consequence, if CP is conserved, only the K_1^0, the CP eigenstate with the eigenvalue $CP = +1$, can decay into 2π.

Let us now consider the neutral 3π systems. The case of $3\pi^0$ is easy. We have

$$CP(\pi^0\pi^0\pi^0) = \left[CP(\pi^0)\right]^3 = (-1)^3 = -1. \tag{8.5}$$

The state $\pi^+\pi^-\pi^0$ requires more work. Let us call l the angular momentum of the two-pion $\pi^+\pi^-$ system in their centre of mass reference and L the π^0 angular momentum relative to the two-pion system in the overall centre of mass frame. The total angular momentum of the 3π system is the sum of the two and must be zero, namely $J = l \otimes L = 0$, implying $l = L$. Therefore, the parity is $P =$

$P^3(\pi)(-1)^l(-1)^L = -1$. As for the charge conjugation we have $C(\pi^0) = +1$ and $C(\pi^+\pi^-) = (-1)^l$. In total we have

$$CP(\pi^+\pi^-\pi^0) = (-1)^{l+1}.$$

We now take into account the fact that the difference between the K mass and the mass of three pions is small, $m(K) - 3m(\pi) = 80$ MeV, and therefore that the phase space volume in the decay is very small. This strongly favours the S wave, namely $l = 0$ and then $CP = -1$. In principle the $CP = +1$ decays might occur, but with minimum angular momenta $l = L = 1$; in practice their kinematic suppression is so large that they do not exist and we have

$$CP(\pi^+\pi^-\pi^0) = (-1)^{l+1} = -1. \tag{8.6}$$

In conclusion, if CP is conserved, only the CP eigenstate with the eigenvalue $CP = -1$, the K_2^0, can decay into 3π. Summing up, if CP is conserved we have

$$K_1^0 \to 2\pi \qquad K_2^0 \not\to 2\pi, \qquad K_1^0 \not\to 3\pi \qquad K_2^0 \to 3\pi. \tag{8.7}$$

If CP were absolutely conserved, K_1^0 and K_2^0 would be the states of definite mass and lifetime. As we shall see, CP is very slightly violated and therefore the states of definite mass and lifetime, called K_S and K_L ('K short' and 'K long' respectively) are not exactly K_1^0 and K_2^0. However, the difference is very small, and we shall neglect it for the time being.

Experimentally, the lifetime of the (short) state decaying into 2π, τ_S, is about 580 times shorter than the lifetime of the (long) state decaying into 3π, τ_L. The values are

$$\tau_S = 89.53 \pm 0.05 \text{ ps} \qquad \tau_L = 51.14 \pm 0.21 \text{ ns}. \tag{8.8}$$

The long life of K_L is due to the fact that its decay into 2π is forbidden by CP while its CP conserving decay into 3π is hindered by the small Q-value of the decay. This very fact shows that the CP violation by weak interactions is small, if any.

Let us also look at the widths and at their difference. From (8.8) we have

$$\Gamma_S = \frac{1}{\tau_S} = 7.4 \ \mu\text{eV} \qquad \Gamma_L = \frac{1}{\tau_L} = 0.013 \ \mu\text{eV} \tag{8.9}$$

$$\Delta\Gamma \equiv \Gamma_L - \Gamma_S \approx -\Gamma_S = -7.4 \ \mu\text{eV} = -11.2 \text{ ns}^{-1}.$$

Two other related quantities are

$$c\tau_S = 2.67 \text{ cm} \qquad c\tau_L = 15.5 \text{ m}. \tag{8.10}$$

Suppose we produce a neutral K beam by sending a proton beam extracted from an accelerator onto a target. Initially it contains both K_S and K_L. However, the beam composition varies with the distance from the target. Take for example a K beam momentum of 5 GeV, corresponding to the Lorentz factor $\gamma \approx 1$. In a lifetime the K_S travel $\gamma c \tau_S = 27$ cm. Therefore, at a distance of a few metres (in vacuum) we have a pure K_L beam, for whatever initial composition.

Let us now consider the masses, which, we recall, are defined for the states K_L and K_S. It happens that their difference is extremely small, so small that it cannot be measured directly. The measured average value of the neutral K masses is

$$m_{K^0} = 497.648 \pm 0.022 \text{ MeV.} \tag{8.11}$$

The mass difference is indirectly measured from the strangeness oscillation period, which we shall see in the next section. Its value is

$$\Delta m \equiv m_L - m_S = 3.48 \pm 0.006 \,\mu\text{eV} = 5.292 \pm 0.009 \text{ ns}^{-1}, \tag{8.12}$$

which in relative terms is only 7×10^{-15} of m_{K^0}. Notice that $\Delta m_{K^0} > 0$, which is not a consequence of the definitions, but means that the larger mass K^0 lives longer.

8.2 Strangeness oscillations

In 1955 Gell-Mann and Pais (Gell-Mann & Pais 1955) pointed out that a peculiar phenomenon, strangeness oscillations, should happen in an initially pure K^0 beam, prepared for example using the reaction $\pi^- p \to K^0 \Lambda$. Let us find the probability of finding a K^0 and that of finding a $\bar{K}^0$ as functions of the proper time t. From the experimental point of view, the time corresponds to the distance from the target. The states of definite mass m_i and definite lifetime, or equivalently definite width Γ_i, have the time dependence $\exp[-i(m_i - i\Gamma_i/2)]t$.

These are neither the K^0 nor the $\bar{K}^0$ but, provided $\mathcal{CP}$ is conserved, they are $\mathcal{CP}$ eigenstates

$$\left|K_1^0\right\rangle = \frac{1}{\sqrt{2}}\left(\left|K^0\right\rangle + \left|\bar{K}^0\right\rangle\right) \qquad \left|K_2^0\right\rangle = \frac{1}{\sqrt{2}}\left(\left|K^0\right\rangle - \left|\bar{K}^0\right\rangle\right). \tag{8.13}$$

The K^0 is a superposition of these, namely

$$\left|K^0\right\rangle = \left(\left|K_1^0\right\rangle + \left|\bar{K}_2^0\right\rangle\right)/\sqrt{2}. \tag{8.14}$$

Therefore, the temporal evolution of the wave function (the suffix 0 is to remind us that at $t=0$ the state is K^0, as opposed to $\bar{K}^0$) is

$$\Psi_0(t) = \frac{1}{2}\left[\left(K^0 + \bar{K}^0\right)e^{-im_S t - \frac{\Gamma_S}{2}t} + \left(K^0 - \bar{K}^0\right)e^{-im_L t - \frac{\Gamma_L}{2}t}\right]. \tag{8.15}$$

To understand the phenomenon better, assume for the time being that the mesons are stable, $\Gamma_S = \Gamma_L = 0$. Expression (8.15) becomes

$$\Psi_0(t) = \frac{1}{2}\left[e^{-im_S t} + e^{-im_L t}\right]K^0 + \frac{1}{2}\left[e^{-im_S t} - e^{-im_L t}\right]\bar{K}^0. \qquad (8.16)$$

The probability of finding a K^0 in the beam at time t is

$$\left|\langle K^0|\Psi_0(t)\rangle\right|^2 = \frac{1}{4}\left|e^{-im_S t} + e^{-im_L t}\right|^2 = \frac{1}{2}[1 + \cos(\Delta m \cdot t)] = \cos^2\left(\frac{\Delta m}{2}t\right).$$
$$(8.17)$$

A correlated feature is the appearance in time of $\bar{K}^0$s in the initially pure K^0 beam. The probability of finding a $\bar{K}^0$ is

$$\left|\langle \bar{K}^0|\Psi_0(t)\rangle\right|^2 = \frac{1}{4}\left|e^{-im_S t} - e^{-im_L t}\right|^2 = \frac{1}{2}[1 - \cos(\Delta m \cdot t)] = \sin^2\left(\frac{\Delta m}{2}t\right). \quad (8.18)$$

Summing up, the probabilities of finding a K^0 and, respectively, a $\bar{K}^0$ are initially one and zero. As time passes, the former decreases, the latter increases, so much so that at time $T/2$, the probability of finding a K^0 becomes zero, that of finding a $\bar{K}^0$ is one. Then the process continues with inverted roles. The two-state quantum system 'oscillates' between the two opposite flavour states. It is a 'beat' phenomenon between the monochromatic waves corresponding to the two eigenstates. In NU the two angular frequencies are equal to the masses, as seen in (8.18). Therefore, the oscillation period is $T = 2\pi/|\Delta m| \approx 1.2$ ns. As anticipated, the measurement of the period gives the mass difference but, notice, only in absolute value.

To appreciate the order of magnitude, consider a beam energy of 10 GeV. The first oscillation maximum is at the distance $\gamma c T/2 = 3.6$ m.

As for the sign of Δm, we give only the following hint. If the K^0 beam travels in a medium its refractive index is different from that in vacuum, as happens for photons. Since the index depends on Δm in magnitude and sign, the latter can be determined. The result is that $\Delta m > 0$.

We talked above of the probability of observing a K^0 or a $\bar{K}^0$, but how can we distinguish them? We cannot do that by observing the 2π or 3π decay, because these channels select the states with definite CP, not those of definite strangeness.

To select definite strangeness states we must observe their semileptonic decays. These decays obey the '$\Delta S = \Delta Q$ rule' which reads: 'the difference between the strangeness of the hadrons in the final and initial states is equal to the difference of their electric charges'. The rule, which was established experimentally, is a consequence of the quark contents of the states

$$
\begin{array}{llll}
K^0 = \bar{s}d & \bar{s} \to \bar{u}l^+\nu_l & \Rightarrow K^0 \to \pi^- l^+ \nu_l & K^0 \not\to \pi^+ l^- \bar{\nu}_l \\
\bar{K}^0 = s\bar{d} & s \to ul^- \bar{\nu}_l & \Rightarrow \bar{K}^0 \to \pi^+ l^- \bar{\nu}_l & \bar{K}^0 \not\to \pi^- l^+ \nu_l.
\end{array}
\tag{8.19}
$$

We see that the sign of the charged lepton flags the strangeness of the K. The semileptonic decays are called K^0_{e3} and $K^0_{\mu3}$ depending on the final charged lepton. It is easy to observe them due to their large branching ratios, namely

$$
\mathrm{BR}\left(K^0_{e3}\right) \approx 39\% \qquad \mathrm{BR}\left(K^0_{\mu3}\right) \approx 27\%.
\tag{8.20}
$$

Let us now call $P^\pm(t)$ the probabilities of observing a $+$ and a $-$ lepton respectively, at time t. These are the survival probability of the initial flavour and the appearance probability of the other flavour. Considering unstable kaons, the probabilities are

$$
P^+(t) = \left|\langle K^0|\Psi_0(t)\rangle\right|^2 = \frac{1}{4}\left[e^{-\Gamma_S t} + e^{-\Gamma_L t} + 2e^{-\frac{\Gamma_S+\Gamma_L}{2}t}\cos(\Delta m \cdot t)\right]
\tag{8.21a}
$$

$$
P^-(t) = \left|\langle \bar{K}^0|\Psi_0(t)\rangle\right|^2 = \frac{1}{4}\left[e^{-\Gamma_S t} + e^{-\Gamma_L t} - 2e^{-\frac{\Gamma_S+\Gamma_L}{2}t}\cos(\Delta m \cdot t)\right].
\tag{8.21b}
$$

Both expressions are the sums of two decreasing exponentials and a damped oscillating term. The damping is dominated by the smaller lifetime $\tau_S = 90$ ps. Therefore, the phenomenon is observable only within a few τ_S. Over such short times we can consider the term $e^{-\Gamma_L t}$ as a constant (remember that $\tau_L = 51.7$ ns). Observe finally that τ_S is much smaller than the oscillation period $T \approx 1.2$ ns. Therefore the damping is strong. Figure 8.1 shows the two probabilities.

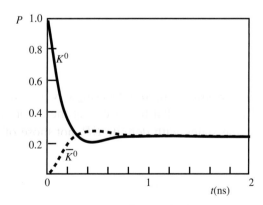

Fig. 8.1. Probabilities of observing K^0 and $\bar{K}^0$ in a beam initially pure in K^0.

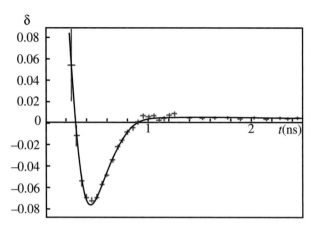

Fig. 8.2.   Charge asymmetry. (From Gjesdal *et al.* 1974)

Experimentally one measures the charge asymmetry, namely the difference between the numbers of observed $K^0 \to \pi^- l^+ \nu_l$ events and $\bar{K}^0 \to \pi^+ l^- \bar{\nu}_l$ events. We see from (8.21) that this is a damped oscillation

$$\delta(t) \equiv P^+(t) - P^-(t) = e^{-\frac{\Gamma_S}{2}t} \cos(\Delta m \cdot t). \qquad (8.22)$$

The experimental results are shown in Fig. 8.2.

The interpolation of the experimental points gives us Γ_S and $|\Delta m|$. We have already given their values.

Let us look more carefully at the data. We see that at very late times ($t \gg \tau_S$) when only K_L survive, the asymmetry does not go to zero as it should, according to (8.22). This implies that the two components K^0 and $\bar{K}^0$ did not become equal and consequently that the long-life state is not a CP eigenstate. The wave function of the eigenstate contains a small 'impurity' with the 'wrong' CP. We shall come back to CP violation in Section 8.4.

8.3 Regeneration

The decisive test of the Gell-Mann and Pais theory discussed in the previous section was proposed by Pais and Piccioni in 1955 (Pais & Piccioni 1955) and performed by Piccioni and collaborators in 1960 (Muller *et al.* 1960).

Figure 8.3 shows an idealised scheme of the experiment. A π^- beam bombards the thin target A producing K^0 by the reaction $\pi^- p \to K^0 \Lambda$. The K^0 state is the mixture (8.14) of K_1^0 and K_2^0. The former component decays mainly into 2π and does so at short distances, the latter survives for longer times and does not decay into 2π, provided CP is conserved. We observe the 2π decays immediately after

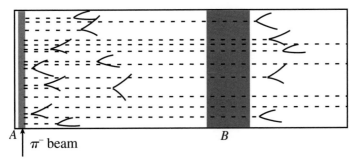

A | π^- beam B

Fig. 8.3. Logical scheme of the Pais–Piccioni experiment.

the target, with a decreasing frequency as we move farther away. When the short component has disappeared for all practical purposes we are left with a pure K_2^0 beam with half of the original intensity. If we insert here a second target B, the surviving neutral K mesons interact with the nuclei in this target by strong interactions.

Strong interactions distinguish between the states of different strangeness, namely between K^0 and $\bar{K}^0$. Indeed, if the energy is as low as we suppose, the only inelastic reaction of the K^0 is the charge exchange, whilst the $\bar{K}^0$ can also undergo reactions with hyperon production such as

$$\bar{K}^0 + p \rightarrow \Lambda + \pi^+ \qquad \bar{K}^0 + n \rightarrow \Lambda + \pi^0 \qquad (8.23)$$

and similarly with a Σ hyperon in place of the Λ. Therefore, the total inelastic cross section is much higher for $\bar{K}^0$ than for K^0 and B preferentially absorbs the former, provided its thickness is large enough. To simplify the discussion, let us consider an idealised absorber that completely absorbs the $\bar{K}^0$ while transmitting the K^0 component without attenuation. After B we then have again a pure K^0 beam with intensity exactly 1/4 of the original one. After the absorber we observe the reappearance of 2π decays. The absorber has regenerated the short-lifetime component. The phenomenon is very similar to those exhibited by polarised light. Its observation established the nature of the short- and long-lifetime neutral kaons as coherent superpositions of states of opposite strangeness.

Question 8.2 Consider two pairs of mutually perpendicular linear polarisation states of light rotated by 45° one to the other. Consider the following analogy: let the K^0, $\bar{K}^0$ system be analogous to the first pair of axes, and let the K_1^0, K_2^0 be analogous to the second pair. Is this analogy correct? Design an experiment analogous to the Pais–Piccioni experiment, using linear polarisers.

8.4 CP violation

Violations of the CP symmetry have been observed in weak interaction processes only, specifically in the decays of the neutral K and B mesons. There are three kinds of CP violation.

1. Violation in the wave function. This happens if the wave functions of the free Hamiltonian are not CP eigenstates. It has been observed only in the neutral K meson system. It is a small, but important, effect. The short-lifetime state is not exactly K_1^0 (the CP eigenstate with eigenvalue $+1$), but contains a small K_2^0 component (the CP eigenstate with eigenvalue -1); symmetrically, the long-lifetime state is not exactly K_2^0 but contains a bit of K_1^0. We shall discuss this phenomenon in this section.
2. Violation in decays. Let M be a meson and f the final state of one of its decays. Let $\bar{M}$ be its antimeson and $\bar{f}$ the conjugate state of f. If CP is conserved, the two decay amplitudes are equal, namely $A(M \rightarrow f) = A(\bar{M} \rightarrow \bar{f})$. The equality holds both for the absolute values, namely for the decay probabilities, and for the phases. The phase is detectable by the interference between different amplitudes contributing to the matrix element. In principle this violation might appear in the decays both of charged and of neutral mesons. However, up to now, it has been observed only in the K^0 and B^0 systems. We shall discuss this type of violation in Section 8.6.
3. Violation in the interference with the oscillations. This may happen for neutral meson decays into a final state f that is a CP eigenstate. The phenomenon has been observed in the K^0 and B^0 systems, as we shall discuss, in an example, in Section 8.5.

Historically, the first CP violation to be discovered was by J. Christenson, J. Cronin, V. Fitch and R. Turlay in 1964 (Christenson *et al.* 1964). Specifically, they observed that the long-lifetime neutral K mesons decay, in a few cases per thousand, into 2π.

The first element of the experiment is the neutral beam, containing the K mesons, obtained by steering the proton beam extracted from a proton synchrotron (the AGS of the Brookhaven National Laboratory) onto a target. A dipole magnet deflects the charged particles produced in the target, while the neutral ones travel undeflected. A collimator located beyond the magnet selects the neutral component. After a few metres, this contains the K_L long-life mesons and no K_S. To this must be added an unavoidable contamination of neutrons and gammas.

The experiment aims to establish whether the CP violating decay

$$K_L \rightarrow \pi^+ + \pi^- \tag{8.24}$$

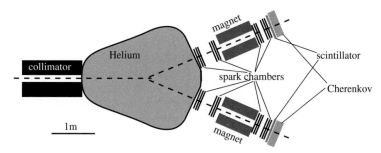

Fig. 8.4. Schematic view of the Christenson *et al.* experiment. (© Nobel Foundation 1980)

exists. Experimentally the topology of the event consists of two opposite tracks. This decay, if it does exist, is expected to be much rarer than other decays with the same topology, K_{e3}^0, $K_{\mu3}^0$ and

$$K_L \rightarrow \pi^+ + \pi^- + \pi^0. \tag{8.25}$$

However, the latter are three-body decays containing a non-observed neutral particle. One takes advantage of this kinematic property to select the events (8.24).

Figure 8.4 is a drawing of the experiment. The volume in which the decays are expected, a few metres long, should ideally be empty, to avoid K_S^0 regeneration and interactions of beam particles simulating the decay. In practice, it is filled with helium gas that, with its light atoms, acts as a 'cheap vacuum'. The measuring apparatus is a two-arm spectrometer, adjusted to accept the kinematics of the decay (8.24). Each arm has two pairs of spark chambers located before and after a bending magnet. In this way the momentum and charge of each particle are measured. The spark chambers are photographed like bubble chambers, but, unlike those, can be triggered by an electronic signal. The trigger signal originated in two Cherenkov counters at the ends of the arms.

In the data analysis, the three-body events are suppressed imposing two conditions: (1) the angle θ between the direction of the sum of the momenta of the two tracks and the beam direction should be compatible with zero; (2) the mass $m(\pi^+\pi^-)$ of the two-particle system should be compatible with the K mass.

Figure 8.5 shows three $\cos\theta$ distributions. Part (b) is for the events with $m(\pi^+\pi^-)$ near to the K mass. The panels (a) and (c) are for two control zones with $m(\pi^+\pi^-)$ immediately below and above the K mass. In the central panel, and only in it, a clear peak is visible at $\theta=0$ above the background. This is the evidence that the long-lifetime neutral K also decays into $\pi^+\pi^-$, a state with $CP=+1$.

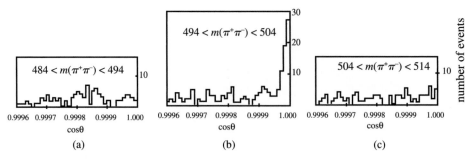

Fig. 8.5.   Distribution of $\cos\theta$ (see text for definition) in three different ranges of $m(\pi^+\pi^-)$: (a) below the K mass; (b) around the K mass; (c) above the K mass. (Christenson *et al.* 1964 © Nobel Foundation 1980)

The measured value of the branching ratio in the CP violating channel is

$$\mathrm{BR}\left(K_L \to \pi^+\pi^-\right) = 2\times 10^{-3}. \tag{8.26}$$

Summarising, the experiment shows that the two CP eigenstates K_1^0 and K_2^0 are not the states with definite mass and lifetime. The latter can be written as

$$
\begin{aligned}
|K_S\rangle &= \frac{1}{\sqrt{1+|\varepsilon|^2}}\left(|K_1^0\rangle + \varepsilon|K_2^0\rangle\right) \\[2mm]
|K_L\rangle &= \frac{1}{\sqrt{1+|\varepsilon|^2}}\left(\varepsilon|K_1^0\rangle + |K_2^0\rangle\right).
\end{aligned}
\tag{8.27}
$$

The ε parameter measures the small impurity of the wrong CP. The Fitch and Cronin experiment is sensitive to the absolute value of this complex parameter. Let us now define the ratio of the transition amplitudes of K_L and K_S into $\pi^+\pi^-$ as

$$\eta_{+-} \equiv \left|\eta_{+-}\right|e^{i\phi^{+-}} \equiv \frac{A\left(K_L \to \pi^+\pi^-\right)}{A\left(K_S \to \pi^+\pi^-\right)}. \tag{8.28}$$

Its absolute value is the ratio of the decay rates. If CP violation is only due to the wave function impurity, one finds that

$$|\varepsilon|^2 \equiv \left|\eta_{+-}\right|^2 = \frac{\Gamma\left(K_L \to \pi^+\pi^-\right)}{\Gamma\left(K_S \to \pi^+\pi^-\right)}. \tag{8.29}$$

We have just seen how the numerator was measured. The denominator is easily determined, being the main decay of the K_S. The present value of $|\varepsilon|$ is (Yao *et al.* 2006)

$$|\varepsilon| = \left|\eta_{+-}\right| = (2.284 \pm 0.014)\times 10^{-3}. \tag{8.30}$$

We now go back to the observation we made at the end of the previous section. Figure 8.2 shows that at late times, when only K_Ls survive, they decay through $K_L \to \pi^- l^+ \nu_l$ a little more frequently than through the CP conjugate channel $K_L \to \pi^+ l^- \bar{\nu}_l$. To be quantitative, the measurement gives

$$\delta_L \equiv \frac{N(K_L \to \pi^- l^+ \nu_l) - N(K_L \to \pi^+ l^- \bar{\nu}_l)}{N(K_L \to \pi^- l^+ \nu_l) + N(K_L \to \pi^+ l^- \bar{\nu}_l)} = (3.27 \pm 0.12) \times 10^{-3}. \quad (8.31)$$

This shows, again and independently, that matter and antimatter are somewhat different. Let us suppose that we wish to tell an extraterrestrial being what we mean by matter and by antimatter. We do not know whether his world is made of the former or the latter. We can tell him: 'prepare a neutral K meson beam and go far enough from the production point to be sure to have been left only with the long-lifetime component.' At this point he is left with K_L mesons, independently of the matter or antimatter constitution of his world. We continue: 'count the decays with a lepton of one or the other charge and call positive the charge of the sample that is about three per thousand larger. Humans call matter the one that has positive nuclei.' If, after a while, our correspondent answers that his nuclei have the opposite charge, and comes to meet you, be careful, apologise, but do not shake his hand.

The measurement of the charge asymmetry (8.31) determines the real part of ε. Actually we can write the K_L wave function as

$$|K_L\rangle = \frac{1}{\sqrt{1 + |\varepsilon|^2}} \left(\varepsilon |K_1^0\rangle + |K_2^0\rangle \right) \simeq \varepsilon |K_1^0\rangle + |K_2^0\rangle$$
$$= \frac{1}{\sqrt{2}} (1 + \varepsilon) |K^0\rangle - \frac{1}{\sqrt{2}} (1 - \varepsilon) |\bar{K}^0\rangle. \quad (8.32)$$

Consequently

$$\delta_L = \frac{|1 + \varepsilon|^2 - |1 - \varepsilon|^2}{|1 + \varepsilon|^2 + |1 - \varepsilon|^2} = 2 \frac{\mathrm{Re}\varepsilon}{1 + |\varepsilon|^2} \simeq 2\mathrm{Re}\varepsilon. \quad (8.33)$$

The present value is (Yao *et al.* 2006)

$$\mathrm{Re}\,\varepsilon = (1.657 \pm 0.021) \times 10^{-3}. \quad (8.34)$$

Comparing with (8.30) we see that the ε phase is about $\pi/4$. Its measured value is (Yao *et al.* 2006)

$$\phi^{+-} = (43.4° \pm 0.7°). \quad (8.35)$$

8.5 Oscillation and CP violation in the neutral B system

In this section we shall discuss two phenomena in the B^0 system, the beauty oscillations and the CP violation in the interference between decays without and with oscillation. Both phenomena have been discovered at the 'beauty factories' by the BELLE and BABAR experiments. In this example we shall see how CP violation is originated by the presence of the phase factor in the quark mixing matrix.

Let us now start by considering the CKM matrix (7.91) and performing the two products. We obtain

$$
V = \begin{pmatrix} V_{ud} & V_{us} & V_{ub} \\ V_{cd} & V_{cs} & V_{cb} \\ V_{td} & V_{ts} & V_{tb} \end{pmatrix}
$$

$$
= \begin{pmatrix} c_{12}s_{13} & s_{12}c_{13} & s_{13}e^{-i\delta_{13}} \\ -s_{12}c_{23} - c_{12}s_{23}s_{13}e^{i\delta_{13}} & c_{12}c_{23} - s_{12}s_{23}s_{13}e^{i\delta_{13}} & s_{23}c_{13} \\ s_{12}s_{23} - c_{12}c_{23}s_{13}e^{i\delta_{13}} & -c_{12}s_{23} - s_{12}c_{23}s_{13}e^{i\delta_{13}} & c_{23}c_{13} \end{pmatrix}.
\tag{8.36}
$$

The phase factor appears in five elements, the remaining four being real. In all cases the phase factor is multiplied by $\sin\theta_{13}$, the smallest of the mixing angles. This explains why CP violation effects are so small. In three elements a second sine or more is present making them almost real. With a good approximation we can consider only V_{ub} and V_{td} as being complex.

We shall now discuss the measurement of the phase β of V_{td}, which we define as

$$
V_{td} \equiv |V_{td}| e^{i\beta}.
\tag{8.37}
$$

We mention in passing, that the precise definition of β is

$$
\beta \equiv \arg\left(-\frac{V_{cd}V_{cb}^*}{V_{td}V_{tb}^*}\right)
$$

where all the factors but V_{td} are close to real or are real.

The neutral B system behaves very similarly to the neutral K system. However, we shall describe its evolution in time with a slightly different formalism. The formalism in the K case would be

$$
|K_S\rangle = p|K^0\rangle + q|\bar{K}^0\rangle
$$
$$
|K_L\rangle = p|K^0\rangle - q|\bar{K}^0\rangle
$$

where p and q are two complex numbers satisfying the normalisation condition $\sqrt{p^2 + q^2} = 1$. Moreover, $\arg(p/q^*)$ is a phase common to K_S and K_L and does not have a physical meaning.

The formalism of the previous section is recovered with

$$p = (1 + \varepsilon)/\sqrt{2} \qquad q = (1 - \varepsilon)/\sqrt{2}. \qquad (8.38)$$

There are two important differences between the B and the K systems. The first is that the lifetimes of the two Bs are equal within the errors (see Table 8.1 at the end of the section). The reason is that the Q-values of the decays of both particles are large and not one large and one small. We label the two eigenstates according to their larger and smaller masses as B_H and B_L (heavy and light). Their mass difference is

$$\Delta m_B \equiv m_H - m_L > 0 \qquad (8.39)$$

that is positive by definition. We shall call Γ_B the common value of the widths

$$\Gamma_B = \Gamma_{B_H} = \Gamma_{B_L} \simeq 0.43 \, \text{meV}. \qquad (8.40)$$

The second difference is the suppression of the common decay channels of B^0 and $\bar{B}^0$ due to the smallness of the corresponding mixing elements. The important consequence is that $|p/q| \approx 1$, namely the expected CP violation in the mixing is small. There is however a probability amplitude for transitions between B^0 and $\bar{B}^0$ given at the lowest order by the 'box' diagrams shown in Fig. 8.6.

The box diagrams with u or c quarks replacing one or two t quarks should also be considered. However, the contribution of a quark internal line is proportional to the square of its mass. Consequently, the diagrams with quarks different from top are negligible. The Standard Model gives the rules to compute the mass difference from the box diagram. In particular, the product $|V_{td}|^2 |V_{tb}|^2$ is proportional to Δm_B and can be determined by measuring the latter from the oscillation period as anticipated in Section 7.9.

Question 8.3 Evaluate the distance between the vertices of the diagram in Fig. 8.6.

We now describe the neutral B system starting from the expressions

$$|B_L\rangle = p|B^0\rangle + q|\bar{B}^0\rangle$$
$$|B_H\rangle = p|B^0\rangle - q|\bar{B}^0\rangle. \qquad (8.41)$$

Fig. 8.6.   The dominant 'box' diagrams of the neutral B system.

Setting

$$m \equiv (m_H + m_L)/2 \tag{8.42}$$

the evolution in time of the L and H eigenstates is given by the factors $e^{-\frac{\Gamma_B}{2}t}e^{-imt}e^{+i\frac{\Delta m_B}{2}t}$ and $e^{-\frac{\Gamma_B}{2}t}e^{-imt}e^{-i\frac{\Delta m_B}{2}t}$ respectively, where t is the proper time.

Let us call $\Psi_0(t)$ and $\Psi_{\bar{0}}(t)$ the wave functions of the states that are purely B^0 and purely $\bar{B}^0$ at $t=0$ respectively. A calculation similar to that we made in Section 8.2 leads to the following expressions

$$\Psi_0(t) = h_+(t)B^0 + \frac{q}{p}h_-(t)\bar{B}^0 \tag{8.43}$$

and

$$\Psi_{\bar{0}}(t) = \frac{p}{q}h_-(t)B^0 + h_+(t)\bar{B}^0 \tag{8.44}$$

where (notice the imaginary unit in the second expression)

$$h_+(t) = e^{-\frac{\Gamma_B}{2}t}e^{-imt}\cos\left(\frac{\Delta m_B}{2}t\right); \quad h_-(t) = ie^{-\frac{\Gamma_B}{2}t}e^{-imt}\sin\left(\frac{\Delta m_B}{2}t\right). \tag{8.45}$$

If at $t=0$ we have a pure B^0 state, the probability of finding a B^0 at a generic t is

$$\left|\langle B^0|\Psi_0(t)\rangle\right|^2 = |h_+(t)|^2 = e^{-\Gamma_B t}\cos^2\left(\frac{\Delta m_B}{2}t\right) = \frac{1}{2}e^{-\Gamma_B t}(1 + \cos\Delta m_B t) \tag{8.46}$$

and the probability of finding a $\bar{B}^0$ is

$$\left|\langle \bar{B}^0|\Psi_0(t)\rangle\right|^2 = |h_-(t)|^2 = \left|\frac{q}{p}\right|^2 e^{-\Gamma_B t}\sin^2\left(\frac{\Delta m_B}{2}t\right)$$
$$= e^{-\Gamma_B t}\sin^2\left(\frac{\Delta m_B}{2}t\right) = \frac{1}{2}e^{-\Gamma_B t}(1 - \cos\Delta m_B t) \tag{8.47}$$

in the approximation $|p/q| = 1$. Similar expressions are valid starting from a pure $\bar{B}^0$ state, i.e. with the wave function $\Psi_{\bar{0}}(t)$. The difference between the probabilities of observing opposite-flavour and same-flavour decays, normalised to their sum, called flavour asymmetry

$$\frac{P_{\mathrm{OF}} - P_{\mathrm{SF}}}{P_{\mathrm{OF}} + P_{\mathrm{SF}}} \propto \frac{1}{2}e^{-\Gamma_B t}\cos\Delta m_B t \tag{8.48}$$

is measurable as a function of time, as we shall see. This determines Δm_B.

To measure the phase of p/q we need a second phase to use as a reference. Only phase differences have a physical meaning. Consider to this purpose a CP eigenstate f of eigenvalue η_f into which both B^0 and $\bar{B}^0$ can decay. Let A_f be the amplitude for $B^0 \to f$ and $\bar{A}_f$ the amplitude for $\bar{B}^0 \to f$. If $A_f \neq \bar{A}_f$, CP is violated. If $|A_f| \neq |\bar{A}_f|$ we observe the violation as a difference between the two decay rates. However, in the important case that we shall discuss, the absolute values of the two amplitudes are equal, and the CP violation is due to the phase difference between the amplitudes. We shall see how to measure the relative phase between the complex numbers p/q and $A_f/\bar{A}_f$ or, more precisely, the observable

$$\lambda_f \equiv \eta_f \frac{p\,A_f}{q\,\bar{A}_f}. \tag{8.49}$$

Notice that $|\lambda_f| = 1$.

The amplitudes for the decay into the final state f of $\Psi_0(t)$ and of $\Psi_{\bar{0}}(t)$ are respectively

$$\langle f|\Psi_0(t)\rangle = A_f h_+(t) + \frac{q}{p}\bar{A}_f h_-(t)$$
$$= \frac{A_f}{\lambda_f} e^{-imt} e^{-\frac{\Gamma_B t}{2}}\left[\lambda_f \cos\left(\frac{\Delta m_B}{2}t\right) + i\sin\left(\frac{\Delta m_B}{2}t\right)\right]$$

and

$$\langle f|\Psi_{\bar{0}}(t)\rangle = \frac{p}{q}A_f h_-(t) + \bar{A}_f h_+(t)$$
$$= \bar{A}_f e^{-imt} e^{-\frac{\Gamma_B t}{2}}\left[i\lambda_f \sin\left(\frac{\Delta m_B}{2}t\right) + \cos\left(\frac{\Delta m_B}{2}t\right)\right].$$

The CP violating observable is the ratio between the difference and the sum of the two probabilities. After a few passages, taking into account that $|A_f| = |\bar{A}_f|$ and that $|\lambda_f| = 1$, we obtain

$$|\langle f|\Psi_0(t)\rangle|^2 + |\langle f|\Psi_{\bar{0}}(t)\rangle|^2 = 2|A_f|^2 e^{-\Gamma_B t}$$

and

$$|\langle f|\Psi_0(t)\rangle|^2 - |\langle f|\Psi_{\bar{0}}(t)\rangle|^2 = |A_f|^2 e^{-\Gamma_B t} 2\eta_f \operatorname{Im}\lambda_f \sin(\Delta m_B \cdot t)$$

and finally

$$a_{fCP} = \frac{|\langle f|\Psi_0(t)\rangle|^2 - |\langle f|\Psi_{\bar{0}}(t)\rangle|^2}{|\langle f|\Psi_0(t)\rangle|^2 + |\langle f|\Psi_{\bar{0}}(t)\rangle|^2} = \eta_f \operatorname{Im}\lambda_f \sin(\Delta m_B \cdot t). \tag{8.50}$$

We see that CP is violated if $\operatorname{Im}\lambda_f \neq 0$.

We now consider the measurements of the mass difference Δm_{B^0} and the CP asymmetry a_{fCP} at the beauty factories. We recall here that these high luminosity e^+e^- colliders provide hundreds of millions of $B^0\bar{B}^0$ pairs in a pure $J^{PC}=1^{--}$ state. The factories operate at the $\Upsilon(4^1S_3)$ resonance that is only 20 MeV above $m_{B^0}+m_{\bar{B}^0}$. Consequently, the Bs move slowly in the centre of mass frame and their decay vertices cannot be resolved in this frame. The beauty factories are consequently built 'asymmetric', meaning that the energies of the two beams are not equal in order to have the centre of mass moving in the laboratory. The electron and positron momenta are $p(e^-)=9$ GeV and $p(e^+)=$ 3.1 GeV in PEP2 and $p(e^-)=8$ GeV and $p(e^+)=3.5$ GeV in KEKB, corresponding to an average Lorentz factor for the Bs of $\langle\beta\gamma\rangle=0.56$ and $\langle\beta\gamma\rangle=0.425$. The average distance between the production and the decay vertices is $\Delta z\approx 200\,\mu$m. It is measured by surrounding the collision point with a 'vertex detector'. The vertex detector is made up of several layers of silicon-microstrip tracking devices assembled with high mechanical accuracy. The accuracy in the vertex reconstruction is typically 80–120 μm, corresponding to about one-half of the flight length in a lifetime. The proper time, the variable that appears in the above written expressions, is the distance measured in the laboratory divided by $c\langle\beta\gamma\rangle$.

We know that the two neutral Bs produced in any e^+e^- annihilation are one B^0 and one $\bar{B}^0$, but we do not know which is which. The time evolution of the two-state system is given by a single wave function that describes both particles. In other words, the phase difference between the B^0 and the $\bar{B}^0$ does not vary in time. However, one of the Bs can identify itself as a particle or an antiparticle when and if it decays semileptonically. Similarly to what we have discussed for the K mesons, we have

$$
\begin{aligned}
B^0 &= \bar{b}d & \bar{b} &\to \bar{c}l^+\nu_l & &\Rightarrow B^0 \to D^-l^+\nu_l \\
\bar{B}^0 &= b\bar{d} & b &\to cl^-\bar{\nu}_l & &\Rightarrow \bar{B}^0 \to D^+l^-\bar{\nu}_l.
\end{aligned}
\tag{8.51}
$$

Consequently, by observing the sign of the lepton or by reconstructing the D we 'tag' the neutral B as a B^0 or $\bar{B}^0$. We measure the time of this decay relative to the time of production by measuring the distance between production and decay vertices, and the velocity of the particle by measuring the momenta of its daughters. We then take as $t=0$ the time of the tagging decay. If the tagging B is a $\bar{B}^0$, the companion is a B^0 at that time, and its wave function evolves as $\Psi_0(t)$, and vice versa. Strange as it may appear, its evolution is given by $\Psi_0(t)$ even before the tag decay, namely for $t<0$. Indeed, the evolution of the wave functions is completely deterministic in quantum mechanics. Once known at an instant, the wave function is known at any time.

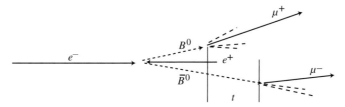

Fig. 8.7. Kinematics for $e^+ e^- \to B^0 \bar{B}^0$ followed by the decays $B^0 \to \mu^+ + \cdots$ and $\bar{B}^0 \to \mu^+ + \cdots$.

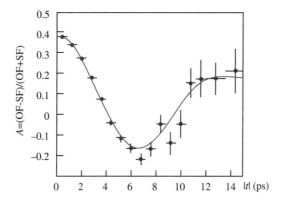

Fig. 8.8. Flavour asymmetry as a function of time (absolute value). (From Abe *et al.* 2005)

Let us now consider firstly the beauty oscillation, which is very similar to that for the *K*. We consider the cases in which the flavours of both *B*s can be identified by the final states of their decays, for example by the signs of the leptons, as sketched in Fig. 8.7. The times between the production of the $B\bar{B}$ pair and each of the two decays are measured, obtaining the time interval *t* between the decays. Since the decay time of one or the other *B* can be taken as $t = 0$ indifferently, *t* is known in absolute value only. Figure 8.8 shows the flavour asymmetry (8.48) measured by the BELLE experiment (Abe *et al.* 2005) (BABAR has similar results) as a function of $|t|$. Notice that the lifetime is about one order of magnitude smaller than the oscillation period. Consequently, the number of events per unit time decreases at longer times and the error bars increase accordingly.

By fitting to the data expression (8.48), corrected to take into account the presence of background, the decay width and the mass difference are obtained. The value of the latter averaged over all the experiments is (Yao *et al.* 2006)

$$\Delta m_B = 0.507 \pm 0.005 \, \text{ps}^{-1} = 0.3337 \pm 0.0033 \, \text{meV}. \tag{8.52}$$

From this measurement we can extract (Ceccucci *et al.* 2006)

$$|V_{td}||V_{tb}| = (7.4 \pm 0.8) \times 10^{-3}. \tag{8.53}$$

The CP-asymmetry a_{CP} has been observed in different channels. We shall consider only the final CP eigenstates $f = J/\psi + K_S$ and $f = J/\psi + K_L$. The final orbital momentum is $L=1$ for angular momentum and parity conservation. As a consequence, the CP eigenvalues are $\eta_{J/\psi+K_L} = +1$ and $\eta_{J/\psi+K_S} = -1$. Notice that the branching ratios are small, about 0.9×10^{-3}. The peak luminosity of the beauty factories is larger than 10^{34} cm^{-2} s^{-1} corresponding to the production of 10^6 $B\bar{B}$ pairs a day. BELLE and BABAR have collected about 5×10^8 events.

The experiments tag the events as we described and select the cases in which the companion B decays into one of the CP eigenstates. An example is sketched in Fig. 8.9.

The Standard Model gives a very clean prediction for Im λ_f in both cases. We shall give a plausibility argument but not a proof. To be concrete, consider the decay of the tagged state $\Psi_{\bar{0}}(t)$. It may decay directly as $\bar{B}^0$, with the diagram of Fig. 8.10(a), or oscillate into a B^0 and then decay, with the diagram of Fig. 8.10(b). The two amplitudes do not interfere at the level of the diagrams shown in the figure, because in one case there is a K^0, in the other a $\bar{K}^0$, which can be distinguished. However, if the K decays as a CP eigenstate, namely as a K_1^0 (or a K_2^0) the final states are identical and the two amplitudes do interfere. In

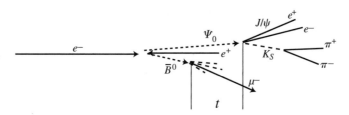

Fig. 8.9.   Kinematics for $e^+e^- \to B^0\bar{B}^0$. One neutral B is tagged as $\bar{B}^0$ by its decay $\bar{B}^0 \to \mu^- + \ldots$. This instant is defined as the origin of the time. The other B decays into $J/\psi + K_S$ at time t (which may have both signs).

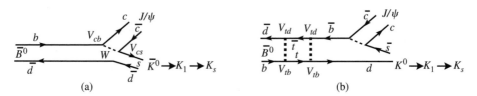

Fig. 8.10.   (a) Feynman diagram at the quark level for the decay $\bar{B}^0 \to J/\psi + \bar{K}^0$ without oscillation; (b) same with oscillation.

the present discussion the difference between K_1^0 and K_S (K_2^0 and K_L) can be safely neglected.

We have reported in the figure the relevant elements of the mixing matrix. All of them, and those relative to the transition to K_1^0 that are not shown, are real except V_{td}. This appears twice, hence squared, in the amplitude. Moreover, the other element in the box is $V_{tb} \approx 1$ to a very good approximation. Taking everything into account, one finds

$$\lambda_{J/\psi+K_S} = \frac{p}{q}\frac{A_{J/\psi+K}}{\bar{A}_{J/\psi+K}} = e^{2i\beta} \tag{8.54}$$

and

$$\mathrm{Im}\,\lambda_{J/\psi+K_S} = 2\mathrm{Im}\,V_{td} = \sin(2\beta) \tag{8.55}$$

and

$$\mathrm{Im}\,\lambda_{J/\psi+K_L} = -\sin(2\beta). \tag{8.56}$$

In conclusion we expect the observables $a_{CP,J/\psi+K_S}(t)$ and $a_{CP,J/\psi+K_L}(t)$ to be two sinusoidal functions of time with the same period, the same amplitude and opposite phases.

Figure 8.11 shows the 'raw' asymmetry as measured by BABAR (the result of BELLE is similar). Indeed the measured asymmetry is not the ideal one for three principal experimental reasons: (1) the presence of backgrounds; (2) the experimental resolution in the measurement of time (1–1.5 ps); (3) the presence of mis-tags, meaning B^0 wrongly tagged as $\bar{B}^0$ and vice versa. Notice that the background is larger in the case of the K_L because this particle does not decay in

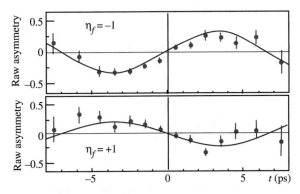

Fig. 8.11.   Asymmetry a_{CP} (before correction for experimental effects) as measured by BABAR for $J/\psi K_S$ and $J/\psi K_L$ final states. (Adapted from Aubert et al. 2006)

the detector due to its long lifetime. The K_L is detected when it interacts in the calorimeter as a hadronic shower.

All these effects reduce the amplitude of the sinusoidal time dependence. After having applied all the necessary corrections, the average value of the two experiments is

$$\sin(2\beta) = 0.685 \pm 0.032. \tag{8.57}$$

Question 8.4 Why are the amplitudes of the two sines of Fig. 8.11 different?

In Section 7.9 we examined three tests of the unitarity of the CKM matrix based on the absolute values of its elements. We saw also that in the Standard Model CP is violated by the complex nature of the matrix. The measurement of β allows a unitarity test at this level. This test is particularly sensitive to possible contributions of physics beyond the Standard Model. We recall that all the elements are real or close to being such, except V_{ub} and V_{tb}. We then consider the unitarity condition in the product of the first and third lines, which contain these elements, namely

$$V_{ud}V_{ub}^* + V_{cd}V_{cb}^* + V_{td}V_{tb}^* = 0. \tag{8.58}$$

We can consider each term in this expression as a vector in the complex plane and read (8.58) as stating that the sum of the three vectors should be zero. Geometrically this means that the vectors make up a triangle, called the 'unitary triangle'.

Since we know $|V_{cd}|, |V_{cb}|, |V_{ud}|$ and $|V_{ub}|$ given in (7.92) and $|V_{td}||V_{tb}|$ given by (8.53), we have the lengths of the vectors. We can safely neglect the imaginary part of $V_{cd}V_{cb}^*$ and set $V_{cd}V_{cb}^* = |V_{cd}||V_{cb}|$. We then divide the three vectors by this quantity, as shown in Fig. 8.12. In this way we fix the positions of the two vertices on the real axis in (0,0) and (1,0). Knowing the lengths of the other two sides we can check if they can close the triangle. Moreover, a further constraint is given by $\sin 2\beta$ that determines β within the four-fold ambiguity $\beta, \beta + \pi, \pm\pi/2 - \beta$. Another constraint is given by the ε CP violating parameter in the K system. All the measurements overlap consistently and define, by a global fitting procedure, the small shaded area shown in Fig. 8.12 (Bona *et al.*

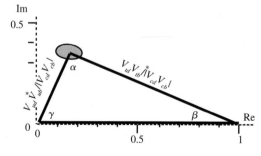

Fig. 8.12. The 'unitary triangle'. The shaded area has 95% confidence level.

Table 8.1 *Lifetimes, total widths and mass differences of the pseudoscalar neutral flavoured mesons*

	τ (ps)	$c\tau$ (μm)	Γ (ps^{-1})	Γ (meV)	Δm (ps^{-1})	Δm (meV)
K_L	$51.14 \pm 0.21 \times 10^3$	15.3×10^6	2.0×10^{-5}	1.3×10^{-5}	$0.005\,292 \pm 0.000\,009$	$0.003\,483 \pm 0.000\,006$
K_S	89.53 ± 0.05	2.67×10^4	0.011	7.4×10^{-3}		
D_H	0.410 ± 0.002	123	2.4	1.61	$0.0209^{+0.0072}_{-0.0082}$	$0.0140^{+0.0055}_{-0.0048}$
D_L	0.410 ± 0.002					
B_H	1.530 ± 0.009	459	0.65	0.43	0.507 ± 0.005	0.3337 ± 0.0033
B_L	1.530 ± 0.009					
B_{sH}	1.466 ± 0.059	439	0.86	0.57	17.77 ± 0.12	11.6 ± 0.7
B_{sL}	1.466 ± 0.059					

2006, also Charles *et al.* 2006). Notice that if the expression (7.91) is valid for the CKM matrix, the CP violating phase δ_{13} is the angle γ.

The high-precision work at the beauty factories and elsewhere continues to measure with increasing accuracy the CKM elements and all the angles.

Before concluding this section we summarise in Table 8.1 the values of the lifetimes (Yao *et al.* 2006), widths and mass differences for the four flavoured pseudoscalar meson pairs that we have partially discussed in this chapter. In particular, the $B_s^0 \bar{B}_s^0$ oscillation has been recently discovered by the CDF experiment at the Tevatron collider (Abulencia *et al.* 2006). This observation was made difficult by the large value of Δm_{B_s} reported in the table, corresponding to an extremely high oscillation frequency, about three times in a picosecond. The D^0 oscillation has been observed recently by the BABAR (Aubert *et al.* 2007) and BELLE (Starich *et al.* 2007) experiments. Indeed the beauty factories are also powerful sources of charm, because B mesons decay preferentially into charmed mesons. The mass difference value reported in Table 8.1 is from a global fit by Asner *et al.* (2007). In this case the lifetime is much shorter than the period.

Question 8.5   Compute the ratio between oscillation period and lifetime in the four cases.

8.6 CP violation in meson decays

CP violation in the decays has been observed both in the K^0 and in the B^0 systems.

We start with the latter that is larger and simpler to describe. Both BABAR and BELLE have searched for differences between the rates of pairs of charge-conjugated decays both for charged and for neutral B mesons. Actually, B mesons can decay in a huge number of different channels, due to their large masses. Consequently, the branching ratios are small, typically of the order of 10^{-5}, and even with the samples of several 10^8 $B\bar{B}$ pairs provided by the beauty factories a few thousand events per channel are available. This gives statistical sensitivities to asymmetries of the order of 10^{-2}.

Both BABAR (Aubert *et al.* 2004) and BELLE (Chao *et al.* 2004) observed a CP violating asymmetry (only) in the channel $B^0 \to K^{\mp}\pi^{\pm}$. The two measurements agree. Their average is

$$A_{K^+\pi^-} \equiv \frac{\Gamma(\bar{B}^0 \to K^-\pi^+) - \Gamma(B^0 \to K^+\pi^-)}{\Gamma(\bar{B}^0 \to K^-\pi^+) + \Gamma(B^0 \to K^+\pi^-)} = -0.115 \pm 0.018. \qquad (8.59)$$

In the K^0 system the CP violation in the decay is much smaller, of the order of 10^{-6}. Consequently, it cannot be directly observed as a difference between the rates of two charge-conjugated decays. Actually the phenomenon has been observed by comparing four decay rates, namely those of the two neutral kaons into $\pi^+\pi^-$ and into $\pi^0\pi^0$. Since the states are CP eigenstates, both CP violation in the mixing and in the decay are present, something that complicates the issue. We shall now try to elucidate how CP violation effects in the decay are extracted. In doing so we will not enter the mathematical details, but give only the results.

Since the two pions from a kaon decay are in a spatially symmetric state, their isospin wave function must be symmetric, i.e. the total isospin can be $I=0$ or $I=2$. Let us consider the amplitudes for the decay of the K^0 into states of definite isospin, A_0 and A_2. Taking into account the Clebsch–Gordan coefficients, the amplitudes for the (weak) decays into the two charge states are

$$A_W\left(K^0 \to \pi^+\pi^-\right) = \frac{1}{\sqrt{3}}\left(A_2 + \sqrt{2}A_0\right)$$
$$A_W\left(K^0 \to \pi^0\pi^0\right) = \frac{1}{\sqrt{3}}\left(\sqrt{2}A_2 - A_0\right). \tag{8.60}$$

We state here without proof that the CPT invariance requires that the corresponding amplitudes for the $\bar{K}^0$ are

$$A_W\left(\bar{K}^0 \to \pi^+\pi^-\right) = -\frac{1}{\sqrt{3}}\left(A_2^* + \sqrt{2}A_0^*\right)$$
$$A_W\left(\bar{K}^0 \to \pi^0\pi^0\right) = \frac{1}{\sqrt{3}}\left(-\sqrt{2}A_2^* + A_0^*\right). \tag{8.61}$$

We search for a difference between the phases of A_2 and A_0, present only if CP is violated. For this we need another phase, which indeed is present, due to strong interactions.

We can consider the strong interaction between the two pions in the final state as a scattering. Since strong interactions conserve the isospin, the absolute values of the probability amplitudes of both the $I=0$ and the $I=2$ two-pion states are 1. Consequently the scattering amplitudes are pure phase factors, say $e^{i\delta_0}$ and $e^{i\delta_2}$. Hence, the complete transition amplitudes are

$$A\left(K^0 \to \pi^+\pi^-\right) = \frac{1}{\sqrt{3}}\left(e^{i\delta_2}A_2 + \sqrt{2}e^{i\delta_0}A_0\right)$$
$$A\left(K^0 \to \pi^0\pi^0\right) = \frac{1}{\sqrt{3}}\left(\sqrt{2}e^{i\delta_2}A_2 - e^{i\delta_0}A_0\right)$$
$$A\left(\bar{K}^0 \to \pi^+\pi^-\right) = -\frac{1}{\sqrt{3}}\left(e^{i\delta_2}A_2^* + \sqrt{2}e^{i\delta_0}A_0^*\right)$$
$$A\left(\bar{K}^0 \to \pi^0\pi^0\right) = \frac{1}{\sqrt{3}}\left(-\sqrt{2}e^{i\delta_2}A_2^* + e^{i\delta_0}A_0^*\right). \tag{8.62}$$

Since strong interactions conserve CP the 'strong phases' have the same sign in a decay and in its charge conjugate.

It has been found experimentally that the A_0 amplitude dominates, namely that $|A_2|/|A_0| \approx 1/22$. Indeed the measured ratio (Yao *et al.* 2006)

$$\Gamma(K_S \to \pi^+\pi^-)/\Gamma(K_S \to \pi^0\pi^0) = 2.25 \pm 0.04$$

is close to the value predicted for a pure $I=0$ final state (taking into account the small mass difference between pion masses). As for the phases of A_0 and A_2, only their difference has a physical meaning. We follow the suggestion of Wu and Yang (Wu & Yang 1964) and choose the arbitrary phase in such a way that the large amplitude A_0 is a real positive number.

The next step is to express the observables in terms of the real numbers A_0, δ_0 and δ_2 and the complex number A_2. The observables are the two complex amplitude ratios

$$\eta_{+-} \equiv |\eta_{+-}|e^{i\phi^{+-}} \equiv A(K_L \to \pi^+\pi^-)/A(K_S \to \pi^+\pi^-) \tag{8.63}$$

which we have already met in (8.28), and

$$\eta_{00} \equiv |\eta_{00}|e^{i\phi^{00}} \equiv A(K_L \to \pi^0\pi^0)/A(K_S \to \pi^0\pi^0). \tag{8.64}$$

The calculation is not difficult but long. The result, obtained by neglecting terms of order higher than the first in Re ε, Im ε and $|A_2/A_0|$, where ε is the CP violating parameter discussed in Section 8.4, is

$$\eta_{+-} = \varepsilon + \varepsilon' \qquad \eta_{00} = \varepsilon - 2\varepsilon' \tag{8.65}$$

and

$$\varepsilon' = \frac{1}{\sqrt{2}} \frac{\mathrm{Im}A_2}{A_0} e^{i\left(\frac{\pi}{2} + \delta_2 - \delta_0\right)}. \tag{8.66}$$

We see that CP is violated only if A_2 is not zero and not real. Moreover, since, as we shall see, the experimentally accessible observable is Re ε', a non-zero strong phase difference is needed. Actually, from scattering experiments we know that $\pi/2 + \delta_2 - \delta_0 = 42.3 \pm 1.5°$.

In the absence of CP violation in the decays $\eta_{+-} = \eta_{00}$. Consequently, we may think to search for a difference in their absolute values or in their phases. We have already mentioned the measurements of η_{+-} in Section 8.4. Comparing with similar measurements for η_{00} we obtain (Yao *et al.* 2006)

$$|\eta_{00}/\eta_{+-}| = 0.9950 \pm 0.008 \qquad \phi^{00} - \phi^{+-} = (-0.022 \pm 0.020)°. \tag{8.67}$$

The measurements are very precise, but not precise enough to show a difference. Let us then try another way, namely to seek, for example, for a possible difference between $\Gamma(K^0 \to \pi^+\pi^-)$ and $\Gamma(\bar{K}^0 \to \pi^+\pi^-)$. Neglecting powers higher than the first of the real and imaginary parts of ε and ε', their ratio is given by (see Problem 8.9)

$$\left| A(\bar{K}^0 \to \pi^+\pi^-)/A(K^0 \to \pi^+\pi^-) \right| \approx 1 - 2\mathrm{Re}\,\varepsilon'. \tag{8.68}$$

Consider next the corresponding ratio for the $\pi^0\pi^0$ final state, i.e.

$$\left| A(\bar{K}^0 \to \pi^0\pi^0)/A(K^0 \to \pi^0\pi^0) \right| \approx 1 + 4\mathrm{Re}\,\varepsilon' \tag{8.69}$$

and the difference between the two ratios,

$$\begin{aligned} \left| A(\bar{K}^0 \to \pi^0\pi^0)/A(K^0 \to \pi^0\pi^0) \right| \\ - \left| A(\bar{K}^0 \to \pi^+\pi^-)/A(K^0 \to \pi^+\pi^-) \right| = 6\mathrm{Re}\,\varepsilon'. \end{aligned} \tag{8.70}$$

However, when measuring decay rates we deal with the free Hamiltonian eigenstates, namely with K_S and K_L. The observable that is directly related to the difference (8.70) is the 'double ratio', i.e. the ratio between the ratios of the decay rates into $\pi^+\pi^-$ and $\pi^0\pi^0$ of K_S and K_L. It is easy to show that

$$\begin{aligned} \mathrm{Re}\left(\frac{\varepsilon'}{\varepsilon}\right) &= \frac{1}{6}\left(1 - \frac{|\eta_{00}|^2}{|\eta_{+-}|^2}\right) \\ &= \frac{1}{6}\left[1 - \frac{\Gamma(K_L \to \pi^0\pi^0)\Gamma(K_S \to \pi^+\pi^-)}{\Gamma(K_L \to \pi^+\pi^-)\Gamma(K_S \to \pi^0\pi^0)}\right]. \end{aligned} \tag{8.71}$$

Obviously the value of $\mathrm{Re}(\varepsilon'/\varepsilon)$ was not initially known. The experimental search started in the 1970s when the sensitivity was of the order of 10^{-2}. The struggle to reduce the systematic and statistical uncertainties continued both at CERN and at Fermilab until the sensitivity reached a few parts in ten thousand and the effect was discovered. We shall only mention here the principal experimental difficulties and the results.

The first difficulty is the rareness of the CP violating decays of the K_L, $K_L \to \pi^+\pi^-$ (with a branching ratio of 2×10^{-3}) and $K_L \to \pi^0\pi^0$ (1×10^{-3}). Moreover, the latter decay suffers from possible contamination from the 200 times more frequent $K_L \to \pi^0\pi^0\pi^0$ decay. Another problem is the large difference in the average decay paths of the two kaons, which for example at 110 GeV are $\beta\gamma\tau_L \simeq 3.4\,\mathrm{km}$ and $\beta\gamma\tau_S \simeq 6\,\mathrm{m}$, while the two decay distributions along the detector should be as similar as possible to avoid instrumental asymmetries.

In practice the experiments have used the following procedures:

- contemporary detection of $\pi^+\pi^-$ and $\pi^0\pi^0$ in order to cancel the uncertainty of the incident fluxes in the ratio;
- two beams of K_L and of K_S simultaneously in the detector. The beams must have energy spectra as equal as possible, have the same direction and produce spatial distributions of the decays that are as similar as possible;
- good spatial resolution and outstanding energy resolution to reduce the contamination from other decay channels.

The CP violation in the decay was discovered by the NA31 experiment at CERN (Barr *et al.* 1993) on a sample of 428 000 $K_L \rightarrow \pi^0\pi^0$ decays. In 1993, NA31 published the result $\text{Re}(\varepsilon'/\varepsilon) = (2.30 \pm 0.65) \times 10^{-3}$. This value is at 3.5 standard deviations from zero; however, the contemporary experiment at Fermilab, E731, with similar statistics, obtained (Gibbons *et al.* 1993) $\text{Re}(\varepsilon'/\varepsilon) = (0.74 \pm 0.56) \times 10^{-3}$, which is compatible with zero. The issue was solved by the next generation of experiments with of the order of 10^7 $K_L \rightarrow \pi^0\pi^0$ decays and improved systematic accuracy. KTeV at Fermilab (Alavi-Harati *et al.* 2003) obtained $\text{Re}(\varepsilon'/\varepsilon) = (2.07 \pm 0.28) \times 10^{-3}$ and NA48 at CERN (Batlay *et al.* 2002) obtained $\text{Re}(\varepsilon'/\varepsilon) = (1.47 \pm 0.22) \times 10^{-3}$. The two values agree. The weighted average of the four measurements gives (Yao *et al.* 2006)

$$\text{Re}(\varepsilon'/\varepsilon) = (1.66 \pm 0.23) \times 10^{-3}. \tag{8.72}$$

Problems

8.1. The DAΦNE ϕ-factory at Frascati is an e^+e^- collider at the centre of mass energy equal to the ϕ mass. Calculate the ratio between the annihilation rates into K^+K^- and $K^0\bar{K}^0$ neglecting the mass difference between charged and neutral kaons. Is this a good approximation? Considering the case $K^0\bar{K}^0$ calculate the relative frequency of $K_1^0K_1^0$, $K_1^0K_2^0$ and $K_2^0K_2^0$.

8.2. From which of the $\bar{p}p$ initial states $^1S_0$, $^3S_1$, $^1P_1$, $^3P_0$, $^3P_1$, $^3P_2$ can each of the following reactions, $\bar{p}p \rightarrow K^+K^-$, $\bar{p}p \rightarrow K_1^0K_1^0$ and $\bar{p}p \rightarrow K_1^0K_2^0$, proceed?

8.3. A π^- is sent onto a target producing neutral K mesons and Λ hyperons. Consider the component of the resulting K beam with momentum $p = 10$ GeV. What is the ratio between K_S and K_L at the production point? What is it at $l = 10$ m from the production point? Determine the fraction of decays into 2π that would be observed in the absence of CP violation.

8.4. An experiment needs an almost monochromatic K^+ beam with momentum $p = 2$ GeV. We obtain it by building a magnetic spectrometer and a system of slits. However, the total length of the beam is limited by the lifetime of the K^+. At what distance is the K^+ intensity reduced to 10% of the initial value?

8.5. Consider the reactions $\pi^- p \to K^0 + X$ and $\pi^- p \to \bar{K}^0 + Y$ and establish the minimum masses of the states X and Y that are compatible with the conservation laws and the two corresponding energy thresholds.

8.6. Consider a neutral K meson beam with momentum $p_K = 400$ MeV impinging on a liquid hydrogen target and determine the reaction channels open for each component K^0 and $\bar{K}^0$. Estimate which has the larger cross section.

8.7. An asymmetric beauty factory operates at the $\Upsilon(4^1S_3)$, namely at $\sqrt{s} = 10\,580$ MeV, to study the process $e^+e^- \to B^0\bar{B}^0$. 'Asymmetric' means that the centre of mass moves in the reference frame of the collider. Consider an event in which both mesons are produced with the Lorentz factor $\beta\gamma = 0.56$. The decay vertex of one of them is at 120 μm from the principal vertex; amongst the B decay products there is a μ^+. How many lifetimes did the particle live? What can we say about the two flavours? The second B decays at 0.5 mm, again with a μ in its final state. How many lifetimes did it live? Can the μ be positive? Why?

8.8. Consider the sample corresponding to an integrated luminosity of 100 eV/fb of the BABAR experiment. In the laboratory frame the centre of mass moves with the average Lorentz factor $\beta\gamma = 0.56$. How many seconds at a luminosity $L = 10^{34}$ cm^{-2} s^{-1} would it take to collect such a sample? Assuming the value $\Delta R = 3$ at the $\Upsilon(4S)$ resonance, how many $B^0\bar{B}^0$ pairs will have been collected? What is the average separation between production and decay vertices?

8.9. Prove expression (8.68) neglecting terms of orders above the first in ε and ε'.

8.10. Prove expression (8.69) neglecting terms of orders above the first in ε and ε'.

8.11. Prove expression (8.71) neglecting terms of orders above the first in ε and ε'.

Further reading

Cronin, J. W. (1980); Nobel Lecture, *CP Symmetry Violation. The Search for its Origin* http://nobelprize.org/nobel_prizes/physics/laureates/1980/cronin-lecture.pdf

Fitch, V. L. (1980); Nobel Lecture, *The Discovery of Charge-Conjugation Parity Asymmetry* http://nobelprize.org/nobel_prizes/physics/laureates/1980/fitch-lecture.pdf

9

The Standard Model

9.1 The electroweak interaction

The development of the theoretical model that led to the electroweak unification started at the end of the 1960s. In this model, a single gauge theory, with the symmetry group $SU(2) \otimes U(1)$, includes the electromagnetic and weak interactions, both neutral current (NC) and charged current (CC). In particular, the electromagnetic and weak coupling constants are not independent but correlated by the theory. On the other hand, electroweak theory and QCD, both being gauge theories, are unified by the theoretical framework while their coupling constants are independent. Electroweak theory and QCD together form the Standard Model of fundamental interactions.

In the first part of this chapter we shall introduce the electroweak theory, as usual without any theoretical rigour. The unification characteristics appear mainly in the NC processes. The transition probabilities of all these processes are predicted by the theory with a single free parameter, the electroweak mixing angle. We shall discuss an example of its determination.

A crucial prediction of the theory is the existence of three vector bosons, W^+, W^- and Z^0, together with predictions of their masses, widths and branching ratios in all their decay channels. All these predictions have been experimentally verified with high accuracy. We shall finally see the experimental proof of the fact that the vector bosons have weak charges themselves and that consequently they interact directly.

In the electroweak theory the photon and the massive vector bosons, which mediate the weak interactions, are initially introduced together, as gauge massless fields. The logical construction of the theory, which we shall not discuss, proceeds by introducing a spontaneous symmetry breaking mechanism, which gives mass to $W^\pm$ and Z^0, while leaving the photon massless.

The fundamental representation of $SU(2) \otimes U(1)$ hosts three and one gauge fields. A quantity called *weak isospin*, which we shall indicate by I_W, corresponds to $SU(2)$. From now on we shall call it simply isospin. The quantity corresponding to $U(1)$ is called *weak hypercharge* or simply hypercharge, Y_W. All the members of the same isospin multiplet have the same hypercharge.

Hypercharge can be defined in two equivalent ways, as twice the average electric charge of the multiplet or as

$$Y_W \equiv 2(Q - I_{Wz}). \tag{9.1}$$

Notice that weak isospin and hypercharge here have nothing to do with those of the hadrons.

Let us call $W = (W_1, W_2, W_3)$ the triplet of fields corresponding to $SU(2)$. Clearly, W has $I_W = 1$ and $Y_W = 0$. It interacts with the isospin of the particles.

Let us call B the field corresponding to $U(1)$. Its isospin, its electric charge and its hypercharge are zero. It interacts with the hypercharge of the particles.

These four fields are not the physical fields that mediate the interactions. The weak CC interactions are mediated, as we shall see immediately, by W^+ and W^-, which are linear combinations of W_1 and W_2, while the mediators of the electromagnetic and weak NC interactions, the photon and the Z, are linear combinations of W_3 and B.

The experiments we have discussed in Chapter 7 (and many others) showed that the charged W, the mediator of the CC weak interactions, couples to the left component of leptons and quarks and to the right component of antileptons and antiquarks. We must take this into account in assigning isospin and hypercharge to the particles.

Let us start with the leptons. There are two left leptons in every family; we assume them to be in the same isospin doublet ($I_W = 1/2$) as in the equations

$$\begin{pmatrix} I_{Wz} = +1/2 \\ I_{Wz} = -1/2 \end{pmatrix} = \begin{pmatrix} \nu_{eL} \\ e_L^- \end{pmatrix}, \begin{pmatrix} \nu_{\mu L} \\ \mu_L^- \end{pmatrix}, \begin{pmatrix} \nu_{\tau L} \\ \tau_L^- \end{pmatrix}. \tag{9.2}$$

Unlike the charged current, the neutral current also interacts with right charged fermions, with different couplings, but not with right neutrinos. The charged right lepton of each family is an isospin singlet ($I_W = 0$)

$$e_R^-, \quad \mu_R^-, \quad \tau_R^-. \tag{9.3}$$

Right neutrinos do not exist. More precisely, if they existed, since they would have zero isospin and hypercharge, they would not interact by any known interaction except gravitation.

The situation for the quarks is similar, provided we take mixing into account, i.e. that the W couples universally to the rotated quark states d', s' and b'. For every colour, there are three isospin doublets, one for each family (nine in total)

$$\begin{pmatrix} I_{Wz} = +1/2 \\ I_{Wz} = -1/2 \end{pmatrix} = \begin{pmatrix} u_L \\ d'_L \end{pmatrix}, \begin{pmatrix} c_L \\ s'_L \end{pmatrix}, \begin{pmatrix} t_L \\ b'_L \end{pmatrix}. \tag{9.4}$$

and the singlets (18 in total)

$$d_R, \quad u_R, \quad s_R, \quad c_R, \quad b_R, \quad t_R. \tag{9.5}$$

Quark mixing is irrelevant for the NC weak interactions. So, we can write it in terms of the 'rotated' quarks or of the quarks of definite flavours, with the same result. Indeed, what we have observed for two families at the end of Section 7.8 is also true for three families.

Notice that the weak isospin of the left quark is, by chance, equal to its flavour isospin.

All the quantum numbers of the antiparticles are equal and opposite to those of the corresponding particles. In the CC sector we deal with right antileptons (the 'anti' component of the left bi-spinor). They belong to three doublets

$$\begin{pmatrix} I_{Wz} = +1/2 \\ I_{Wz} = -1/2 \end{pmatrix} = \begin{pmatrix} e^+_R \\ \bar{v}_{eR} \end{pmatrix}, \begin{pmatrix} \mu^+_R \\ \bar{v}_{\mu R} \end{pmatrix}, \begin{pmatrix} \tau^+_R \\ \bar{v}_{\tau R} \end{pmatrix}. \tag{9.6}$$

The left antileptons, which appear in the NC sector, are isospin singlets

$$e^+_L, \quad \mu^+_L, \quad \tau^+_L. \tag{9.7}$$

Left antineutrinos do not exist.

The antiquark doublets are

$$\begin{pmatrix} I_{Wz} = +1/2 \\ I_{Wz} = -1/2 \end{pmatrix} = \begin{pmatrix} \bar{d}'_R \\ \bar{u}_R \end{pmatrix}, \begin{pmatrix} \bar{s}'_R \\ \bar{c}_R \end{pmatrix}, \begin{pmatrix} \bar{b}'_R \\ \bar{t}_R \end{pmatrix}. \tag{9.8}$$

Their singlets are

$$\bar{d}_L, \quad \bar{u}_L, \quad \bar{s}_L, \quad \bar{c}_L, \quad \bar{b}_L, \quad \bar{t}_L. \tag{9.9}$$

Table 9.1 summarises the values of isospin, hypercharge and electric charge of the fundamental fermions. The values are identical for every colour.

Example 9.1 Establish whether any of the following processes exist: $W^- \rightarrow e^-_L + \bar{v}_{eR}$, $W^- \rightarrow d'_L + \bar{u}_R$, $Z^0 \rightarrow \bar{u}_R + u_R$, $W^+ \rightarrow \bar{d}'_R + u_L$, $Z^0 \rightarrow \bar{u}_R + u_L$, $Z^0 \rightarrow \bar{u}_L + u_L$.

Electric charge and hypercharge are absolutely conserved quantities. The former conservation is satisfied by all the above processes. Let us check hypercharge

Table 9.1 *Fermion isospin, hypercharge, electric charge and 'Z-charge factors'* $c_Z = I_{Wz} - s^2 Q$ *(see later in Section 9.3)*

	I_W	I_{Wz}	Q	Y_W	c_Z
ν_{lL}	1/2	+1/2	0	-1	1/2
l_L^-	1/2	$-1/2$	-1	-1	$-1/2+s^2$
l_R^-	0	0	-1	-2	s^2
u_L	1/2	+1/2	2/3	1/3	$1/2-(2/3)s^2$
d_L'	1/2	$-1/2$	$-1/3$	1/3	$-1/2+(1/3)s^2$
u_R	0	0	2/3	4/3	$-(2/3)s^2$
d_R'	0	0	$-1/3$	$-2/3$	$(1/3)s^2$
$\bar{\nu}_{lR}$	1/2	$-1/2$	0	1	$-1/2$
l_R^+	1/2	+1/2	$+1$	1	$1/2-s^2$
l_L^+	0	0	$+1$	2	$-s^2$
$\bar{u}_R$	1/2	$-1/2$	$-2/3$	$-1/3$	$-1/2+(2/3)s^2$
$\bar{d}_R'$	1/2	+1/2	1/3	$-1/3$	$1/2-(1/3)s^2$
$\bar{u}_L$	0	0	$-2/3$	$-4/3$	$(2/3)s^2$
$\bar{d}_L'$	0	0	1/3	2/3	$-(1/3)s^2$

conservation. Consider $W^- \to e_L^- + \bar{\nu}_{eR}$. The initial W^- has $Y=0$ as all the gauge bosons, the left electron has $Y=-1$, the right antineutrino has $Y=+1$, synthetically $0 \to -1+1$. The process exists. For $W^- \to d_L' + \bar{u}_R$ we have $0 \to 1/3 - 1/3$, OK. For $Z^0 \to \bar{u}_R + u_R$ we have $0 \to 1/3 - 4/3$; hypercharge is not conserved, the process does not exist. For $W^+ \to \bar{d}_R' + u_L$ we have $0 \to 1/3 - 1/3$, OK. For $Z^0 \to \bar{u}_R + u_L$ we have $0 \to -1/3 + 1/3$, OK. For $Z^0 \to \bar{u}_L + u_L$ we have $0 \to -4/3 + 1/3$; does not exist.

9.2 Structure of the weak neutral currents

The NC weak interactions are mediated by the Z boson. They have two important characteristics.

1. Neutral current couples each fermion with itself only, for example ee and not $e\mu$. If they are quarks, they must have the same colour: $^Bu^Bu$, not $^Ru^Bu$, because the Z, like the W, does not carry any colour. Figure 9.1 shows four non-existent vertices.

Example 9.2 Consider the couplings $u_R \to Z^0 + u_R$ and $u_R \to Z^0 + u_L$. Are they possible? Electric charge is conserved in both cases. The former process is possible because the hypercharge balance is $4/3 \to 0 + 4/3$, the latter process is not because $4/3 \to 0 + 1/3$.

2. Neutral currents do not have the space-time $V-A$ structure.

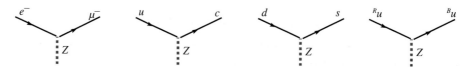

Fig. 9.1. Four couplings that do not exist.

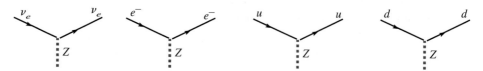

Fig. 9.2. The existing couplings for the first family.

For every family there are seven currents coupling Z to every fermion: six for the left and right charged fermions, and one for neutrinos, which are only left. We write down the seven currents of the first family in Eqs. (9.10) and draw the vertices in Fig. 9.2.

$$g_L^{\nu_e} \bar{\nu}_e \gamma_\mu (1 - \gamma_5) \nu_e = g_L^{\nu_e} \bar{\nu}_{eL} \gamma_\mu \nu_{eL}$$

$$g_L^e \bar{e} \gamma_\mu (1 - \gamma_5) e + g_R^e \bar{e} \gamma_\mu (1 + \gamma_5) e = g_L^e \bar{e}_L \gamma_\mu e_L + g_R^e \bar{e}_R \gamma_\mu e_R$$

$$g_L^u \bar{u} \gamma_\mu (1 - \gamma_5) u + g_R^u \bar{u} \gamma_\mu (1 + \gamma_5) u = g_L^u \bar{u}_L \gamma_\mu u_L + g_R^u \bar{u}_R \gamma_\mu u_R$$

$$g_L^d \bar{d} \gamma_\mu (1 - \gamma_5) d + g_R^u \bar{d} \gamma_\mu (1 + \gamma_5) d = g_L^d \bar{d}_L \gamma_\mu d_L + g_R^u \bar{d}_R \gamma_\mu d_R.$$

$$(9.10)$$

Notice that the first term in every row has the structure of the CC, namely $V - A$ coupling with left fermions. The second terms couple with right fermions. Every term corresponds to a different physical process and the corresponding coupling constant might be, a priori, different. Therefore, we have used different symbols. With three families, we have in total 27 NC 'weak charges'. The power of the electroweak theory is to give all these charges in terms of two constants, the elementary electric charge and the electroweak mixing angle θ_W, which we shall meet soon.

The coupling of the Z is a universal function of the charge Q and the third isospin component I_z of the particle, as we shall see in the next section. We immediately establish that the Z:

- couples to both left and right fermions;
- couples to the Ws;
- also couples to electrically neutral particles, provided they have $I_z \neq 0$, such as left neutrinos;
- couples to the γ and itself.

9.3 Electroweak unification

The electroweak model was developed mainly by S. Glashow (Glashow 1961), A. Salam (Salam & Ward 1964) and S. Weinberg (Weinberg 1967). The Feynman rules and the procedures needed for the renormalisation of the theory were developed by 't Hooft ('t Hooft 1971) and by Veltman. Let us now see the relationships between the fields W and B and the physical fields $W^\pm$, Z and γ together with their couplings to fermions.

The field $W^\mu \equiv \left(W_1^\mu, W_2^\mu, W_3^\mu \right)$ is a four-vector in space-time (index μ) and a vector in isotopic space. The fields of the physical charged bosons are

$$W^\pm = \frac{1}{\sqrt{2}} (W_1 \pm i W_2). \tag{9.11}$$

For every fermion doublet, there is a space-time four-vector, isospin vector, called the 'weak current' $j_\mu \equiv (j_{1\mu}, j_{2\mu}, j_{3\mu})$. The field W^μ couples to j_μ as $g W^\mu j_\mu$ with the dimensionless coupling constant g. The charged currents are linear combinations of two components of the current

$$j^\pm = \frac{1}{\sqrt{2}} (j_1 \pm i j_2). \tag{9.12}$$

Considering for example the doublet $\begin{pmatrix} v_{eL} \\ e_L^- \end{pmatrix}$, the corresponding charged currents are

$$j_{e\mu}^- = \bar{v}_{eL} \gamma_\mu e_L^- \qquad j_{e\mu}^+ = \bar{e}_L^- \gamma_\mu v_{eL}. \tag{9.13}$$

The field B^μ is a space-time four-vector, isospin scalar. It couples with the hypercharge current j_μ^Y, which is also four-vector and isoscalar. The coupling constant is g'. The hypercharge current is twice the difference between the electromagnetic current j_μ^{EM} and the neutral component of the weak NC, in agreement with Eq. (9.1)

$$j_\mu^Y = 2 j_\mu^{EM} - 2 j_{3\mu}. \tag{9.14}$$

The first term is the electromagnetic current that we know, which, for the charged fermion f is

$$j_{f\mu}^{EM} = \bar{f} \gamma_\mu f. \tag{9.15}$$

Chirality is not specified because the electromagnetic interaction does not depend on it. The second term in (9.14) corresponds to the currents (9.10).

Let us call A and Z the physical fields that mediate the electromagnetic and the weak NC interactions respectively. They are two mutually orthogonal linear superpositions of W_3 and B. We shall determine them by imposing that the photon does not couple to neutral particles, while the Z^0 does. The transformation is expressed in terms of the two coupling constants g and g' or, equivalently, as a rotation through an angle θ_W, called the weak mixing angle

$$\begin{pmatrix} Z^0 \\ A \end{pmatrix} = \frac{1}{\sqrt{g^2 + g'^2}} \begin{pmatrix} g & -g' \\ g' & g \end{pmatrix} \begin{pmatrix} W_3 \\ B \end{pmatrix} = \begin{pmatrix} \cos\theta_W & -\sin\theta_W \\ \sin\theta_W & \cos\theta_W \end{pmatrix} \begin{pmatrix} W_3 \\ B \end{pmatrix}. \quad (9.16)$$

The weak mixing angle is defined by the relationship

$$\theta_W \equiv \tan^{-1}\frac{g'}{g}. \quad (9.17)$$

The rotation is not small, $\theta_W \approx 29°$, as we shall see. The interaction Lagrangian, being symmetrical under the gauge group, is an isoscalar, namely

$$L = g\left(j_\mu^1 W_1^\mu + j_\mu^2 W_2^\mu + j_\mu^3 W_3^\mu\right) + \frac{g'}{2}j_\mu^Y B^\mu. \quad (9.18)$$

We can write this expression as

$$L = \frac{g}{\sqrt{2}}\left(j_\mu^- W_+^\mu + j_\mu^+ W_-^\mu\right) + j_\mu^3\left(gW_3^\mu - g'B^\mu\right) + g'j_\mu^{EM} B^\mu.$$

Also introducing the neutral physical fields and grouping terms, we obtain

$$L = \frac{g}{\sqrt{2}}\left(j_\mu^- W_+^\mu + j_\mu^+ W_-^\mu\right) + \frac{g}{\cos\theta_W}\left(j_\mu^3 - \sin^2\theta_W\, j_\mu^{EM}\right)Z^\mu + g\sin\theta_W\, j_\mu^{EM} A^\mu. \quad (9.19)$$

Let us examine this fundamental expression. Its terms are, in order: the CC weak interaction, the NC weak interaction and the electromagnetic interaction.

The last term is the electromagnetic interaction. Consequently the constant in front of it must be proportional to the electric charge, assuming that the photon does not couple to neutral particles. Actually, the relationship with the elementary electric charge is

$$g\sin\theta_W = \frac{q_e}{\sqrt{\varepsilon_0 \hbar c}} = \sqrt{4\pi\alpha}. \quad (9.20)$$

This expression unifies the weak and the electric charges. As anticipated, all the interactions mediated by the four-vector bosons are expressed in terms of two constants, the electric charge q_e and the weak angle θ_W. However, the model does not predict the values of the two fundamental parameters. They must be determined experimentally.

From (9.17) and (9.20) we immediately have the relationship between the coupling constant of $U(1)$ and the electric charge

$$g' \cos \theta_W = \sqrt{4\pi a}. \tag{9.21}$$

From (9.20) and (9.21) we have also

$$\frac{1}{a} = \frac{4\pi}{g'^2} + \frac{4\pi}{g^2} \tag{9.22}$$

which shows how the couplings of both gauge groups contribute to $1/a$. At low energies where $1/a \approx 137$, with $\sin^2 \theta_W \approx 0.232$, we have

$$4\pi/g'^2 = 105.2 \quad \text{and} \quad 4\pi/g^2 = 31.8. \tag{9.23}$$

The second term of (9.19) gives the coupling of Z with fermions. We see that it is universal in the sense that it is a universal function of the charge and of the third isospin component

$$g_Z \equiv \frac{g}{\cos \theta_W} \left(I_{Wz} - Q \sin^2 \theta_W \right)$$

$$= \frac{\sqrt{4\pi a}}{\sin \theta_W \cos \theta_W} \left(I_{Wz} - Q \sin^2 \theta_W \right) = \frac{g}{\cos \theta_W} c_Z. \tag{9.24}$$

In the last member we introduced the 'Z-charge factor' c_Z

$$c_Z \equiv I_{Wz} - Q \sin^2 \theta_W. \tag{9.25}$$

The structure of c_Z is determined by the gauge group $SU(2) \otimes U(1)$, as the colour factors $\kappa_\lambda^{c_i c_j}$ are determined by $SU(3)$. We gave the fermion Z-charge factors in Table 9.1.

The first term of Eq. (9.19) describes the CC weak processes we discussed in Chapter 7. As we know, the coupling constant g is given in terms of the Fermi constant and of the W mass by (7.9b), which we repeat here

$$G_F = \frac{\sqrt{2}g^2}{8M_W^2}. \tag{9.26}$$

Using Eq. (9.20) we have the prediction of the W mass as a function of the fine structure constant, the Fermi constant and the weak angle

$$M_W = \left(\frac{g^2\sqrt{2}}{8G_F}\right)^{1/2} = \sqrt{\frac{\pi a}{\sqrt{2}G_F}}\frac{1}{\sin\theta_W} = \frac{37.3}{\sin\theta_W}\,\text{GeV}. \qquad (9.27)$$

In this model, a measurement of the weak angle gives the W mass.

We shall give only a few hints on one fundamental part of the theory, namely the origin of the masses, in Section 9.11. We anticipate here that the problem arises because the gauge theories describe infinite range forces, i.e. massless mediators, such as the photon and the gluons. However, the weak interaction mediators are massive. Moreover, all fermions are massless too in the 'unperturbed' theory. The mechanism that allows the masses to be introduced, without destroying the renormalisation property of the theory, was found by Higgs. The mechanism introduces, in particular, a space-time scalar particle called the Higgs boson. In its simplest form, which is the choice made in the Standard Model, the Higgs boson is an isospin doublet. The model is unable to predict the Higgs boson mass.

The Standard Model does give a precise prediction of the vector boson masses. Beyond (9.27), we have, at the lowest perturbative order, the ratio of the masses.

$$M_W/M_Z = \cos\theta_W. \qquad (9.28)$$

We shall discuss the (indirect) measurement of the weak angle in Section 9.5. Its value is $\sin^2\theta_W = 0.232$ and we have

$$M_W \approx 80\,\text{GeV} \qquad M_Z \approx 90\,\text{GeV}. \qquad (9.29)$$

We shall describe the discovery of the vector bosons in Section 9.7.

In complete generality, we can say that the Standard Model provides a unified description of all the known elementary processes of Nature (but see Chapter 10). It is the most comprehensive theoretical structure ever built by mankind and the most accurately tested one. Its electroweak section, in particular contains:

- The CC weak processes. We have studied a few examples at low energies, where the Standard Model coincides with the four-fermion interaction.
- The NC weak processes. These are the processes in which unification appears directly, especially at energies comparable to the mediator masses. We shall see a few experimental tests in this chapter.
- The direct interaction between mediators. It was tested precisely at LEP, as we shall see in Section 9.10.
- The 'mass generation' by the Higgs mechanism. This sector has been tested experimentally only in an indirect way. We do not yet know whether the Higgs

boson exists or not. The search for it is one of the principal goals of the new LHC at CERN. We shall give a few hints in Section 9.10.

9.4 Determination of the electroweak angle

Electroweak unification appears in the NC weak processes, where the 'weak charges' are predicted in terms of $\sin^2\theta_W$. This parameter cannot be measured directly; however, if the theory is correct, its values as extracted from measurements of cross sections or decay rates of different processes must agree. The extraction itself is made according to the prescriptions of the theory, which always imply the calculation of Feynman diagrams. In the case of precise measurements, calculations must go beyond the tree-level, including radiative corrections at the level needed to reach the required precision.

A long series of high-accuracy experiments has tested the universality of the interaction in a wide energy range, from keV to hundreds of GeV and for many different couplings. We shall only mention the main experiments here, without entering into details, and come back to the most precise determinations in Section 9.8.

- The gauge boson masses have been measured with high accuracy, as we shall see. Their ratio gives the most precise value of the weak angle.
- Parity violation in atoms. The atomic electrons are bound to the nucleus not only by the electromagnetic interaction, exchanging a photon with its quarks, but also by the weak NC one, exchanging a Z with them. The effect of the latter is extremely small and not observable as a shift of the levels. However, the interference between the amplitudes is observable (Zel'dovich 1959) as a parity violating effect (of the order of a part per million, ppm), providing a test of the theory at the several keV energy scale (Noecker *et al.* 1988, Grossman *et al.* 2005).
- Polarised-electron elastic scattering on deuterium. Both photon exchange and Z exchange contribute to the process. The latter contribution is too small to be observable in the cross section. However, the interference between the two amplitudes can be measured as a small difference (asymmetry) between the differential cross sections of the two electron polarisation states (of the order of a few ppm). Indeed, the elementary process to be measured would be the neutrino–quark scattering. However, quarks are inside nucleons that, in turn, are inside the deuterium nucleus. Consequently, the accuracy in the determination of the weak angle is limited by the theoretical uncertainties in the QCD calculations. The first experiment was done at SLAC at $|Q^2| = 1.6\,\text{GeV}^2$ (Prescott *et al.* 1978).

- Electron–electron (Møller) elastic scattering with polarised beam on unpolarised target. The process is similar to the previous one, but it is purely leptonic. Consequently, the extraction of the weak angle from the measured asymmetry A_{PV} is free of hadronic uncertainties. The experiment is more difficult because the asymmetry is very small, only a fraction of a ppm, and because the Møller events must be separated from the more frequent electron–proton scatterings. The first experiment was done at SLAC in 2004 on a 50 GeV high-intensity polarised electron beam at $|Q^2| = 0.026\,\text{GeV}^2$, measuring A_{PV} with 20% precision (Anthony *et al.* 2004).

- Forward–backward asymmetry in the differential cross section of the electron–positron annihilation into a fermion pair $e^+e^- \to f^+f^-$. The effect is due, again, to the interference between photon and Z exchanges, which violates parity resulting in an asymmetry around 90°. The asymmetry has been measured in a wide energy range between 10 GeV and 200 GeV. It is large at energies comparable to M_Z.

- Deeply inelastic scattering of ν_μ and $\bar{\nu}_\mu$ on nuclei. In this case, since the probe is a neutrino, which only has the weak interaction, we can determine the weak angle directly from the measurement of the cross sections. However, the use of a complex hadronic target limits the accuracy that can be reached in the weak angle determination. Typical momentum transfer values are of several GeV.

- Scattering of ν_μ and $\bar{\nu}_\mu$ on electrons. This is a purely leptonic process, free from the problem just mentioned. The measurement of its cross section consequently provides a clean means of determining the weak angle without theoretical uncertainties. There is, of course, a price to be paid: the neutrino–electron cross sections are four orders of magnitude smaller than the neutrino–nucleus ones. The most precise experiment of this type was the CHARM2 experiment performed at CERN in the 1980s and 1990s, which we shall now describe.

The aim of CHARM2 was to measure the ratio of the total cross sections of the two elastic scattering processes

$$\nu_\mu e^- \to \nu_\mu e^- \qquad \bar{\nu}_\mu e^- \to \bar{\nu}_\mu e^-. \tag{9.30}$$

These are similar to (7.23), with the Z as a mediator instead of the W, as shown in Fig. 9.3 for the first of them. Even at neutrino beam energies of a hundred GeV, the momentum transfers are small compared to M_Z. Therefore, as discussed in relation to Eq. (7.24), the cross sections are proportional to s, namely to the product of neutrino energy and target mass, hence to $G_F^2 m_e E_\nu$. As the electron mass is several thousand times smaller than the mass of a nucleus, neutrino–electron cross sections are, as anticipated, four orders of magnitude smaller than those on light nuclei.

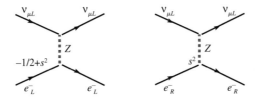

Fig. 9.3. Diagrams for the $\nu_\mu e^-$ scattering. The Z-charge factors are shown at the lower vertex.

Let us now find how the ratio of the cross sections of the processes (9.30) depends on the weak angle. Even if we do not have the theoretical instruments for the calculation, we can obtain the result by physical arguments. Since we are interested in the ratio, we can ignore common factors. Since the energies are very high we shall consider the electrons as massless.

Let us start with the first reaction to which the two diagrams in Fig. 9.3 contribute, namely the diffusion of the left neutrino (the only one that exists) on left electrons and on right electrons. The two contributions can in principle be distinguished, by measuring the helicity, and therefore do not interfere. We take the sum of the squares of the two amplitudes.

Since the upper vertex is the same, the two contributions are proportional to the square of the Z-charge factors, c_Z, of the lower vertices. These are written in the figure, with the notation $s^2 = \sin^2\theta_W$.

Moving on to the second reaction, we observe that its cross section is equal, for $\mathcal{CPT}$ invariance, to the cross section of $\nu_\mu e^+ \to \nu_\mu e^+$. The corresponding diagrams, shown in Fig. 9.4, have the same upper vertex as those in Fig. 9.3. Therefore, in the ratio we only have to consider the lower vertices, summing the squares of the Z-charge factors.

We now observe that, for both processes, the two contributions are a scattering of a left fermion on a left fermion $(L+L)$ and a left one on a right one $(L+R)$. The two cases are different. Let us see why, with reference to neutrino scattering (the antineutrino case is similar).

We analyse the two contributions with reference to Fig. 9.5. In the $(L+L)$ term, shown in Fig. 9.5(a), the angular momenta of both the incoming and outgoing pairs are zero. Therefore, since the interaction is point-like, all the diffusion angles have the same probability, namely the angular cross section is constant. In the $(L+R)$ case of Fig. 9.5(b) the total angular momentum is $\mathbf{J}=1$ with third component $J_z = -1$ (the quantisation axis is the neutrino flight line). Therefore, only one out of the $2J+1=3$ a-priori possible spin states is allowed. In conclusion, there is a 1/3 factor in the $(L+R)$ cross section.

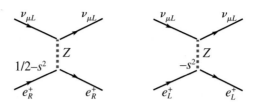

Fig. 9.4.   Diagrams for the $\nu_\mu e^+$ scattering. The Z-charge factors are shown at the lower vertex.

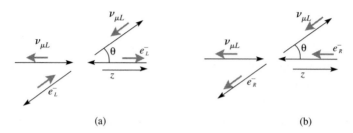

(a)   (b)

Fig. 9.5.   (a) The three-momenta and the spins in the scattering $\nu_{\mu L} + e_L^- \to \nu_{\mu L} + e_L^-$ in the centre of mass frame; (b) same for the scattering $\nu_{\mu L} + e_R^- \to \nu_{\mu L} + e_R^-$.

Before going on we observe that the conclusion just reached is general and that it can be written in a form valid for all the neutral currents as: the total NC cross sections between fermions $L+R$ and $R+L$ are, when considered at the same energy, three times larger than the corresponding $L+L$ and $R+R$.

Returning to our calculation, and summing up the results, we have, aside from a common constant (which, for the curious, is $2/\pi$)

$$\sigma_{\nu_\mu e} \propto G_F^2 m_e E_\nu \left[\left(-\frac{1}{2} + \sin^2 \theta_W \right)^2 + \frac{1}{3} \sin^4 \theta_W \right]$$

$$\sigma_{\bar\nu_\mu e} \propto G_F^2 m_e E_{\bar\nu} \left[\frac{1}{3} \left(-\frac{1}{2} + \sin^2 \theta_W \right)^2 + \sin^4 \theta_W \right].$$

Taking the ratio, at the same energy for neutrinos and antineutrinos, we have

$$R = \frac{\sigma_{\nu_\mu e}}{\sigma_{\bar\nu_\mu e}} = 3 \frac{1 - 4\sin^2 \theta_W + \frac{16}{3}\sin^4 \theta_W}{1 - 4\sin^2 \theta_W + 16\sin^4 \theta_W}. \tag{9.31}$$

To measure the ratio, we expose the detector to both a neutrino and an anti-neutrino beam. Let us call $N(\nu_\mu e)$ and $N(\bar\nu_\mu e)$ the numbers of neutrino–electron scattering events obtained in the two exposures. The incident neutrino energy is not well defined, rather the beams have a wide energy spread. Since the cross sections are proportional to the energy, the observed numbers of events are normalised to the ratio F of the energy-weighted fluxes

$$F \equiv \int \Phi_{\bar{v}_\mu}(E_{\bar{v}})E_{\bar{v}}\,dE_{\bar{v}} \bigg/ \int \Phi_{v_\mu}(E_v)E_v\,dE_v \qquad (9.32)$$

obtaining the empirical ratio

$$R_{\text{exp}} = \frac{N(v_\mu e)}{N(\bar{v}_\mu e)}F. \qquad (9.33)$$

Let us now see the main characteristics the detector needs to have. First of all, its sensitive mass must be large, given the smallness of the cross sections. In practice, with the available neutrino beam intensities, it is of the order of hundreds of tons. Secondly, the detector must visualise the tracks of the events and measure their energy. At the same time, it must provide the target for neutrino interactions. In practice a 'fine grain calorimeter' must be built.

The third problem is the background. Neutrinos interact both with the nuclei and with the electrons of the detector, but the latter process, which we are interested in, happens only once in ten thousand events. Moreover, the 'signal' is

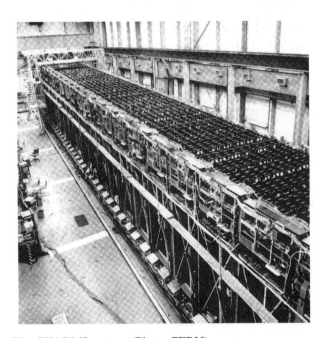

Fig. 9.6.   The CHARM2 set-up. (Photo CERN)

a single-electron track, a topology that can be easily simulated by background events. There are two principal types of background:

1. The muon neutrino beam contains an unavoidable contamination of electron neutrinos. It is small, about 1%, but the probability of the process

$$\nu_e + N \rightarrow e + X \qquad (9.34)$$

   is ten thousand times that of the elastic scattering, which is our signal. Consequently, the background to signal ratio is of the order of one hundred.
2. The CC neutrino interactions with the nuclei are recognised by the μ track, which is straight and deeply penetrating. However, this is not the case for the NC interactions. Consider in particular

$$\nu_\mu + N \rightarrow \nu_\mu + \pi^0 + X \qquad \pi^0 \rightarrow \gamma\gamma. \qquad (9.35)$$

Sometimes the hadronic part, called X, does not have enough energy and escapes detection; sometimes one of the photons materialises in a positron–electron pair that is confused with a single electron, simulating the signal.

Both backgrounds can be discriminated on kinematic grounds. Indeed, the electron hit by a neutrino maintains its direction within a very small angle, because its mass is small. Let us see.

The collision kinematic is depicted in Fig. 9.7. Let us write the energy and momentum conservation

$$E_i + m_e = E_e + E_\nu \qquad 0 = E_\nu \sin\theta_\nu + E_e \sin\theta_e \qquad E_i = E_\nu \cos\theta_\nu + E_e \cos\theta_e. \qquad (9.36)$$

The last equation can be written in the form

$$E_i = E_\nu + E_e - E_\nu(1 - \cos\theta_\nu) - E_e(1 - \cos\theta_e).$$

Using the first equation in (9.36) we have

$$E_i = E_i + m_e - E_\nu(1 - \cos\theta_\nu) - E_e(1 - \cos\theta_e)$$

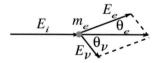

Fig. 9.7.   Kinematic of the neutrino–electron scattering.

and in conclusion

$$E_e(1 - \cos\theta_e) = m_e - E_v(1 - \cos\theta_v) \leq m_e \quad \Rightarrow 1 - \cos\theta_e \leq \frac{m_e}{E_v}.$$

Finally, with a very good approximation

$$E_e\theta_e^2 \leq 2m_e. \tag{9.37}$$

We see that in the elastic scattering events the product of the angle of the electron with the beam direction and the square of its energy is extremely small. We conclude that our detector must have a good energy resolution and, in order to measure the electron direction with high accuracy, a very good spatial resolution. The latter condition implies a low atomic mass medium in order to minimise multiple scattering. A further condition is a good granularity to distinguish electrons from π^0s.

The detector has a modular structure, as shown schematically in Fig. 9.8. Its mass is 792 t, with a 4 m × 4 m section and 33 m length. Each module is made of a glass (a low Z material) slab 48 cm thick, followed by a pair of tracking chambers, each measuring one coordinate, and an array of scintillation counters. These have two functions, to trigger the read-out electronics if an event takes place and to measure the energy.

Figure 9.9 shows the $E_e\theta_e^2$ distributions for both neutrino and antineutrino exposures. The peak close to 0 is the signal of neutrino–electron scattering, above the background. The contributions of the above-mentioned two principal background sources are evaluated and their sum compared to the experimental data at large angles, where the signal is absent. Having found agreement, the background function is extrapolated in the peak region and its contribution subtracted, to obtain the size of the signal.

The final result is (Vilain *et al.* 1994)

$$\sin^2\theta_W = 0.2324 \pm 0.0083. \tag{9.38}$$

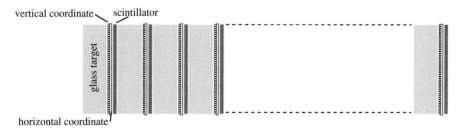

vertical coordinate   scintillator

glass target

horizontal coordinate

Fig. 9.8. Schematic of the CHARM2 experiment.

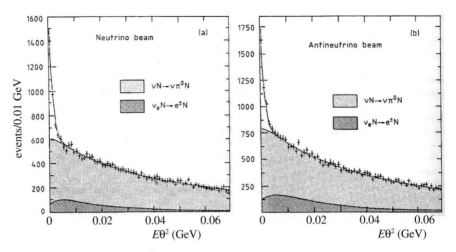

Fig. 9.9. CHARM2. $E\theta^2$ distribution for neutrinos and antineutrinos. (Geiregat *et al.* 1991)

9.5 The intermediate vector bosons

A crucial prediction of the electroweak theory is the existence of the vector mesons W and Z and of their characteristics: masses and total and partial widths. Let us look at them in detail. As we saw in Section 9.3, the predicted values of the masses are

$$M_W \approx 80\,\text{GeV} \qquad M_Z \approx 90\,\text{GeV}. \qquad (9.39)$$

As we know, the couplings of W and Z to all the leptons and quarks are predicted by the theory. The calculation of the partial widths requires knowledge of quantum field theory and cannot be done here. However, once we know one of them we can calculate the others with elementary arguments. We start from the partial width of the decay

$$W^- \to l^- + \nu_l \qquad (9.40)$$

that is

$$\Gamma_{e\nu} = \Gamma_{\mu\nu} = \Gamma_{\tau\nu} = \left(\frac{g}{\sqrt{2}}\right)^2 \frac{M_W}{24\pi} = \frac{1}{2}\frac{G_F M_W^3}{3\sqrt{2\pi}} \approx 225\,\text{MeV}. \qquad (9.41)$$

Notice that the factor $g/\sqrt{2}$, in evidence in the above expression, is simply the constant in the CC term in the Lagrangian (9.19). We have also given its expression in terms of the Fermi constant, using Eq. (7.9).

We now assume that the phase space factors are equal, because the lepton masses are negligible in comparison to the energies in the W decay. As for the decays into quark–antiquark pairs, we must take into account the mixing on one side and the existence of three colours on the other, namely that there are three possibilities for each decay channel.

Charge conservation implies that the quarks of the pair must be one of up-type and one of down-type. Not all the channels are open, namely the decays into $\bar{t}+d$, $\bar{t}+s$ and $\bar{t}+b$ do not exist because $M_W < m_t$ (historically, this was not known at the time of the W discovery).

The quark and the antiquark of the pair may be in the same or different families. Given the smallness of the mixing angles, the partial widths in the latter channels are small. Neglecting the quark masses and recalling the colour factor 3, we have

$$\Gamma_{us} \equiv \Gamma(W \to \bar{u}s) = 3 \times |V_{us}|^2 \Gamma_{ev} = 3 \times 0.224^2 \times \Gamma_{ev} \approx 35\,\text{MeV} \qquad (9.42)$$

and

$$\Gamma_{cd} \equiv \Gamma(W \to \bar{c}d) = 3 \times |V_{cd}|^2 \Gamma_{ev} = 3 \times 0.22^2 \times \Gamma_{ev} \approx 33\,\text{MeV}. \qquad (9.43)$$

Γ_{ub} and Γ_{cb} are very small.

The widths in a quark–antiquark pair of the same family are

$$\begin{aligned} \Gamma_{ud} \equiv \Gamma(W \to \bar{u}d) &= 3 \times |V_{ud}|^2 \Gamma_{ev} \\ &= 3 \times 0.974^2 \times \Gamma_{ev} = 2.84 \times \Gamma_{ev} \approx 640\,\text{MeV} \end{aligned} \qquad (9.44)$$

and

$$\Gamma_{cs} \equiv \Gamma(W \to \bar{c}s) = 3 \times |V_{cs}|^2 \Gamma_{ev} = 3 \times 0.99^2 \times \Gamma_{ev} \approx 660\,\text{MeV}. \qquad (9.45)$$

We obtain the total width by summing the partial ones

$$\Gamma_W \approx 2.04\,\text{GeV}. \qquad (9.46)$$

The couplings of the Z are proportional to the Z-charge factors c_Z (Table 9.1), as we see by rewriting Eq. (9.19)

$$g_Z \equiv \frac{g}{\cos\theta_W}\left(I_3^W - Q\sin^2\theta_W\right) = \frac{g}{\cos\theta_W}c_Z. \qquad (9.47)$$

Let us start with the neutrino–antineutrino channels. We can obtain their expression from (9.41), with the constant $g/\cos\theta_W$ in place of $g/\sqrt{2}$, because this is the factor of the NC term in the Lagrangian (9.19), and with M_Z in place of M_W. We have

$$\Gamma_\nu \equiv \Gamma(Z \to \nu_l\bar{\nu}_l) = \left(\frac{g}{\cos\theta_W}\right)^2 \frac{M_Z}{24\pi}\left(\frac{1}{2}\right)^2 = \frac{G_F M_W^2 M_Z}{\cos^2\theta_W 3\sqrt{2}\pi}\left(\frac{1}{2}\right)^2.$$

We now use Eq. (9.27) to eliminate M_W, obtaining

$$\Gamma_\nu = \frac{G_F M_Z^3}{3\sqrt{2}\pi}\left(\frac{1}{2}\right)^2 \approx 660 \times \frac{1}{4}\,\text{MeV} = 165\,\text{MeV}. \tag{9.48}$$

The two-neutrino final states are not observable. Considering that other invisible particles might exist, one defines as 'invisible width' the total width in the invisible channels. If the only contribution to this is given by three neutrinos, the width is

$$\Gamma_{\text{inv}} = 3\Gamma_\nu \approx 495\,\text{MeV}. \tag{9.49}$$

The measurement of Γ_{inv} provides a way of testing whether there are more 'light' neutrinos, i.e. with masses smaller than $M_Z/2$ and if there are other invisible particles.

Going now to the charged leptons and setting $s^2 = \sin^2\theta_W$, we have

$$\Gamma_l = \Gamma_e = \Gamma_\mu = \Gamma_\tau = \frac{G_F M_Z^3}{3\sqrt{2}\pi}\left[\left(-\frac{1}{2}+s^2\right)^2 + s^4\right] \approx 660 \times 0.125 \approx 83\,\text{MeV}. \tag{9.50}$$

For the quark–antiquark decays we do not need to worry about mixing but we must remember the three colours. The $t\bar{t}$ channel is closed. For the other two up-type pairs, neglecting the quark masses, we have

$$\Gamma_u = \Gamma_c = 3\frac{G_F M_Z^3}{3\sqrt{2}\pi}\left[\left(\frac{1}{2}-\frac{2}{3}s^2\right)^2 + \left(-\frac{2}{3}s^2\right)^2\right] \approx 660 \times 0.42 \approx 280\,\text{MeV}. \tag{9.51}$$

Finally, for the three down-type pairs, we obtain

$$\Gamma_d = \Gamma_s = \Gamma_b = 3\frac{G_F M_Z^3}{3\sqrt{2}\pi}\left[\left(-\frac{1}{2}+\frac{1}{3}s^2\right)^2 + \left(\frac{1}{3}s^2\right)^2\right] \approx 660 \times 0.555 \approx 370\,\text{MeV}. \tag{9.52}$$

From the experimental point of view it is not generally possible to distinguish the different quark–antiquark channels. Indeed this is only possible, in some instances, for the $c\bar{c}$ and $b\bar{b}$ channels. Therefore, the total hadronic cross section is measured. The predicted value is

$$\Gamma_h = 2\Gamma_u + 3\Gamma_d \approx 1.67\,\text{GeV}. \tag{9.53}$$

Summing up, the total width is

$$\Gamma_Z = \Gamma_{\text{inv}} + 3\Gamma_l + \Gamma_h \approx 2.42 \,\text{GeV}. \tag{9.54}$$

We still need other predictions, namely those of the vector meson production cross sections. To be precise, let us consider the W and Z formation experiments. Both vector mesons can be obtained by a quark–antiquark annihilation and they were discovered this way at the 'quark–antiquark collider' at CERN, which was, of course, the proton–antiproton collider. A second possibility for Z formation is given by the e^+e^- colliders. Two such colliders were built for precision studies following this discovery, the storage ring LEP at CERN and the linear collider SLC at SLAC. Let us evaluate the cross sections.

In a proton–antiproton collision it may happen that a quark and an antiquark come very close to each other and annihilate into a Z or a W. The probability of this process, which is weak, is very small compared to the much more frequent strong reactions, even at resonance. Let us call x_q the momentum fraction carried by the quark and $x_{\bar{q}}$ the momentum fraction carried by the antiquark. The collision is observed in the proton–antiproton centre of mass system, which is not in general the quark–antiquark centre of mass system. If $\sqrt{s}$ is the centre of mass energy of the proton–antiproton collision, the quark–antiquark centre of mass energy is

$$\sqrt{\hat{s}} = x_q x_{\bar{q}} \sqrt{s}. \tag{9.55}$$

Let us start by considering the process

$$\bar{u} + d \to e^- + \bar{\nu}_e \tag{9.56}$$

in the neighbourhood of the resonance, namely for $\sqrt{\hat{s}} \approx M_W$. The dominant diagram is given in Fig. 9.10.

The situation is analogous to that of $e^+ + e^- \to e^+ + e^-$ near a resonance. We can use, in a first approximation, the Breit–Wigner expression of the cross section with two spin 1/2 particles both in the initial and in the final states, through an intermediate vector state. The expression is Eq. (4.67), taking into account that the two quarks must have the same colour. For a given colour, this happens one

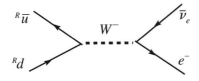

Fig. 9.10. The upper index labels the colour.

time out of nine. On the other hand, we have already taken into account that there are three colours when evaluating the partial width. Therefore, the cross section, summed over the colours, is

$$\sigma(\bar{u}d \to e^- \bar{v}_e) = \frac{1}{9} \frac{3\pi}{\hat{s}} \frac{\Gamma_{ud}\Gamma_{ev}}{\left(\sqrt{\hat{s}} - M_W\right)^2 + (\Gamma_W/2)^2}. \tag{9.57}$$

At the resonance peak $\sqrt{\hat{s}} = M_W$, using the values we have just computed for the widths, we have

$$\sigma_{\max}(\bar{u}d \to e^- \bar{v}_e) = \frac{4\pi}{3} \frac{1}{M_W^2} \frac{\Gamma_{ud}\Gamma_{ev}}{\Gamma_W^2} = \frac{4\pi}{3} \frac{1}{81^2} \frac{0.64 \times 0.225}{2.04^2}$$

$$\times 388 \left[\mu b/\text{GeV}^{-2}\right] \approx 8.8 \text{ nb}. \tag{9.58}$$

Obviously the charge conjugated process $u + \bar{d} \to e^+ + v_e$ contributes to the W^+ with an equal cross section.

We now consider the Z production followed by its decay into $e^+ e^-$. Two processes contribute

$$\bar{u} + u \to e^- + e^+ \qquad \bar{d} + d \to e^- + e^+. \tag{9.59}$$

Their cross sections in resonance are

$$\sigma_{\max}(\bar{u}u \to e^- e^+) = \frac{4\pi}{3} \frac{1}{M_Z^2} \frac{\Gamma_u \Gamma_e}{\Gamma_Z^2} = \frac{4\pi}{3} \frac{1}{91^2} \frac{0.280 \times 0.083}{2.42^2} \times 388 \,\mu b \approx 0.8 \text{ nb} \tag{9.60}$$

and

$$\sigma_{\max}(\bar{d}d \to e^- e^+) = \frac{4\pi}{3} \frac{1}{M_Z^2} \frac{\Gamma_d \Gamma_e}{\Gamma_Z^2} \approx 1 \text{ nb}. \tag{9.61}$$

Notice that the cross sections for the Z are almost an order of magnitude smaller than those for the W. This is due to the fact that the Z partial widths are smaller and the mass is larger.

9.6 The UA1 experiment

In 1976 C. Rubbia, D. Cline and P. McIntyre (Rubbia *et al.* 1976) proposed transforming the CERN Super Proton Synchrotron (SPS) into a storage ring in which protons and antiprotons would counter-rotate and collide head-on, as we have already discussed in Section 1.10. In this way, with 270 GeV per beam, the energy needed to create the W and the Z could be reached. To this aim a large number of antiprotons had to be produced, concentrated in a dense beam and collided with an intense proton beam. Let us evaluate the necessary luminosity.

We can think of the proton and the antiprotons as two groups of partons, quarks, antiquarks and gluons, travelling in parallel directions, as shown in Fig. 9.11, neglecting, in a first approximation, the transverse momentum of the partons. Let us consider the valence quarks and antiquarks respectively. They carry the largest fraction of the total momentum, about 1/6 on average, with a rather broad distribution (see Fig. 6.14). It is important to notice that the width of the $\sqrt{\hat{s}}$ distribution is much larger than the widths of the W and Z resonances. Therefore, the W and Z production cross sections grow with collision energy because the larger $\sqrt{s}$ the greater the probability of finding a quark–antiquark pair with $\sqrt{\hat{s}}$ close to resonance. In conclusion, the higher the energy the better. The initial design centre of mass energy at CERN was $\sqrt{s} = 540$ GeV, to reach 630 GeV later on.

The calculation of the proton–antiproton cross sections starts from those at the quark level and takes into account the quark distribution functions and the effects of the colour field. The evaluation made in the design phase gave the values

$$\sigma(\bar{p}p \to W \to ev_e) \approx 530\,\text{pb} \qquad \sigma(\bar{p}p \to Z \to ee) \approx 35\,\text{pb}. \qquad (9.62)$$

To be precise, both the valence and the sea quarks contribute to the process, however at $\sqrt{s} = 540$ GeV the average momentum fraction at the W and Z resonances is $\langle x \rangle_W / \sqrt{s} \approx 0.15$. Therefore, the process is dominated by the valence quarks, while the sea quarks have momentum fractions that are too small. We thus know that the annihilating quark is in the proton, the antiquark in the antiproton. This information is lost at higher collision energies.

As we have mentioned in Section 1.10, the stochastic cooling technique had been developed at CERN to increase the density of particles within bunches at the collision point. Starting from this experience, an advanced accelerator physics programme was launched, under the guidance of S. Van der Meer, which made it

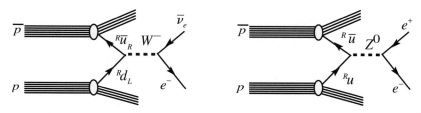

Fig. 9.11. W and Z production in a $\bar{p}p$ collider. Upper left indices label the colour.

possible to reach the luminosity $\mathcal{L} = 10^{32}$ m^{-2} s^{-1}, large enough to search for W and Z, in 1983.

Example 9.3   How many $W \to ev$ events and how many $Z \to e^+e^-$ events are observed in one year with the luminosity $\mathcal{L} = 10^{32}$ m^{-2} s^{-1} and 50% detection efficiency? We apply the mnemonic rule that one year $= \pi \times 10^7$ s. Taking into account the time needed to fill the machine, for maintenance, etc. we take 10^7 s. With the above-mentioned cross sections we have $\sim$25 $W \to ev$ and $\sim$2 $Z \to ee$. Actually the W was discovered several months before the Z.

The production of weak vector bosons is a rare event. Indeed the cross sections in (9.62) are eight and nine orders of magnitude smaller than the total proton–antiproton cross section, which is 60 mb at the energies we are considering. Weak interactions are weak indeed! Consequently, the detector must be able to detect the interesting events with a discriminating power of at least 10^{10}. This is the reason why we considered above only the leptonic channels, which can be discriminated. The hadronic channels $W \to \bar{q}q'$, $Z \to \bar{q}q$ are more frequent but are submerged in a huge background due to strong interaction processes, such as

$$gg \to gg \qquad gq \to gq \qquad g\bar{q} \to g\bar{q} \qquad q\bar{q} \to q\bar{q}. \qquad (9.63)$$

The leptonic channels are

$$\bar{p}p \to W \to ev_e \qquad \bar{p}p \to W \to \mu v_\mu \qquad \bar{p}p \to W \to \tau v_\tau \qquad (9.64)$$

and

$$\bar{p}p \to Z \to ee \qquad \bar{p}p \to Z \to \mu\mu \qquad \bar{p}p \to Z \to \tau\tau. \qquad (9.65)$$

Leptons can be present in the strong interaction processes too, being produced indirectly by hadron decays, but they can be discriminated. The crucial variable is the 'transverse momentum', p_T, namely the momentum component perpendicular to the colliding beams. In the largest fraction of cases the proton–antiproton collision is soft, namely it gives rise to low transverse momentum hadrons. Consider one of them, for example a charm, which decays into a charged lepton. The latter might simulate one of the (9.64) or (9.65) processes. However, in the rest frame of the decaying particle the lepton momentum is a fraction of the charm mass, less than a GeV. The Lorentz transformation to the laboratory frame does not alter the lepton component normal to the charm velocity, which is about that of the beams. The transverse momenta of the kaons are even smaller, while those of the beauties are somewhat larger.

However, there are cases in which two partons come very close to each other and collide violently, namely with a large momentum transfer, by one of the processes (9.63). These events are rare, because, as discussed in Chapter 6, the QCD coupling constant is small at high momentum transfer. The hit parton appears as a jet at high transverse momentum. A possible semileptonic decay of a hadron produces a high p_T lepton. However, these leptons are inside a jet, while those from the W and Z decays are not. In conclusion, we search for leptons that have a high p_T and are 'isolated', namely without other particles in a properly defined cone around its direction.

The same criteria also apply to the neutrino, in the case of the W. Even if neutrinos cannot be detected, we can infer their presence indirectly. To this aim we must build a hermetic detector, which completely surrounds the interaction point with homogeneous calorimeters, in order to intercept all the hadrons and charged leptons and to measure their energies. Moreover, the calorimeters are divided into cells, in order to measure also the direction of the particles. However, the energy of the high-energy muons cannot be measured with calorimetric means because these particles cannot be absorbed in a reasonable length. We solve the problem by determining their momenta by measuring their trajectories in a magnetic field. With this information, we check if the vector sum of all the momenta is compatible with zero or not. In the presence of one (or more) neutrinos we find an imbalance and we say that the 'missing momentum' is the momentum of the neutrino(s). This is possible even if the detector cannot be closed at small angles with the beams, where the physical elements needed to drive the beam itself are located, because we only need the *transverse* component of the missing momentum, p_T^{miss}, to which the undetected particles at small angles make a negligible contribution.

Summarising, the principal channels for the W and Z search and the corresponding topologies are

$$W \rightarrow e^{\pm} \nu_e \qquad \text{isolated electron at high } p_T \text{ and high } p_T^{miss} \qquad (9.66a)$$

$$W \rightarrow \mu^{\pm} \nu_\mu \qquad \text{isolated muon at high } p_T \text{ and high } p_T^{miss} \qquad (9.66b)$$

$$Z \rightarrow e^+ e^- \qquad \text{two isolated electrons, opposite sign, at high } p_T \qquad (9.67a)$$

$$Z \rightarrow \mu^+ \mu^- \qquad \text{two isolated muons, opposite sign, at high } p_T. \qquad (9.67b)$$

This discussion determines the main specifications of the experimental apparatus.

Two experiments were built at the CERN proton–antiproton collider, called UA1 and UA2. The W and the Z were observed by UA1 (Arnison *et al.* 1983a, 1983b) first and immediately afterwards by UA2 (Banner *et al.* 1983, Bagnaia *et al.* 1983). The results of the two experiments are in perfect agreement and of the same quality. We shall describe here those of UA1.

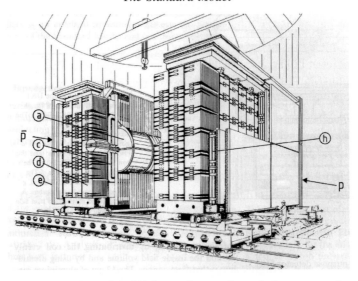

Fig. 9.12. Artist's view of the UA1 experiment, shown in its open configuration. The labels indicate the components: (a) tracking central detector, (c) magnetic field coil, (d) hadronic calorimeters, (e) drift chambers for μ detection, (h) Fe absorber. (Albajar *et al.* 1989)

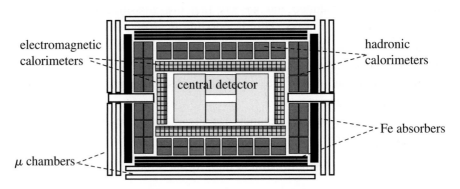

Fig. 9.13. Simplified horizontal cross section of UA1.

Figure 9.12 shows an artist's view of the UA1 experiment, when open. Figure 9.13 shows the UA1 logic structure. The two beams travelling in the vacuum pipe enter the detector from the left and the right respectively, colliding at the centre of the detector. A particle produced in the collision meets in series the following elements:

1. The central detector, which is a large cylindrical time-projection chamber providing electronic images of the charged tracks, and is immersed in a horizontal magnetic field in the plane of the drawing, perpendicular to the beams.
2. The electromagnetic calorimeters, made up of a sandwich of lead plates alternated with plastic scintillator plates. In the calorimeter electrons and photons lose all their energy, which is measured.

3. The other particles penetrate the hadronic calorimeter, which is a sandwich of iron and plastic scintillator plates. The iron plates on the left and right sides of the beams also act as the yoke of the magnet driving the magnetic return flux. In the calorimeter the hadrons lose all (or almost all) of their energy, which is measured.
4. In practice the highest-energy hadronic showers, especially in the forward directions, are not completely contained in the calorimeters, as ideally they should be. They are absorbed in iron absorbers.
5. The particles that survive after the iron absorbers are neutrinos and muons. Large tracking drift and streamer chambers detect the latter.

The detector is hermetic but at small angles with the beams; the response of the calorimeters is made as homogeneous as possible.

9.7 The discovery of W and Z

Figure 9.14 shows the reconstruction of one of the first $W \to e\nu$ events observed by UA1. We observe many tracks that make the picture somewhat confused. These are particles pertaining to the 'rest of the event', i.e. coming from the interaction of partons different from those that produced the W. They are soft, because the strong coupling constant is large at small momentum transfers, and can be easily eliminated simply by neglecting all tracks with p_T smaller than a few times λ_{QCD}, in practice with $p_T < 1$ GeV, as shown in Fig. 9.14(b).

With this simple 'cut' we are left with a clean picture of a single charged track with the characteristics of an electron. Its momentum, measured from its curvature, and its energy, measured in the calorimeter, are equal within the errors. We also find that the transverse momentum is not balanced. The transverse missing momentum is shown in Fig. 9.14(a).

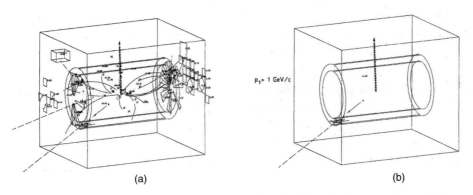

(a)   (b)

Fig. 9.14.   A $W \to e\nu$ event. (a) The tracks, the hit calorimeter cells and the missing transverse momentum are shown; (b) only tracks with $p_T > 1$ GeV. (Rubbia 1985 © Nobel Foundation 1984)

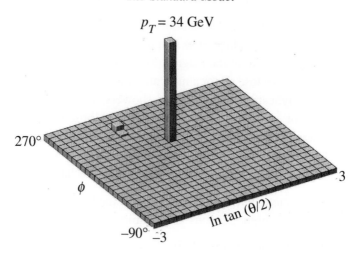

$p_T = 34\ \text{GeV}$

270°

ϕ

3

ln tan (θ/2)

−90° −3

Fig. 9.15. Lego plot of a $W \to e\nu$ event in the electromagnetic calorimeter. ϕ azimuth, θ anomaly to the beam direction. (Adapted from Rubbia 1985)

The calorimeters give a complementary view of the events, namely they show the energy flow from the collision point as a function of the angles. Figure 9.15 shows such a view for a $W \to e\nu$ event. The cells of the diagram, called a 'lego plot', correspond to the physical electromagnetic calorimeter cells. The two coordinates are the azimuth ϕ and a function of the anomaly θ with the beam direction as polar axis. Since the frequency of tracks in the forward directions, namely for $\theta = 0$ and $\theta = \pi$, is very high, the function ln tan $\theta/2$ is used to obtain a smooth distribution. In this event there is in practice a single, large, localised energy deposit. This is how the calorimeter sees the electron.

Summarising, we see that simple kinematic selection criteria allow unambiguous identification of the very rare cases in which a W is produced. It subsequently decays into $e\nu$.

The situation is similar for the decays into $\mu\nu$ and into $\tau\nu$, which we shall not discuss. We mention, however, that the comparison of the three cross sections gives a test of lepton universality, namely

$$g_\mu/g_e = 1.00 \pm 0.07(\text{stat}) \pm 0.04(\text{syst})$$
$$g_\tau/g_e = 1.01 \pm 0.10(\text{stat}) \pm 0.06(\text{syst}).$$

$$(9.68)$$

Let us now consider the measurement of the W mass. As the calorimetric measurement of the electron energy is more precise than the muon momentum measurement, we choose the $e\nu$ channel. We cannot reconstruct the electron-neutrino mass because only the transverse component of the neutrino

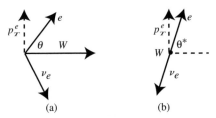

Fig. 9.16.   The momenta (a) in the laboratory frame and (b) in the centre of mass frame of the W.

momentum is known. However, we can measure M_W with the 'Jacobian peak' method.

Figure 9.16(a) gives a scheme of the W decay kinematic in the laboratory frame. The W momentum component transverse to the beam is very small in general. Neglecting it in a first approximation, the flight direction of the W is the direction of the beams. Consider the electron momentum, which is measured. Its component normal to the W motion, p_T is equal in the laboratory frame and in the centre of mass frame (Fig. 9.16(b))

$$p_T = \frac{M_W}{2} \sin \theta^*. \tag{9.69}$$

Let $dn/d\theta^*$ be the decay angular distribution in the rest frame of the W. The transverse momentum distribution is then given by

$$\frac{dn}{dp_T} = \frac{dn}{d\theta^*} \frac{d\theta^*}{dp_T}. \tag{9.70}$$

The quantity $d\theta^*/dp_T$ is called the Jacobian of the variable transformation. Its expression is

$$\frac{dn}{dp_T} = \frac{1}{\sqrt{(M_W/2)^2 - p_T^2}} \frac{dn}{d\theta^*}. \tag{9.71}$$

The essential point is that the Jacobian diverges for

$$p_T = M_W/2. \tag{9.72}$$

Consequently, the p_T distribution has a sharp maximum at $M_W/2$. Notice that the conclusion does not depend on the longitudinal momentum of the W, which may be large. The position of the maximum does, on the other hand, depend on the transverse momentum of the W, which, as we have said, is small but not completely negligible. Its effect is a certain broadening of the peak.

The electron transverse momentum distribution for the W events is shown in Fig. 9.17, where the Jacobian peak is clearly seen. From this distribution UA1 measured $M_W = 83$ GeV, with ± 3 GeV uncertainty, substantially determined by the systematic uncertainty on the energy calibration. UA2 measured $M_W = 80$ GeV with an uncertainty of ± 1.5 GeV.

A further test of the electroweak theory is the measurement of the electron helicity in the decay $W \rightarrow e\nu$. Consider the process in the W rest frame as in Fig. 9.18.

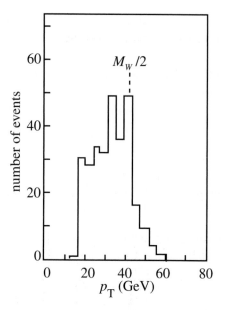

Fig. 9.17. Electron p_T distribution for W events. (Adapted from Albajar *et al.* 1989)

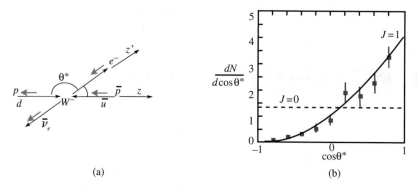

(a)   (b)

Fig. 9.18. (a) Kinematics of the W production and decay. (b) Angular distribution measured by UA1. (Adapted from Albajar *et al.* 1989)

For the $V - A$ structure of the CC weak interactions the leptons are left, and, if their energy is much larger than their masses as in the present case, their helicity is -1; the antileptons are right, with helicity $+1$. We take z, the direction of the beams, as the quantisation axis for the angular momenta in the initial state, as in Fig. 9.18. The total angular momentum is $J = 1$. As already seen, since the W production is due to valence quarks, we know that the initial quark has the direction of the proton, the antiquark that of the antiproton. Therefore, the third component of the angular momentum is $J_z = -1$.

We take the electron direction z' as the quantisation axis in the final state. By the same token the third component is $J_{z'} = -1$. Therefore, the angular dependence of the differential cross section is given by

$$\frac{d\sigma}{d\Omega} \propto \left[d^1_{-1,-1} \right]^2 = \left[\frac{1}{2}(1 + \cos \theta^*) \right]^2. \tag{9.73}$$

The distribution measured by UA1 is shown in Fig. 9.18(b); the curve is Eq. (9.73), which is in perfect agreement with the data. The dotted line is the prediction for W spin $J = 0$. In this way we measure the W spin.

Notice that the observed asymmetry shows that parity is violated but does not prove that the CC structure is $V - A$. The $V + A$ structure predicts the same angular distribution. Only polarisation measurements can distinguish the two cases.

Question 9.1 Prove the last statement.

We now consider the discovery of the Z. Figure 9.19 shows the UA1 tracking view of a typical $Z \to e^- e^+$ event. Again, the confused view becomes clear with

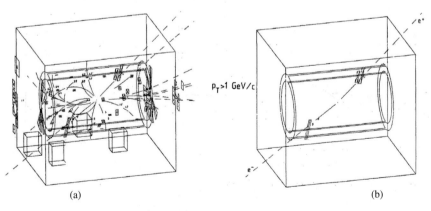

(a)   (b)

Fig. 9.19. (a) A $Z \to e^- e^+$ event; (b) only tracks with $p_T > 1$ GeV. (Rubbia 1985 © Nobel Foundation 1984)

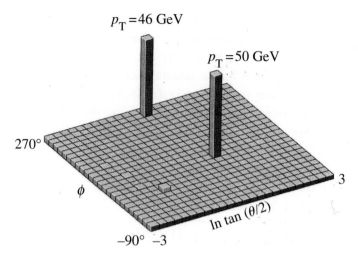

Fig. 9.20.   Lego plot of a $Z \to ee$ event in the UA1 electromagnetic calorimeter.

the selection $p_T > 1$ GeV. Only two tracks remain. One of them is positive, the other negative; for both, the energy as measured in the electronic calorimeter is equal to the momentum measured from curvature.

Figure 9.20 shows the calorimetric view of a $Z \to e^- e^+$ event: two localised, isolated energy deposits appear in the electromagnetic calorimeter.

The mass of Z is obtained by measuring the energies of both electrons in the electromagnetic calorimeters and the angle between their tracks in the central detector. Figure 9.21 is the M_Z distribution of the first 24 UA1 events. The average is $M_Z = 93$ GeV with a systematic uncertainty of ± 3 GeV; the UA2 measurement gave $M_Z = 91.5$ GeV with a systematic uncertainty of ± 1.7 GeV.

In conclusion, by 1983 the UA1 and UA2 experiments had confirmed that the vector mesons predicted by the electroweak theory exist and have exactly the predicted characteristics.

Particularly important is the ratio of the two masses, experimentally because it is not affected by the energy scale calibration and theoretically because it directly provides the weak angle. Indeed, Eq. (9.28) valid at the tree-level, gives

$$\cos^2 \theta_W = 1 - (M_W/M_Z)^2. \tag{9.74}$$

The ratio of the masses measured by UA1 and UA2 gives

$$\text{UA1}: \sin^2 \theta_W = 0.211 \pm 0.025 \qquad \text{UA2}: \sin^2 \theta_W = 0.232 \pm 0.027. \tag{9.75}$$

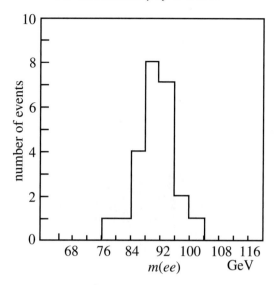

Fig. 9.21. Distribution of m (e^+e^-) for the first 24 UA1 events. (Adapted from Albajar *et al.* 1989)

These values are in agreement with the low-energy measurements we mentioned in Section 9.4. We shall come back to this point in Section 9.8 for a more accurate discussion.

Question 9.2 Does the Z decay into two equal pseudoscalar mesons? And into two scalar mesons?

Question 9.3 In their first data-taking period, UA1 and UA2 collected ~300 W and ~30 Z each. What is the principal source of uncertainty on the W mass? On the Z mass? On their ratio?

Before closing this section, let us see how the quarks appear in a hadronic collider. As we know, to observe a quark we should not try to break a nucleon in order to extract one of them, rather we must observe the hadronic energy flux in a high-energy collision at high momentum transfer. One of the first observations of UA2 (Banner *et al.* 1982) and UA1 was that of events with two hadronic jets in back-to-back directions. They are violent collisions between two quarks, which in the final state hadronise into jets. More rarely a third jet was observed, due to the radiation of a gluon.

The lego plot of a two-jet event as seen in the UA1 calorimeter is shown in Fig. 9.22. Comparing it with Fig. 9.20 we see that the two quarks, as seen in the calorimeter, are very similar to electrons, with some differences: the peaks are

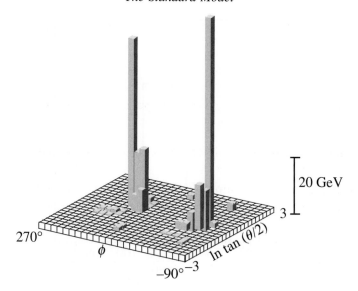

Fig. 9.22. Two quark jets in the UA1 hadronic calorimeter. (Adapted from Albajar *et al.* 1987)

wider and more activity is present outside them, two features that are well understood by thinking of the antiscreening QCD phenomenon.

9.8 The evolution of $\sin^2 \theta_W$

As already stated, accurate comparison of the weak angle values extracted from different physical processes requires a theoretical calculation beyond the tree-level, including higher-order terms. The most important radiative corrections are those to the W mass, corresponding to the diagrams in Fig. 9.23. The correction to M_W due to diagram (a) is proportional to the difference between the squares of the masses of the two quarks in the loop, namely to

$$G_F \left(m_t^2 - m_b^2 \right) \approx G_F m_t^2. \tag{9.76}$$

Therefore, an accurate measurement of M_W allows us to predict the top mass.

The correction due to diagram (b) depends on the Higgs boson mass M_H. The dependence is, however, only logarithmic and consequently the prediction of the Higgs boson mass is less precise. At the end of the 1980s Amaldi *et al.* (1987) and Costa *et al.* (1988) analysed the weak angle values measured at different Q^2 scales and the relevant radiative corrections. The top mass, unknown at that time, was left as a free parameter. The result was that the data were in perfect agreement with the Standard Model, provided the top mass was not too large, namely $m_t < 200 \, \text{GeV}$.

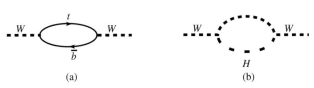

Fig. 9.23. Principal radiative correction diagrams to the W mass.

At that time the limit appeared very difficult to reach, because the larger the top mass, the smaller its production cross section. On this basis a campaign of technological improvements aimed at increasing the Tevatron collider luminosity was planned at Fermilab. The CDF and D0 detectors were gradually modified.

In the same years, as we shall see in the next section, the LEP experiments at CERN measured the vector meson masses with increasing accuracy and, with a number of different methods, the weak angle. These accurate data and the equally accurate calculations of the radiative corrections gave a prediction of the top mass of increasing accuracy, which in 1993, assuming the Higgs boson mass in the range $60 < M_H < 700\,\text{GeV}$, was

$$m_t = 166 \pm 27\,\text{GeV}. \tag{9.77}$$

Finally, two years later CDF at Fermilab discovered the top with the predicted mass, as we saw in Chapter 4.

We studied in Section 5.8 the evolution of the QED 'constant' a, which is proportional to the square of the electric charge, and in Section 6.6 the evolution of the QCD 'constant' a_s, which is proportional to the colour charges squared. The corresponding gauge groups are $U(1)$ and $SU(3)$. The gauge group of the electroweak theory is $SU(2) \otimes U(1)$. The 'electroweak charges' are g for $SU(2)$ and g' for $U(1)$. The tangent of the weak angle is the ratio of the $U(1)$ and $SU(2)$ charges g'/g. These two charges need to be renormalised in the theory in a manner similar to the other charges and consequently are functions of the momentum transfer. Like the other charges, the electroweak charges cannot be measured directly. The quantity that can be extracted from the observables in the most direct way is their ratio, i.e. the weak angle.

The evolution of $\sin^2 \theta_W$ as a function of the momentum transfer $|Q|$ is more complicated than that of a or a_s, because both numerator and denominator vary. It is shown in Fig. 9.24.

First of all notice that the variation of $\sin^2 \theta_W$ is very small, only a small percentage, even in the huge range of Q^2 we are considering. Consequently, only the most precise determinations of those mentioned in Section 9.4 are reported in the figure.

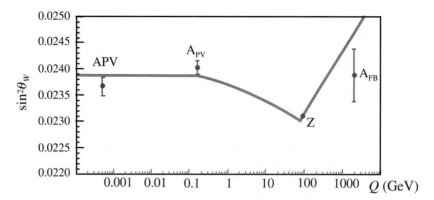

Fig. 9.24. $\sin^2\theta_W$ as a function of $|Q|$. APV: atomic parity violation; A_{PV}: asymmetry in polarised Møller scattering; Z: Z-pole measurements; A_{FB}: forward–backward asymmetry at LEP2. (Adapted from Yao *et al.* 2006 by permission of Particle Data Group and the Institute of Physics)

The curve is the Standard Model prediction. Its change at M_W is due to the following reason. Since the gauge bosons carry weak charges, both fermion and boson loops contribute to the renormalisation of the weak charge, with opposite effects on its slope. This situation is similar to what we have met in QCD. However, different from QCD, the weak gauge bosons are massive. Therefore, only fermionic loops are important at energies below M_W. Here the W loops set in. However, they contribute to the evolution of the $SU(2)$ constant g, not to the evolution of the $U(1)$ constant g'. This inverts the slope of the weak angle evolution.

9.9 Precision tests at LEP

As we have just discussed, at the beginning of the 1990s all the crucial predictions of the electroweak theory had been experimentally verified, with the very important exception of the Higgs boson, a prediction that remains to be tested at the time of writing (2007).

The following steps were the high-precision tests. For these, the ideal instrument is the e^+e^- collider. For this purpose the LEP machine was designed and built at CERN with a 27 km circumference and with energy and luminosity adequate for studying not only all the features of the Z resonance but also the crucial processes $e^+e^- \rightarrow W^+W^-$.

In the same period, B. Richter and collaborators designed at Stanford a novel type of collider. We recall in this context that in a circular machine the electrons continually radiate energy due to their centripetal acceleration. A big problem of the circular e^+e^- colliders is the large amount of radiated power. The electrical power that must be spent just to maintain them circulating in the rings at a constant energy grows with the fourth power of the energy to mass ratio, at fixed orbit radius. Clearly, the construction costs of a machine increase with the length of its tunnel and, above a certain energy, it may become convenient to accelerate electrons and positrons in a linear structure and collide them head-on only once. In such a way one spends more electrical energy on accelerating the particles, which are 'used' only once, but does not spend energy keeping them in orbit. In practice the trade-off is reached around 100–200 GeV.

An added advantage was that at SLAC a linear accelerator already existed. However, several technological developments were necessary in order to produce extremely dense and thin bunches, a few micrometers across at the collision point. The Stanford Linear Collider, SLC, with the Mark II and later with the SLD experiment, started producing physical results at the same time as LEP in 1989. The SLC luminosity was much smaller than that of LEP but its beams could be polarised allowing different tests of the theory, which, however, we shall not discuss. The SLC was also a fundamental step forward from the technological point of view: the next generation electron–positron collider cannot be circular, for the above-mentioned reasons, it will be a linear collider.

The e^+e^- colliders are precision instruments, providing collisions that are in every case, and not just rarely as in a hadron collider, between elementary, point-like objects. This has two consequences: all the events are interesting, not one in a billion or so. Moreover, the events are very clean, and no 'rest of the event' is present as in a hadron collider. Finally, the e^+e^- annihilation leads to a pure quantum state, of definite quantum numbers, $J^{PC} = 1^{--}$.

Four experiments, called ALEPH, DELPHI, L3 and OPAL, were run at LEP from 1989 on the Z peak (collecting 4 000 000 events each) and from 1996 to 2000 at increasing energies up to 209 GeV. Even if rather different in important details, the basic features of all the set-ups are the same. Each of them has a central tracking chamber in a magnetic field oriented in the direction of the beams, electromagnetic and hadronic calorimeters and large muon chambers. Silicon-microstrip detectors are located between the central detector and the beam pipe to provide a close-up image of the vertex region, with ten-micrometer resolution, necessary to look for secondary vertices of charm and beauty decays.

Figure 9.25 shows four events of different types at DELPHI.

Most of the work at LEP was dedicated to the search for small effects that might show violations of the Standard Model and the presence of 'new physics'. These tests require not only extremely accurate experimental work but also high-precision theoretical calculations of radiative corrections, which are beyond our

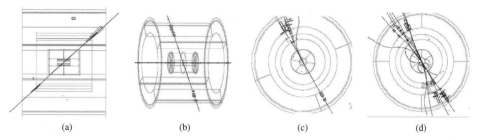

(a) (b) (c) (d)

Fig. 9.25.   DELPHI events. (a) e^+e^- pair, (b) $\mu^+\mu^-$ pair, (c) $\tau^+\tau^-$pair, (d) quark pair. (Images CERN)

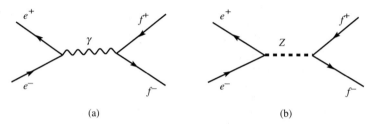

(a) (b)

Fig. 9.26.   Tree-level Feynman diagrams for the process $e^+e^- \rightarrow f^+f^-$ with exchange in the s channel of (a) a photon, (b) a Z boson.

scope here. We shall consequently limit our discussion to the more elementary results.

A fundamental measurement is the shape of the resonance line. In practice, the experiments measure the hadronic cross section, since it is the largest one, as a function of the machine energy. An outstanding experimental effort by the scientists working on the experiments and on the machine led to the astonishing precision of $\Delta\sqrt{s} = \pm 2\,\text{MeV}$, which is about 20 ppm.

Consider the generic f^+f^- final state, different from e^+e^-, to which t-channel exchange also contributes. Two s-channel diagrams are present in general, with γ and Z exchange, as shown in Fig. 9.26. Near the resonance, where the latter dominates, we can express the cross section in the Breit–Wigner approximation as

$$\sigma(e^+e^- \rightarrow f^+f^-) = \frac{3\pi}{s} \frac{\Gamma_e \Gamma_f}{\left(\sqrt{s} - M_Z\right)^2 + \left(\Gamma_Z/2\right)^2}. \tag{9.78}$$

At the peak, namely for $\sqrt{s} = M_Z$, we have

$$\sigma(e^+e^- \rightarrow f^+f^-) = \frac{12\pi}{M_Z^2} \frac{\Gamma_e \Gamma_f}{\Gamma_Z^2}. \tag{9.79}$$

Example 9.4 Calculate the cross section for $e^+e^- \to \mu^+\mu^-$ at the peak. How many Z are produced in this channel at the typical LEP luminosity, $\mathcal{L} = 10^{31}$ cm^{-2}s^{-1}?

$$\sigma(e^+ + e^- \to \mu^+ + \mu^-) = \frac{12\pi}{m_Z^2}\frac{\Gamma_e\Gamma_\mu}{\Gamma^2} = \frac{12\pi}{91^2}\frac{84^2}{2450^2}$$
$$= 5.3 \times 10^{-6}\,\text{GeV}^{-2} \times 388\,\mu\text{b/GeV}^{-2} = 2.1\,\text{nb}.$$

About one Z per minute.

Example 9.5 Repeat the calculation for the hadronic cross section.

$$\sigma(e^+ + e^- \to \text{hadrons}) = \frac{12\pi}{m_Z^2}\frac{\Gamma_e\Gamma_\mu}{\Gamma^2} = \frac{12\pi}{91^2}\frac{84 \times 1690}{2450^2} = 40.2\,\text{nb}.$$

About one thousand Z per hour.

However, as we discussed in Section 4.9, the Breit–Wigner approximation is a rather bad one, as we can see in Fig. 9.27, which shows Eq. (9.78) as a dotted line together with the experimental data. The disagreement is due mainly to the bremsstrahlung of a photon from one of the initial or final particles, as in Fig. 9.28.

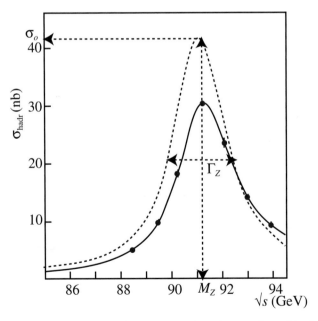

Fig. 9.27.   The Z resonance at LEP. Error bars have been enlarged by 20 to make points visible. (Adapted from LEP & SLD 2006)

Fig. 9.28.   Initial and final bremsstrahlung diagrams.

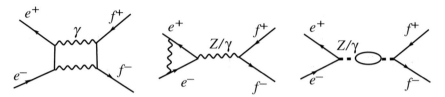

Fig. 9.29.   Higher-order photonic and non-photonic corrections.

Other, smaller but theoretically more interesting, corrections are due to the diagrams in Fig. 9.29.

The analysis starts from the continuous curve in Fig. 9.27 obtained by interpolating the experimental points. The curve is then corrected by taking into account the calculated contribution of the electromagnetic corrections of Fig. 9.28. The result is the dotted curve. From this we extract the mass M_Z, the total width Γ_Z and the peak value σ_0. The most precise values of these observables and of those that we shall now discuss are extracted from an over-constrained fit to all the available data, including the Standard Model radiative corrections. The goodness of the fit provides a precise test of the Standard Model, as we shall see at the end of Section 9.10. We now give the resulting best values. The Z mass (LEP & SLD 2006), with its 23 ppm accuracy, is

$$M_Z = 91.1875 \pm 0.0021 \, \text{GeV}. \tag{9.80}$$

The total width is

$$\Gamma_Z = 2.4952 \pm 0.0023 \, \text{GeV}. \tag{9.81}$$

The peak hadronic cross section is

$$\sigma^0 = 41.540 \pm 0.037 \, \text{nb}. \tag{9.82}$$

We can see that these values are in agreement with our approximate evaluations of Section 9.5.

We now consider the partial widths, obtained by measuring the partial cross sections in the corresponding channels.

The values of the three leptonic widths measured by LEP are equal within 2‰ accuracy, a fact that verifies the universality of the coupling of the leptons to the neutral current, while the tests of Section 7.4 were for the charged current. Their average is

$$\Gamma_l = 83.984 \pm 0.086 \, \text{MeV}. \tag{9.83}$$

The total hadronic width is

$$\Gamma_h = 1744.4 \pm 2.0 \, \text{MeV}. \tag{9.84}$$

It is in general impossible to establish the nature of the quark–antiquark pair in the final state. Exceptions are the charm and beauty cases, in which the presence of a short lifetime particle can be established by observing secondary vertices inside the hadronic jets with the vertex detector. The kinematic reconstruction of the event can then distinguish between the two on statistical grounds. The two partial widths, given as fractions of the hadronic width are, again in agreement with the Standard Model predictions

$$R_c \equiv \Gamma_c/\Gamma_h = 0.1721 \pm 0.0030 \tag{9.85}$$

and

$$R_b \equiv \Gamma_b/\Gamma_h = 0.216\,29 \pm 0.000\,36. \tag{9.86}$$

Example 9.6 Calculate the distances travelled by a D^0 and by a B^0 with 50 GeV energy. [3 mm for the D^0, 4.2 mm for the B^0.]

A fundamental contribution made by the LEP experiments is the precision measurement of the number of 'light neutrino' types, N_ν. The invisible partial width, Γ_{inv} is determined by subtracting from the total width the partial widths in the hadronic and charged leptonic channels, $\Gamma_{\text{inv}} = \Gamma_Z - 3\Gamma_l - \Gamma_h$. We assume the invisible width to be due to N_ν neutrino types each contributing with the partial width as given by the Standard Model $(\Gamma_\nu)_{\text{SM}}$. More precisely, the ratio of the neutrino width to charged lepton width is used, because of its smaller model dependence

$$(\Gamma_\nu/\Gamma_l)_{\text{SM}} = 1.991 \pm 0.001 \tag{9.87}$$

obtaining

$$N_\nu = \frac{\Gamma_{\text{inv}}}{\Gamma_l} \left(\frac{\Gamma_l}{\Gamma_\nu}\right)_{\text{SM}}. \tag{9.88}$$

The combined result from the four LEP experiments and SLD is

$$N_\nu = 2.984 \pm 0.008. \tag{9.89}$$

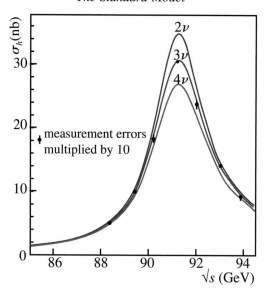

Fig. 9.30. The Z line shape and expectations for different numbers of neutrinos. (Yao *et al.* 2006 by permission of Particle Data Group and the Institute of Physics)

Figure 9.30 shows the resonance curve measured by LEP compared with the curves calculated for two, three and four neutrino flavours.

In conclusion, there are no other neutrinos beyond the three we know, if their mass is smaller than $M_Z/2$. In the hypothesis that the structure of the families is universal, there are no more families beyond those we know.

9.10 The interaction between intermediate bosons

As we have already recalled, LEP was designed to test the Standard Model not only at the Z pole but also at higher energies, above the threshold of the W (pair) production through the reaction processes $e^+e^- \rightarrow W^+W^-$. This study is important because it tests a fundamental aspect of the electroweak theory, namely the fact that the vector mesons self-interact.

Let us start by considering the weak process

$$\nu_\mu + e^- \rightarrow \mu^- + \nu_e. \tag{9.90}$$

As we saw in Section 7.2, its cross sections would grow indefinitely with increasing energy if the interaction was the Fermi point-like interaction represented in Fig. 9.31(a).

The W meson mediating the interaction, as in Fig. 9.31(b), solves the problem.

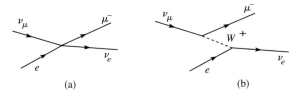

Fig. 9.31. Feynman diagrams for the scattering $\nu_\mu + e^- \rightarrow \mu^- + \nu_e$; (a) in the low-energy point-like Fermi approximation; (b) as mediated by the W boson.

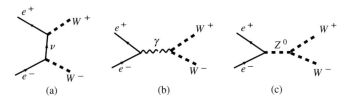

Fig. 9.32. The three tree-level diagrams of $e^+ + e^- \rightarrow W^+ + W^-$.

However, it introduces a new problem. Indeed, if the W exists, then the process

$$e^+ + e^- \rightarrow W^+ + W^- \tag{9.91}$$

exists. It is mediated by a neutrino, as shown in Fig. 9.32(a). Now, computing this diagram we find a diverging cross section, as shown by the curve in Fig. 9.33.

However, we have forgotten that the W is charged. We must include in our calculation the photon exchange shown in the diagram in Fig. 9.32(b). However, the sum of the two diagrams again gives a diverging cross section, the intermediate curve in Fig. 9.33.

In the electroweak theory there is another neutral vector meson beyond the photon, the Z, and it couples directly to the Ws, because the weak vector mesons carry weak charges. Finally, the cross section, calculated also including the diagram of Fig. 9.32(c), does not diverge. It is the continuous curve in Fig. 9.33.

Figure 9.33 shows the cross section of the reaction (9.91) as measured by the LEP experiments up to $\sqrt{s} = 209\,\text{GeV}$. The perfect agreement with the predictions tests another crucial aspect of the theory, namely the weak charge of the weak interaction mediators.

Another important result obtained by measuring the energy dependence of the W production cross section is the accurate determination of the W mass and width. The determination of the energy at which the cross section first becomes different from zero, namely the energy threshold, gives the W mass. The rapidity of the initial growth determines the W width, because if Γ_W is larger the growth is slower.

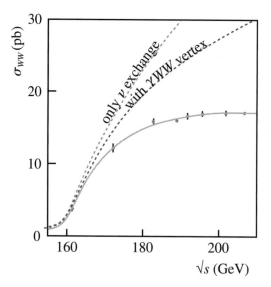

Fig. 9.33. Cross section for $e^+ + e^- \to W^+ + W^-$. (Yao *et al.* 2006 by permission of Particle Data Group and the Institute of Physics)

Other measurements were made at the Tevatron hadron collider in the CDF and D0 experiments with the Jacobian peak method, as we saw when discussing UA1, but with increased precision using, in particular, the Z peak and the value of M_Z of LEP as a calibration of the energy scales of their calorimeters. The result of the global fit is (LEP 2006)

$$M_W = 80.392 \pm 0.029 \, \text{GeV} \qquad (36 \, \text{ppm}) \tag{9.92}$$

and

$$\Gamma_W = 2.1417 \pm 0.060 \, \text{GeV}. \tag{9.93}$$

We now come back to the best-fit procedure used to determine the best values of the fundamental observables. This procedure uses as input the measurements from the experiments at LEP, SLC and Tevatron (including a few that we did not discuss) and theoretical calculations, which take into account the radiative corrections of the Standard Model up to the order that matches the accuracy of the experimental data. There are five free parameters in this fit: the masses of the Z, the top and the Higgs boson, the value of the fine structure constant at m_Z, and the contribution of the strong interactions to the radiative correction of α that cannot be calculated with adequate accuracy. A further input is the Fermi constant G_F, which is known with a 0.8 ppm accuracy. We have already reported the best-fit values of the observables, which we now call generically O^{fit}. They are compared

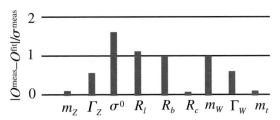

Fig. 9.34.   The pulls of the principal observables of the Standard Model.

with the measured values O^{meas} in Fig. 9.34 (where $R_l = \Gamma_h/\Gamma_l$). The bars are the absolute values of the 'pulls' of each observable, namely $\left|O^{\mathrm{meas}} - O^{\mathrm{fit}}\right|/\sigma^{\mathrm{meas}}$. Notice, in particular, the pulls of m_Z and m_t. They are very small because these observables are measured with high accuracy and consequently are almost fixed in the fit. We see that the fit is good and very cogent, considering the accuracy of the measurements. There is no evidence of physics beyond the Standard Model.

9.11 The search for the Higgs boson

As we saw in Section 9.3, the theoretical mechanism giving rise to the masses in the Standard Model, the Higgs mechanism, has not been experimentally tested yet. We have already said in Section 5.1 that the gauge invariance of the Lagrangian of all the interactions implies that their mediators have zero mass. However, while the photon and the gluons are massless, W and Z are not.

A second issue is the non-zero mass of the fermions. Indeed, for massless particles the Dirac equation separates into two independent equations, one for the left, and one for the right spinor. This property allowed us to consider them as different particles and to classify them, one in an isotopic doublet and one in a singlet. Namely, the two stationary states are invariant under two different gauge transformations. This is impossible if the fermion is massive, because in this case the chirality does not commute with the Hamiltonian.

The Higgs mechanism solves both problems. Without entering into any detail, we say only that the mechanism is a spontaneous symmetry-breaking phenomenon. Let us just give two examples of these processes.

The first example is an example of mechanical instability. Consider a rectangular perfectly symmetric metal plate. Let us lean it vertically with its shorter side on a horizontal plane and let us apply a vertical downwards force in the centre of the other short side. The state is symmetric under the exchange of the left and right faces of the plate. If we now gradually increase the intensity of the force, we observe that at a definite value (which can be calculated from the mechanical characteristics of the system), the plate bows with curvature to the left or to the right. The original symmetry is lost, it has been spontaneously broken.

The second example arises from the spontaneous magnetisation of iron. Consider a piece of iron (or any ferromagnetic material) above its Curie temperature. The atomic magnetic moments are randomly oriented. Consider a microcrystal and its spontaneous magnetisation axis; for the interaction responsible for the ferromagnetism the two directions parallel and antiparallel to the axis are completely equivalent, namely the system is symmetric under their exchange. We now lower the temperature below the Curie point. The Weil domains take shape in the crystal. In each of them the magnetic moments have chosen one of the two directions. Again the symmetry has been spontaneously broken.

In the relativistic quantum field theory the 'vacuum' state is the state of minimum energy. The expectation values both of the fermion and of the boson fields are zero: there are no real particles in the vacuum. The Higgs mechanism (Higgs 1964, Englert & Brout 1964) introduces a field, which respects the symmetry. However, its energy is at a minimum for expectation values of the field different from zero. As a consequence, the equilibrium is unstable, similar to those of the plate and of the magnetic dipoles. The symmetry spontaneously breaks and the Higgs field takes one of these non-zero minimal values. We cannot give any description of the mechanism here and only say that its final result is that the W, the Z and charged fermions, but not neutrinos, acquire masses.

In its simplest form, the mechanism predicts the existence of a neutral scalar particle, the Higgs boson H. The theory does not predict its mass M_H but predicts the Higgs couplings to the fermions and to the intermediate bosons as functions of M_H. Actually the coupling amplitude to fermions $Hf\bar{f}$ is proportional to the fermion mass. Consequently, the decays into larger mass fermions are favoured.

The Standard Model gives an indirect prediction of M_H by calculating its effects on the radiative corrections to M_Z/M_W, a quantity measured with high accuracy. Actually the prediction is similar to that of the top quark mass discussed in Section 9.7.

However, while in the case of the top the diagram in Fig. 9.23(a) gives a contribution proportional to m_t^2, in the case of the Higgs boson the diagram in Fig. 9.23(b), which we reproduce in Fig. 9.35, contributes in proportion to the logarithm of M_H. The prediction is consequently less stringent. The global fit that

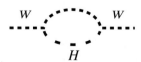

Fig. 9.35.   Higgs loop correction to M_W.

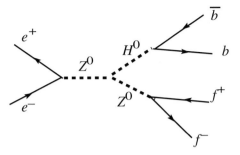

Fig. 9.36.   Higgs production at an e^+e^- collider followed by $\bar{b}b$ decay.

we have mentioned in the previous sections gives the upper limit at 95% confidence level (LEP 2006)

$$M_H < 194\,\text{GeV}. \tag{9.94}$$

The most sensitive searches for the Higgs bosons were carried out in the LEP experiments. The main search channel is shown in Fig. 9.36, with the $H \to \bar{b} + b$ decay favoured by the large b mass. The diagram corresponds to two different situations, depending on the energy. At the Z peak, i.e. for energy $\sqrt{s} \approx M_Z$, the first Z is a real particle, the second is a virtual one, while at higher energies, $\sqrt{s} > M_Z$, the first Z is virtual, the second is real.

At the Z peak, the cross section is appreciable if the virtual Z is not very far from resonance, namely if M_H is rather smaller than M_Z. The search at the Z resonance did not find the Higgs, providing a limit on its mass of that order.

For similar reasons the limit that can be reached on M_H at higher energies is about $\sqrt{s} - m_Z$. The LEP energy was increased as much as possible by installing as many superconductive radiofrequency cavities as could be fitted in the ring to provide the power necessary to compensate for the increasing synchrotron radiation, up to $\sqrt{s} = 209$ GeV, but the Higgs boson was not found. The final limit on the mass at 95% confidence level is (LEP 2006)

$$M_H > 114.4\,\text{GeV}. \tag{9.95}$$

The search for the Higgs particle will be the main task of the LHC (Large Hadron Collider) that will become operational at CERN at $\sqrt{s} = 14$ TeV, together with the two large experiments ATLAS and CMS. The Higgs boson will be discovered for mass values up to $M_H \approx 1000$ GeV, if it exists. Figure 9.37 shows a typical Higgs production channel.

The cross section of any point-like process decreases with increasing energy as $1/s$ just for dimensional reasons. The Higgs boson production cross section does not escape this rule. On the other hand, the total cross section of the collision of

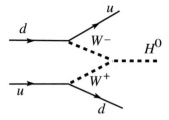

Fig. 9.37.   A Higgs boson production diagram at LHC.

two extended objects, such as protons, does not decrease with increasing energy. Consequently, the fraction of 'interesting' events decreases with increasing energy. At the LHC energy the luminosity must be much higher than the present standard, namely $\mathcal{L} = 10^{33} - 10^{34}$ cm^{-2} s^{-1}. The detectors must be able to resist the huge interaction rate and to select in real time the very rare interesting events. The experimental groups have dedicated great effort in the last fifteen years to design and build ATLAS and CMS, which are now ready to take data.

Problems

9.1. The CHARM2 experiment at CERN studied the reaction $v_\mu e^- \to v_\mu e^-$ using a 'narrow-band beam' with mean energy $\langle E_{v_\mu} \rangle = 24$ GeV. The Super Proton Synchrotron provided two pulses $\Delta t = 6$ ms apart, with a cycle of $T = 14.4$ s. The useful mass of the target detector was $M = 547$ t, the side length of its square useful cross section was $l = 3.2$ m. The target nuclei contained an equal number of protons and neutrons. What was the duty cycle (fraction of time in which interactions take place)? If we want to have an interaction every four pulses on average, how large a neutrino flux Φ is needed? How much intensity I? ($\sigma/E_v = 1.7 \times 10^{-45}$ m^2/GeV)

9.2. Consider an electron of energy $E_e = 20$ GeV detected by the CHARM2 experiment, produced by an elastic $v_\mu e$ scattering. How large can the scattering angle be, at most? Evaluate the accuracy that is necessary in the measurement of the electron direction to verify this to be the case. Can we build the calorimeter using iron?

9.3. To produce a narrow-band v_μ beam, one starts from an almost monochromatic π^+ beam (neglecting the K^+ contamination) and lets the pions decay as $\pi^+ \to \mu^+ + v_\mu$. Assume the pion energy to be $E_\pi = 200$ GeV. Find the neutrino energy in the π^+ rest frame. In the laboratory frame, the neutrino energy depends on the decay angle θ. Find the maximum and minimum neutrino energy. Find the laboratory angle θ of neutrinos emitted in the CM frame at $\theta^* = 50$ mrad.

9.4. Consider the CC cross sections for neutrinos and antineutrinos on nuclei containing the same number of neutrons and protons. Their masses can be neglected in comparison with their energies. Show that the neutrino total cross section is three times larger than that of the antineutrino.

9.5. Give the values of the isospin, its third component and the hypercharge for e_L, $\nu_{\mu R}$, u_L, d_R, and for their antiparticles.

9.6. Give the values of the isospin, its third component and the hypercharge for μ_L^+, τ_R^+, $\bar{t}_L$, $\bar{b}_R$.

9.7. Establish which of the following processes, which might be virtual, are allowed and which forbidden, and give the reasons: $W^- \to d_L + \bar{u}_L$, $W^- \to u_L + \bar{u}_R$, $Z \to W^- + W^+$, $W^+ \to e_R^+ + \bar{\nu}_R$.

9.8. Establish which of the following processes, which might be virtual, are allowed and which forbidden, and give the reasons: $d \to W^- + u_L$, $Z \to e_L^+ + e_L^-$, $W^+ \to Z + W^+$, $W^+ \to e_L^- + \bar{\nu}_R$.

9.9. A 'grand unification' theory assumed the existence at very high energies of a symmetry larger than $SU(2) \otimes U(1)$, namely $SU(5)$. A prediction of the theory, which was experimentally proved to be false, was the value of the weak angle, $\sin^2\theta_W = 3/8$. Find the partial widths of the Z in this hypothesis. Find the value of $\Gamma_{\mu\mu}/\Gamma_h$.

9.10. Calculate the partial and total widths of the Z for $\sin^2\theta_W = 1/4$. Find $\Gamma_{\mu\mu}/\Gamma_h$.

9.11. Evaluate the branching ratio for $W \to e^+\nu_e$.

9.12. Evaluate the ratio g_{Zee}^2/g_{Wev}^2 and the decay rates ratio $\Gamma(Z \to e^+e-)/\Gamma(W \to e^+\nu_e)$.

9.13. Evaluate the ratio $g_{Zuu}^2/g_{Wud'}^2$ and the decay rates ratio $\Gamma(Z \to u\bar{u})/\Gamma(W \to d'\bar{u})$.

9.14. Assume the cross section value $\sigma(\bar{u}d \to e^+\nu_e) = 10\,\text{nb}$ at the W resonance. Evaluate the total $\sigma(\bar{u}d \to \bar{q}q)$ cross section at resonance.

9.15. Assume that the number of neutrinos with mass $\ll M_Z$ is 3, 4 or 5 in turn, without changing anything else. Evaluate for each case the Z branching ratio into $\mu^+\mu^-$ and the ratio $\Gamma_{\mu\mu}/\Gamma_Z$. Evaluate the ratio of the cross sections at the peak for e^+e^- into hadrons for $N = 3$, 4 and 5.

9.16. Calculate the cross section $\sigma(e^+e^- \to \mu^+\mu^-)$ at the Z peak and $\sigma(\bar{u}d \to e^+\nu_e)$ at the W peak.

9.17. A Z is produced in a $\bar{p}p$ collider working at $\sqrt{s} = 540\,\text{GeV}$. The Z moves in the direction of the beams with a momentum $p_Z = 140\,\text{GeV}$. It decays as $Z \to e^+e^-$ with electrons at $90°$ to the beams in the Z rest frame. Calculate the two electron energies in the laboratory frame.

9.18. Consider the Z production at a proton–antiproton collider and its decay channel $Z \to e^+e^-$. The energies of the two electrons as measured by the electromagnetic calorimeters are $E_1 = 60$ GeV and $E_2 = 40$ GeV. The energy resolution is given by $\sigma(E)/E = 0.15/\sqrt{E}$. The measured angle between the tracks is $\theta = 140° \pm 1°$. Find the error on m_Z.

9.19. Consider the mass predictions for the W and Z before their discovery, in round numbers $M_Z = 90$ GeV and $M_W = 80$ GeV. If $\sin^2\theta_W \approx 0.23$ with an uncertainty of 20%, what is the uncertainty on M_W? If we measure M_Z/M_W with 1% uncertainty, what is the uncertainty on $\sin^2\theta_W$?

9.20. At 1 GeV energy the weak charge g is larger than the electric charge $\sqrt{a}$ by $\sqrt{4\pi}/\sin\theta_W \approx 7.4$. Why is the electrostatic force between two electrons at 1 fm distance so large compared to the weak force?

9.21. Consider the $\bar{p}p$ Tevatron collider working at $\sqrt{s} = 2$ TeV. For a Z produced at rest, what are approximately the momentum fractions of the annihilating quark and antiquark? Evaluate in which fraction they are sea quarks. If the Z is produced with a longitudinal momentum of 100 GeV what are approximately the momentum fractions of the quark and of the antiquark?

9.22. LEP2 was designed to study the process $e^+e^- \to W^+W^-$. If the cross section at $\sqrt{s} = 200$ GeV is $\sigma = 17$ pb and the luminosity is $\mathcal{L} = 10^{32} \text{ cm}^{-2}\text{s}^{-1}$, find the number of events produced per day.

9.23. What is the variation, as a percentage, of the Z total width for an additional neutrino type? What is the variation of the peak hadronic cross section?

9.24. Working at the Z with an electron–positron collider and assuming statistical uncertainty only, how many events are needed to exclude at five standard deviations the existence of a fourth neutrino?

9.25. If there were more than three families, the Z would decay into more neutrino–antineutrino channels. Given the existing limits on the masses, however, the charged leptons and quark channels of the new families would be closed. Consequently, the Z width would increase by $\Gamma_{\nu\nu}$ for every extra family. The total width of the W would not increase, because the third family channel $W \to t + \bar{b}$ is already closed (as was established in 1990 when CDF gave the limit $m_t > 90$ GeV). UA1 and UA2 measured the ratio $R = \dfrac{\sigma_W \text{BR}(W \to e\nu_e)}{\sigma_Z \text{BR}(Z \to e^+e^-)}$ simply from the numbers of events observed in the two channels. Both theoretical and experimental systematic uncertainties cancel out in the ratio. Writing the ratio as
$$R = \frac{\Gamma(W \to l\nu_l)}{\Gamma(Z \to \bar{l}l)}\frac{\Gamma_Z}{\Gamma_W}\frac{\sigma_W}{\sigma_Z}$$
one sees that it increases with the number of neutrino types. The experimental upper limit established by joining the UA1 and UA2 data was $R < 10.1$ at 90% confidence level. Evaluate R for 3,

4 and 5 and establish an upper limit for the number of neutrinos. Take $\sigma_W/\sigma_Z = 3.1$.

9.26. Calculate the ratio between the CC cross sections of neutrinos and anti-neutrinos on nuclei with the same numbers of neutrons and protons, considering only the valence quark contributions. Repeat the calculation for NC interactions.

9.27. The largest fraction of matter in our Galaxy is invisible. It might consist of particles similar to neutrinos but much more massive, the 'neutralinos'. Let us indicate them by χ and let m_χ be their mass. According to one theory these particles coincide with their antiparticles. The annihilation processes $\chi + \chi \to \gamma + \gamma$ and $\chi + \chi \to Z^0 + \gamma$ would then take place. Assume initial kinetic energies to be negligible. If a gamma telescope observes a monochromatic signal $E_\gamma = 136\,\text{GeV}$, find m_χ in both hypotheses.

9.28. (1) Consider a pair of quarks with colours R and B and a third quark G. Establish whether the force of the pair on G is attractive or repulsive for each of the combinations $RB + BR$ and $RB - BR$.

Considering quarks of the same colour, establish which of the following processes are allowed or forbidden, giving the reasons: (2) $W^+ \to \bar{b}_L + c_R$; (3) $Z \to \tau_R^+ + \tau_R^-$.

9.29. (1) Establish for each of the following decays whether it is allowed or forbidden, giving the reasons. The left upper label is the colour, the right lower one the chirality: (a) $W^- \to {}^B s_L + {}^B \bar{u}_R$, (b) $W^- \to {}^B d_R + {}^B \bar{u}_L$, (c) $W^- \to {}^R d_L + {}^B \bar{u}_R$, (d) $Z^0 \to {}^G u_L + {}^G \bar{u}_R$, (e) $Z^0 \to {}^G u_R + {}^G \bar{u}_R$, (f) $Z^0 \to {}^G u_R + {}^G \bar{c}_L$, (g) $Z^0 \to {}^G t_L + {}^G \bar{t}_R$.
(2) Is the three-quark status $\frac{1}{\sqrt{6}}[{}^R q({}^B q + {}^G q) + {}^B q({}^G q + {}^R q) + {}^G q({}^R q + {}^B q)]$ bound?

Further reading

Hollik, W. & Duckeck, G. (2000); *Electroweak Precision Tests at LEP*. Springer Tracts on Modern Physics

Rubbia, C. *et al.* (1982); *The search for the intermediate vector bosons. Sci. Am.* March 38

Rubbia, C. (1984); Nobel Lecture, *Experimental Observation of the Intermediate Vector Bosons* http://nobelprize.org/nobel_prizes/physics/laureates/1984/rubbia-lecture.pdf

Salam, A. (1979); Nobel Lecture, *Gauge Unification of Fundamental Forces* http://nobelprize.org/nobel_prizes/physics/laureates/1979/salam-lecture.pdf

Steinberger, J. (1988); Nobel Lecture, *Experiments with High Energy Neutrino Beams* http://nobelprize.org/nobel_prizes/physics/laureates/1988/steinberger-lecture.pdf

Weinberg, S. (1974); *Unified theories of elementary-particle interactions. Sci. Am.* July 50

10

Beyond the Standard Model

10.1 Neutrino mixing

In this chapter we shall discuss the only phenomena that have been discovered beyond the Standard Model. As anticipated in Sections 3.8 and 4.11, neutrinos produced with a certain flavour, v_e, v_μ or v_τ, may be detected at later times with a different flavour. Consequently v_e, v_μ and v_τ are not stationary states with definite mass, which we shall call, v_1, v_2, v_3, but quantum superpositions of them.

Neutrinos change flavour by two mechanisms:

- Oscillation, similar but not identical to the K^0 oscillation. It occurs both in vacuum and in matter. It has been discovered in the v_μ indirectly produced by cosmic rays in the atmosphere. The energies of these neutrinos range from below one GeV to several GeV. The distance from the production to the detection point can be as large as several thousand kilometres.
- Transformation in matter, which is a dynamical phenomenon due to the interaction of the v_es with the electrons, similar to the refractive index of light. The phenomenon can most easily be observed if the flight length is large and if the density is high, as in a star. It has been discovered in the v_es coming from the Sun, which have energies of several MeV.

At the mentioned energy scales both phenomena take place on very long characteristic time scales. The corresponding flight lengths are much larger than those that were available on neutrino beams produced at accelerators. This is why they have been discovered in underground laboratories, designed for the study of spontaneously occurring rare events.

From the historical point of view, in 1957 Bruno Pontecorvo published the idea that the neutrino–antineutrino system might oscillate in analogy with the $K^0 \bar{K}^0$ oscillation (Pontecorvo 1957). At that time only one neutrino species was known, actually it had just been discovered. He returned to the idea in 1967, when the second neutrino flavour was known, advancing the hypothesis of the oscillation between flavours (Pontecorvo 1967). Analysing the experimental data he reached the conclusion that ample room was left for lepton number violation. Discussing possible experiments, he concluded that the ideal neutrino source was the Sun. If the oscillation phenomenon existed, only half of the expected electron neutrino flux on Earth would be observed. At that time Pontecorvo was not aware of two articles published by two Japanese theoretical groups in 1962, contemporary to the discovery of the second neutrino. First the Kyoto group (Katayama *et al.* 1962) and then the Nagoya group (Maki *et al.* 1962) had advanced the hypothesis of neutrino mixing, without however mentioning oscillations.

The neutrino flavour transformation in matter was studied by Wolfenstein in 1978 (Wolfenstein 1978) and by Mikheyev and Smirnov in 1985 (Mikheyev & Smirnov 1985). It is called the MSW effect.

Neither oscillation nor flavour change in matter can happen if neutrinos are massless. Consequently, the two mentioned phenomena contradict the Standard Model in two ways: non-conservation of the lepton flavour and non-zero neutrino masses.

We now summarise the present status of our knowledge, leaving the experimental proofs of these conclusions for the following sections.

The definite flavour states v_e, v_μ and v_τ are obtained from the stationary states v_1, v_2 and v_3 with a transformation, which we assume to be unitary. We indicate the masses, which are defined for the stationary states, by m_1, m_2 and m_3. The transformation, analogous to that of the quarks, is

$$
\begin{pmatrix} v_e \\ v_\mu \\ v_\tau \end{pmatrix} = \begin{pmatrix} U_{e1} & U_{e2} & U_{e3} \\ U_{\mu 1} & U_{\mu 2} & U_{\mu 3} \\ U_{\tau 1} & U_{\tau 2} & U_{\tau 3} \end{pmatrix} \begin{pmatrix} v_1 \\ v_2 \\ v_3 \end{pmatrix}.
\tag{10.1}
$$

We can express the transformation in terms of three rotations, of angles that we shall again call θ_{12}, θ_{23} and θ_{13}, and of phase factors. If neutrinos are Dirac particles, as assumed in the Standard Model, all but one of the phase factors can be absorbed, as in the case of quarks, in the wave functions of the states. However, neutrino and antineutrino might be two states of the same particle, namely 'Majorana particles'. In this case two more phases, which we shall call Majorana phases ϕ_1 and ϕ_2, are physically observable. In conclusion, writing $c_{ij} = \cos\theta_{ij}$, and $s_{ij} = \sin\theta_{ij}$, the transformation matrix is

$$
U = \begin{pmatrix} 1 & 0 & 0 \\ 0 & c_{23} & s_{23} \\ 0 & -s_{23} & c_{23} \end{pmatrix} \begin{pmatrix} c_{13} & 0 & s_{13}e^{-i\delta} \\ 0 & 1 & 0 \\ -s_{13}e^{i\delta} & 0 & c_{13} \end{pmatrix} \begin{pmatrix} c_{12} & -s_{12} & 0 \\ s_{12} & c_{12} & 0 \\ 0 & 0 & 1 \end{pmatrix} \begin{pmatrix} 1 & 0 & 0 \\ 0 & e^{i\phi_1} & 0 \\ 0 & 0 & e^{i\phi_2} \end{pmatrix}.
\tag{10.2}
$$

Majorana phases are irrelevant for the oscillation and matter effects. They are observable in the neutrinoless double beta decay, which we shall mention in Section 10.5.

As we shall discuss in the following sections, a number of experiments have measured different observables, such as fluxes and energy spectra, relevant for the oscillation phenomena. A global fit to these measurements (see e.g. Fogli *et al.* 2006) allows us to extract the mixing angles and the differences between the squares of the masses. Specifically, the information on θ_{12} is mainly due to solar neutrinos and reactor antineutrinos, that on θ_{23} to atmospheric neutrinos, reactor antineutrinos and accelerator neutrinos. Of the third angle, θ_{13}, we know only an upper limit. The values are

$$\theta_{12} = 33.9° \pm 1.6°$$
$$\theta_{23} = 45° \pm 3° \qquad (10.3)$$
$$|\theta_{13}| < 9°.$$

Unlike the case of quarks, the neutrino mixing angles are large. Figure 10.1 shows how different the stationary states are from the flavour states.

The phase factor, provided that $\delta \neq 0$ and $\neq \pi$, induces novel CP violation effects, which have not yet been observed.

We do not know the absolute values of the masses but we have measured the differences between their squares as follows.

The characteristic time of the matter conversion phenomenon, observed in the solar neutrinos, depends on the difference between the squares of the masses of the implied eigenstates, which happen to be ν_1 and ν_2. We consequently obtain, in

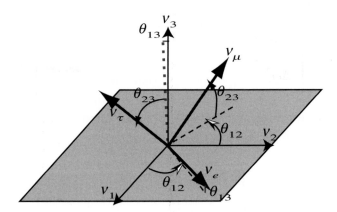

Fig. 10.1.   The rotations of the neutrino mixing.

value and sign, the difference

$$\delta m^2 \equiv m_2^2 - m_1^2. \tag{10.4}$$

In the mass spectrum these two states are very close, while the third, ν_3, is farther away. It is then convenient to define the second relevant square-mass difference as

$$\Delta m^2 \equiv m_3^2 - \frac{m_1^2 + m_2^2}{2}. \tag{10.5}$$

The measured 'atmospheric' oscillation period is inversely proportional to the absolute value of this quantity. The above-mentioned fitting procedure gives the values

$$\delta m^2 = 79 \pm 4\,\text{meV}^2$$
$$\left| \Delta m^2 \right| = 2600 \pm 180\,\text{meV}^2. \tag{10.6}$$

Figure 10.2 shows schematically the neutrino square-mass spectrum, consisting of a singlet and a doublet.

We do not know either the absolute scale or whether the mass of the singlet is larger or smaller than that of the doublet. The so-called hierarchy parameter a is small

$$a \equiv \left| \delta m^2 / \Delta m^2 \right| = 0.03. \tag{10.7}$$

This circumstance decouples the two phenomena, at least within the sensitivity of present experiments.

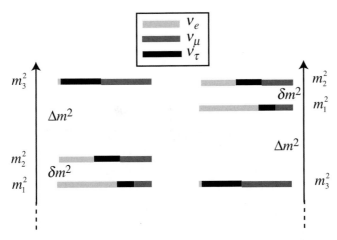

Fig. 10.2. Neutrino square-mass spectrum. The flavour contents of the eigenstates are also shown.

Let us now make explicit a definition that we have already used implicitly. The eigenstate names v_1, v_2 and v_3 are ordered in decreasing order of v_e content. As one can easily see from the values (10.3) of the mixing angles, v_1 is about 70% v_e and for the rest half v_μ and half v_τ. The v_2 contains about one-third of each flavour. The v_3 is almost half v_μ and half v_τ; the fraction of v_e is $|\theta_{13}|^2$, which we know to be small, but not how small.

10.2 Neutrino oscillation

The neutrino oscillation phenomenon is similar to that of neutral kaons, but with two important differences: the K eigenstates are two, those of neutrinos are three; in the K system the flavour mixing is almost maximal while in the neutrino system it is not.

Let us start by considering, for the sake of simplicity, an oscillation between two neutrino species. Actually, this is the situation of the 'atmospheric oscillation', which involves with good approximation only the flavour states v_μ and v_τ and the stationary states v_2 and v_3. The mixing matrix is

$$\begin{pmatrix} v_\mu \\ v_\tau \end{pmatrix} = \begin{pmatrix} U_{\mu 2} & U_{\mu 3} \\ U_{\tau 2} & U_{\tau 3} \end{pmatrix} \begin{pmatrix} v_2 \\ v_3 \end{pmatrix} \tag{10.8}$$

which we can think of as a rotation in a plane, as in the two-quark case

$$\begin{pmatrix} v_\mu \\ v_\tau \end{pmatrix} = \begin{pmatrix} \cos\theta_{23} & \sin\theta_{23} \\ -\sin\theta_{23} & \cos\theta_{23} \end{pmatrix} \begin{pmatrix} v_2 \\ v_3 \end{pmatrix}. \tag{10.9}$$

Figure 10.3(a), which is the same as Fig. 7.19, recalls the Cabibbo rotation, which is of about 13°. In general, a mixing angle can have any value between 0° and 90°. The rotated and non-rotated axes are close to each other both for small angles and for angles near 90°. The difference reaches a maximum at 45°, which is called

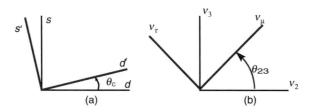

Fig. 10.3.   Flavour rotation for the (1, 2) quark families and for the (2, 3) neutrino families.

maximal mixing. Actually, one of the neutrino mixing angles, the angle involved in the atmospheric oscillation θ_{23}, is equal to $45°$ within the errors, as shown in Fig. 10.3(b).

By explicitly writing Eq. (10.9) we have

$$\begin{aligned}
|v_\mu\rangle &= \cos\theta_{23}|v_2\rangle + \sin\theta_{23}|v_3\rangle \\
|v_\tau\rangle &= -\sin\theta_{23}|v_2\rangle + \cos\theta_{23}|v_3\rangle.
\end{aligned} \tag{10.10}$$

In the case of maximal mixing, $\theta_{23} = 45°$, it becomes

$$\begin{aligned}
|v_\mu\rangle &= \frac{1}{\sqrt{2}}|v_2\rangle + \frac{1}{\sqrt{2}}|v_3\rangle = \frac{1}{\sqrt{2}}(|v_2\rangle + |v_3\rangle) \\
|v_\tau\rangle &= -\frac{1}{\sqrt{2}}|v_2\rangle + \frac{1}{\sqrt{2}}|v_3\rangle = \frac{1}{\sqrt{2}}(-|v_2\rangle + |v_3\rangle).
\end{aligned} \tag{10.11}$$

This transformation is equal to that of the K system neglecting $\mathcal{CP}$ violation, as one can see by recalling Eq. (8.13). However, the analogy with neutrinos ends here because neutrino mixing has nothing to do with $\mathcal{CP}$ violation.

Let us now consider a beam of neutrinos, all with the same momentum **p**. The energies of the two stationary states are not equal due to the difference of the masses. Taking into account that the masses are very small compared to these energies and letting E be the average of the two energies, we can write with very good approximation

$$E_i = \sqrt{p^2 + m_i^2} \approx p + \frac{m_i^2}{2p} \approx p + \frac{m_i^2}{2E}. \tag{10.12}$$

The evolution of the two states in vacuum is given by the Schrödinger equation

$$i\frac{d}{dt}\begin{pmatrix} v_2(t) \\ v_3(t) \end{pmatrix} = H\begin{pmatrix} v_2(t) \\ v_3(t) \end{pmatrix} \tag{10.13}$$

where the Hamiltonian is diagonal

$$H = \begin{pmatrix} E_2 & 0 \\ 0 & E_3 \end{pmatrix} \approx p + \begin{pmatrix} \frac{m_2^2}{2E} & 0 \\ 0 & \frac{m_3^2}{2E} \end{pmatrix}. \tag{10.14}$$

The evolution of the two flavour states is

$$i\frac{d}{dt}\begin{pmatrix} v_\mu(t) \\ v_\tau(t) \end{pmatrix} = H'\begin{pmatrix} v_\mu(t) \\ v_\tau(t) \end{pmatrix} = UHU^+\begin{pmatrix} v_\mu(t) \\ v_\tau(t) \end{pmatrix}. \tag{10.15}$$

We easily find that

$$H' = p + \frac{m_2^2 + m_3^2}{4E} + \frac{\Delta m^2}{4E} \begin{pmatrix} -\cos 2\theta_{23} & \sin 2\theta_{23} \\ \sin 2\theta_{23} & \cos 2\theta_{23} \end{pmatrix}. \tag{10.16}$$

Here we make an observation, which will be useful in the following. The mixing angle, which we call generically θ, is given by the ratio between the non-diagonal element (the two are equal) and the difference between the two diagonal ones, namely by

$$\tan 2\theta = \frac{2H'_{12}}{H'_{22} - H'_{11}}. \tag{10.17}$$

Let $|\nu_\mu(0)\rangle$ and $|\nu_\tau(0)\rangle$ be the amplitudes at the initial time $t=0$ of the definite flavour states. The time evolution of the stationary states is, $|\nu_2(t)\rangle = |\nu_2(0)\rangle e^{-iE_2 t}$ and $|\nu_3(t)\rangle = |\nu_3(0)\rangle e^{-iE_3 t}$. Consequently, writing $c = \cos\theta$ and $s = \sin\theta$, we have

$$
\begin{aligned}
|\nu_\mu(t)\rangle &= c|\nu_2(0)\rangle e^{-iE_2 t} + s|\nu_3(0)\rangle e^{-iE_3 t} \\
&= \left[s^2 e^{-iE_2 t} + c^2 e^{-iE_3 t}\right]|\nu_\mu(0)\rangle + sc\left[e^{-iE_3 t} - e^{-iE_2 t}\right]|\nu_\tau(0)\rangle \\
|\nu_\tau(t)\rangle &= -s|\nu_2(t)\rangle e^{-iE_2 t} + c|\nu_3(0)\rangle e^{-iE_3 t} \\
&= \left[c^2 e^{-iE_3 t} + s^2 e^{-iE_2 t}\right]|\nu_\tau(0)\rangle + sc\left[e^{-iE_3 t} - e^{-iE_2 t}\right]|\nu_\mu(0)\rangle.
\end{aligned}
$$

Let us consider the case, which occurs for cosmic ray and accelerator neutrinos, of an initially pure ν_μ system. At the time t we can observe the 'appearance' of ν_τ with a probability given by

$$
\begin{aligned}
P(\nu_\mu \to \nu_\tau, t) &= \left|\langle \nu_\tau(t)|\nu_\mu(0)\rangle\right|^2 = c^2 s^2 \left|e^{-iE_2 t} - e^{-iE_3 t}\right|^2 \\
&= c^2 s^2 \left|2ie^{-i\frac{E_2 + E_3}{2}t}\sin\frac{E_3 - E_2}{2}t\right|^2.
\end{aligned}
$$

Taking into account that the energy of a particle, if much larger than the mass, is $E \approx p + \frac{2m^2}{E}$, we obtain

$$P(\nu_\mu \to \nu_\tau, t) = 4c^2 s^2 \sin^2\frac{E_3 - E_2}{2}t = \sin^2 2\theta \sin^2\frac{\Delta m^2}{4E}t. \tag{10.18}$$

Notice that the time t we are considering is the proper time, namely the time measured in the neutrino rest frame. In practice, we observe the phenomenon as a function of the distance L travelled by neutrinos in the reference frame of the laboratory. Writing Eq. (10.18) as a function of L, we have, with E in GeV, L in km and Δm in eV

$$P(v_\mu \rightarrow v_\tau, t) = \sin^2 2\theta \sin^2 \left[1.27\Delta m^2 \left(\frac{L}{E} \right) \right]. \qquad (10.19)$$

This is the probability measured by the 'appearance' experiments, which consist of a detector at a distance L from the source capable of detecting the possible presence of the initially non-existent flavour v_τ. 'Disappearance' experiments are another possibility. The initial v_μ flux and energy spectrum must be measured, or at least calculated. The expected v_μ flux and spectrum at the detector, which is at distance L, are then calculated. The ratio of the measured flux to the calculated flux gives the disappearance probability. The survival probability is the complement to unity of the appearance probability

$$P(v_\mu \rightarrow v_\mu, t) = 1 - \sin^2 2\theta \sin^2 \left[1.27\Delta m^2 \left(\frac{L}{E} \right) \right]. \qquad (10.20)$$

An analogy could be useful here. The initially pure v_μ state with definite momentum is the superposition of two monoenergetic stationary states of energies E_2 and E_3, analogous to a dichromatic signal, the sum of two monochromatic ones with angular frequencies, say ω_2 and ω_3. Their mixing is maximal if their amplitudes are equal, a situation shown in Fig. 10.4.

Initially the two monochromatic components are in phase, but their phase difference increases with time and the two components reach phase opposition at $t = 1/|\omega_3 - \omega_2|$, only to return in phase at $2/|\omega_3 - \omega_2|$, etc. The modulated

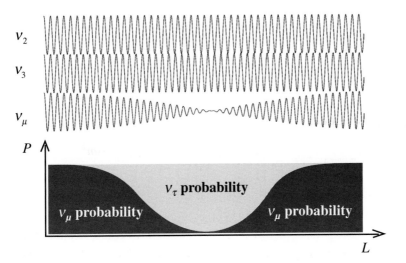

Fig. 10.4.   Summing two monochromatic probability waves.

amplitude varies in time as $\cos\left(\frac{\omega_3 - \omega_2}{2} t\right)$. The probability of observing a ν_μ, which is proportional to the square of the amplitude, namely to $\cos^2\left(\frac{\omega_3 - \omega_2}{2} t\right)$, is 100% initially, decreases to zero (for the first time) at time $t = 1/|\omega_3 - \omega_2|$, then increases again, etc. The probability of observing the new flavours varies as $\sin^2\left(\frac{\omega_3 - \omega_2}{2} t\right)$, according to Eq. (10.19). Notice that the evolution in time of both probabilities depends on the absolute value of the difference between the two angular frequencies, i.e. of the squares of the masses, not on its sign.

If the mixing is not maximal the original flavour never disappears completely and the appearance probability maximum $\sin^2 2\theta_{23}$ is smaller than one. It becomes smaller and smaller as the mixing angle is farther from 45°, either in the first or in the second octant. Indeed, as shown by Eq. (10.19) and Eq. (10.20), the vacuum oscillations do not depend on the sign of $\pi/4 - \theta_{23}$.

In practice no neutrino source is monochromatic. This can be partially compensated by measuring neutrino energy, with a certain energy resolution. In any case neutrino energy is known within a smaller or larger spread. Let us see how a disappearance experiment is affected by the neutrino energy spread.

We try to illustrate that in Fig. 10.5, taking maximal mixing as an example. The survival probabilities of the monochromatic components are in phase at $t = 0$ and remain such at short times, say in the first quarter period, then they gradually dephase and average out at constant value, sooner or later depending on their degree of monochromaticity. This average depends on the mixing angle and is equal to 1/2 in our example of maximal mixing.

Let us now go to the real situation of three neutrinos. We again assume the system to be composed at $t = 0$ purely of neutrinos of the same flavour and all with the same momentum. Let a detector capable of identifying the neutrino flavour be located at distance L. With three possible initial flavours and three possible detected flavours we have in total nine possibilities, which are not all independent due to $\mathcal{CPT}$ invariance. We shall not present here the calculation of

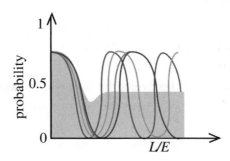

Fig. 10.5. Probability of observing the initial flavour for maximal mixing and with energy spread.

the nine probabilities, which can be easily done starting from the mixing matrix. We only observe that in general there are three oscillation frequencies, corresponding to the three square-mass differences. The oscillation probability between a given pair of flavours is a sum of oscillating terms at these frequencies with maximum excursions that are functions, different in each case, of the three mixing angles.

In practice, two circumstances considerably simplify the situation.

1. Two square-mass differences are equal to all practical effects: $m_3^2 - m_2^2 \approx m_3^2 - m_1^2 \approx \Delta m^2$, and consequently there are only two oscillation periods.
2. The two oscillation periods are very different, as we see from Eq. (10.7). Consequently, the experiments sensitive to the oscillation with shorter period (the atmospheric one) do not see the longer period (solar) oscillation because it has not yet started. On the other hand, the experiments sensitive to the longer period do in practice average the signal on times that are much larger than the shorter period and are not sensitive to the first oscillation.

These approximations are sufficient for the sensitivity of the present experiments. In particular, the discussion made above for two flavours applies directly to atmospheric neutrino oscillation.

Let us now focus on the 'atmospheric' oscillation. It was discovered by Super-Kamiokande in 1998 (Fukuda *et al.* 1998) in the Kamioka underground observatory in Japan. We saw in Section 1.11 that muon and electron neutrinos are present amongst the decay products of the hadrons produced by the cosmic ray collisions with atomic nuclei in the atmosphere. The oscillation probabilities between all flavour pairs have the same dependence on the flight length to energy ratio L/E

$$P(v_x \to v_y, t) = A(v_x \to v_y) \sin^2 \left[1.27 \Delta m^2 (L/E) \right]. \qquad (10.21)$$

The constant $A(v_x \to v_y)$ is the maximum of the probability oscillation between flavours v_x and v_y. Let us see the values of the constants, using our knowledge of the mixing angles, Eq. (10.4).

The specific phenomenon discovered by Super-Kamiokande is the muon neutrino disappearance. For this phenomenon the maximum probability is

$$A(v_\mu \to v_x) = \sin^2(2\theta_{23}) \cos^2(\theta_{13}) \left(1 - \sin^2 \theta_{23} \cos^2 \theta_{13} \right) \approx \frac{1}{2} \qquad (10.22)$$

where, in the last member, we have taken $\cos^2 \theta_{13} \approx 1$ and $\sin^2 \theta_{23} \approx 1/2$.

The missing muon neutrinos appear in part as electron neutrinos, in part as tau neutrinos. The corresponding probabilities at the maximum are

$$A\left(\nu_\mu \to \nu_e\right) = \sin^2(\theta_{23})\sin^2(2\theta_{13}) \approx 2\theta_{13}^2 \tag{10.23}$$

which is very small, and

$$A\left(\nu_\mu \to \nu_\tau\right) = \sin^2(2\theta_{23})\cos^4(\theta_{13}) \approx 1. \tag{10.24}$$

Coming now to the experiment, we recall from Section 1.11, that Super-Kamiokande is a large water Cherenkov detector with 22 500 t fiducial mass, located in the Kamioka underground observatory under the Japanese Alps. Atmospheric neutrinos are detected by their charged-current scattering processes

$$\nu_\mu + N \to \mu + N \qquad \nu_e + N \to e + N. \tag{10.25}$$

In both cases a unique ring signalling the charged lepton is observed. The ring is sharp in the case of the muon corresponding to its straight track, like the ring in Fig. 1.17, while it is diffuse in the case of the electron, which has a track that scatters due to bremsstrahlung. This allows the single-ring events to be classified as 'e-type' or 'μ-type'. Clearly, the charge remains unknown. Neutrino energies range from a few hundred MeV to several GeV. At these values the differential cross sections are strongly forward peaked and consequently the measured final lepton direction is almost the same as the neutrino one. Knowing the incident neutrino direction we also know the distance it has travelled from its production point in the atmosphere, as illustrated in Fig. 10.6. Keep in mind that neutrinos pass through the Earth without absorption.

Calling θ the angle of the neutrino direction with the zenith, the flight length varies from about 10 km for $\theta = 0$ to more than 12 000 km for $\theta = \pi$.

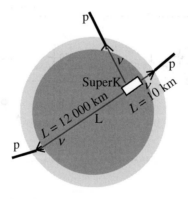

Fig. 10.6.   Flight lengths of atmospheric neutrinos.

The detector gives a rough measurement of the charged lepton energy, which is statistically correlated to the incident neutrino energy. Both e-type and μ-type events are then divided into a low-energy sample, less than about one GeV, and a high-energy sample, up to several GeV.

An essential component of the experiment is the calculation, based on a number of measurements, of the muon and electron neutrino fluxes as expected in the absence of oscillations, as functions of the energy and of the zenith angle.

Having defined four categories of events, 'e-type' and 'μ-type', low and high energy, Fig. 10.7 shows the distributions of the cosine of the zenith angle for each of them. The dotted curves are the predictions in the absence of oscillations; the continuous curves were obtained assuming oscillations and fitting to the data with Δm^2 and θ_{23} as free parameters. The fitting procedure determines their best values.

We observe that the ν_e data do not show signs of oscillations, which implies that θ_{13} is small. Looking at the high-energy ν_μs we see that at small zenith angles,

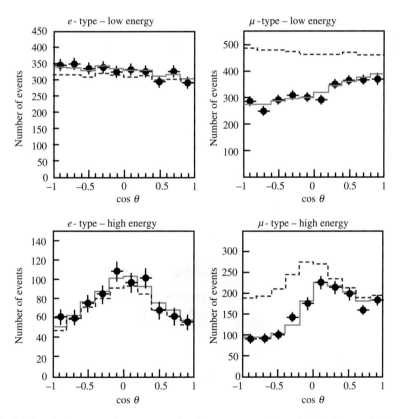

Fig. 10.7.   Electron and muon neutrino fluxes vs. zenith angle. (Ashie *et al.* 2005)

corresponding to short flight lengths, all of them reach the detector; however, above a certain distance the muon neutrino flux is one-half of the expected flux. From Fig. 10.6 we understand that the value of that distance determines Δm^2, while the value 1/2 of the reduction factor says that $\theta_{23} \approx \pi/4$. Finally, the low-energy ν_μs do oscillate even at small distances.

The ν_μ disappearance phenomenon was confirmed by two other experiments on atmospheric neutrinos, MACRO at Gran Sasso and SOUDAN2 in the USA, and by two experiments on accelerator neutrino beams on long base lines. The latter experiments used two detectors, one near to the source to measure the initial neutrino flux and its energy spectrum, and one far away to measure the surviving flux.

The first experiment was K2K with the neutrino source at the proton synchrotron of the KEK laboratory at Tsukuba in Japan, providing a beam with 1.5 GeV average energy. The far detector was Super-Kamiokande at a distance of 250 km (Ahn *et al.* 2003).

In the second experiment the neutrino source was the NuMi beam at the main injector proton accelerator at Fermilab. The far detector was MINOS, which is a tracking calorimeter at a distance of 735 km. Since it is a disappearance experiment, MINOS can work at low neutrino energies, in practice at 2–3 GeV, optimised for being near to the first oscillation maximum (Michael *et al.* 2006).

Now consider θ_{13}, of which we know only that it is small. The best upper limit on θ_{13}^2 is due to the CHOOZ experiment (Apollonio *et al.* 1999) that searched for the disappearance of electron antineutrinos produced by two nuclear power reactors in France. Since the energies are of a few MeV, the oscillation maximum is at a distance of 1–2 km, at which the detector was located. The oscillation probability at its maximum is

$$A(\nu_e \to \nu_x) = \sin^2(2\theta_{13}) \approx 4\theta_{13}^2. \tag{10.26}$$

CHOOZ measured an antineutrino flux equal to the expected flux, as computed from the knowledge of the operational characteristics of the reactors. This led to a limit on $|\theta_{13}|$ or to its difference from 90°. This ambiguity is a consequence of the above-noted fact that the vacuum oscillation is symmetric around $\pi/4$. The matter effects observed in the solar neutrinos chose the first solution. The limit on $|\theta_{13}|$ is given in Eq. (10.3).

All the above arguments would lead to the conclusion that the missing muon neutrinos, which do not appear as electron neutrinos, should appear as tau neutrinos. This fact is implied in Eq. (10.26). However, this conclusion has not yet been experimentally tested. The test is the principal goal of the CERN-INFN project, called CNGS (CERN Neutrinos to Gran Sasso). A new ν_μ beam has been constructed at the CERN Super Proton Synchrotron aimed through the Earth's crust to the Gran Sasso Laboratory at 737 km distance (CNGS 1998). The OPERA detector (OPERA 2000) will search there for the ν_τ appearance, as identified by the reaction

$$\nu_\tau + N \rightarrow \tau + N'. \tag{10.27}$$

The project has been optimised for tau neutrino appearance, implying the following main characteristics. The neutrino energy must be high enough for reaction (10.27) to be substantially above threshold, in practice more than 10 GeV. This implies that the 737 km flight length is small compared to the distance of the oscillation maximum, corresponding to a small expected number of tau neutrinos. Consequently the detector must have a large mass and at the same time an extremely fine granularity to be able to distinguish the production and decay vertices of the τ. In practice, a micrometre-scale resolution over a mass of 2000 t is obtained by OPERA with a combination of emulsion and electronic techniques.

10.3 Flavour transition in matter

The Sun is a main-sequence star in the stable hydrogen-burning stage. Its density is very high in the centre, $\rho_0 = 10^5$ kg/m³, and gradually diminishes towards the surface. The overall reaction that produces 95% of the energy is the fusion of four protons into a helium nucleus

$$4p \rightarrow \text{He}^{++} + 2\nu_e + 2e^+. \tag{10.28}$$

The two positrons immediately annihilate with two electrons. Therefore the energy generation process is

$$4p + 2e^- \rightarrow \text{He}^{++} + 2\nu_e + 26.1\,\text{MeV}. \tag{10.29}$$

The basic elementary reaction is the 'pp fusion'

$$p + p \rightarrow {}^2\text{H} + e^+ + \nu_e. \tag{10.30}$$

The thermonuclear reactions take place in the central part, the core, of the star where the thermal energy is of the order of tens of keV. These energies are much smaller than the Coulomb barriers of the interacting nuclei and consequently the cross sections are very small, but large enough for the above-mentioned reactions to proceed.

Only a small part of the energy released by reaction (10.29) is taken by neutrinos, while the largest fraction is transported by photons. The original MeV-energy photons interact with the solar medium producing other photons of decreasing energies and in increasing number. The energy leaves the surface of the

Sun as light only several thousand years after it was produced. While the light of the Sun and of the stars is a surface phenomenon, neutrinos reach us directly from the centre of the Sun, without any absorption. However, even if the Sun medium is transparent to neutrinos, something happens to them.

Observation of solar neutrinos gives fundamental information both about stellar structure and evolution and about the properties of neutrinos, due to the wide range of matter densities in the Sun and to its large distance from Earth. Clearly, to study neutrino properties, one needs to know their flux and energy spectrum at the source. Today we have a reliable 'solar standard model' (SSM) due mainly to 40 years of work by John Bahcall and collaborators (Bahcall *et al.* 1963, 2005). Figure 10.8

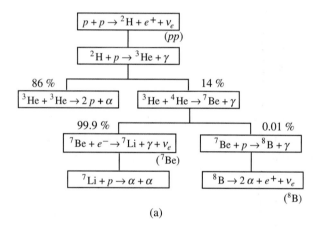

(a)

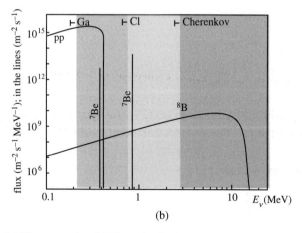

(b)

Fig. 10.8.   (a) The *pp* cycle. (b) The principal components of the neutrino energy spectrum. (From Bahcall *et al.* 2005) The sensitive regions of different experimental techniques are also shown.

shows the principal components of the *pp* cycle and the corresponding contributions to the neutrino energy spectrum, in a simplified form. We have neglected the contribution of the carbon–nitrogen–oxygen (CNO) cycle, which is small in the Sun.

We see in the figure that the largest fraction of the energy flux is due to the elementary reaction (10.30). This component of the flux is obtained from the measured luminous flux and is almost independent of the details of the solar model. However, it is the most difficult experimentally due to its very low neutrino energies, which are below 420 keV.

Two processes produce higher-energy neutrinos. The first process gives the so-called 'beryllium neutrinos' through the reaction

$$^7\text{Be} + e^- \rightarrow\, ^7\text{Li} + \gamma + \nu_e. \tag{10.31}$$

The flux is dichromatic, with the principal line at 0.86 MeV. The second process gives the 'boron neutrinos'

$$^8\text{B} \rightarrow 2a + e^+ + \nu_e. \tag{10.32}$$

This is the highest-energy component with a spectrum reaching 14 MeV (neglecting much weaker components). As such it is the least difficult to detect. Notice that the boron is produced by the beryllium via the reaction

$$^7\text{Be} + p \rightarrow\, ^8\text{B} + \gamma. \tag{10.33}$$

The latter processes make very small contributions to the electromagnetic energy flux we measure and, as a consequence, our knowledge of the corresponding fluxes is heavily based on the solar model and its parameters.

Example 10.1 Knowing that the solar constant, i.e. the flux of electromagnetic energy from the Sun on the Earth's surface, is 1.3 kW/m², evaluate the total neutrino flux.

The energy produced by reaction (10.29) transported by photons is 26.1 MeV for every two neutrinos. The energy per neutrino is $26.1/2 = 13.05$ MeV $= 2.1 \times 10^{-12}$ J.

The neutrino flux is then

$$\Phi_\nu = \frac{\left(1.3 \times 10^3 \text{ J m}^{-2} \text{ s}^{-1}\right)}{\left(2.1 \times 10^{-12} \text{ J}\right)} = 6.2 \times 10^{14} \text{ m}^{-2} \text{ s}^{-1}.$$

We now observe that all the reactions produce electron neutrinos and that they do so in a very high density medium. Neutrinos then cross a decreasing density medium before reaching the surface of the Sun.

The neutrino stationary states in matter are not v_1, v_2 and v_3 and the corresponding mass eigenvalues are not m_1, m_2 and m_3. We shall call them $\tilde{v}_i$ and $\tilde{m}_i$ respectively. The effect is proportional to the scattering amplitude in the forward direction, similarly to the refractive index of light in a medium. As such it is proportional to the Fermi constant and is sizeable, whilst the absorption depends on the cross section that is proportional to the square of the Fermi constant and is negligible. Notice by the way that the commonly given statement that neutrinos cross matter 'without seeing it' is not strictly true. They are not absorbed, but they change speed, just as light does in a medium. A detailed discussion of the physics of neutrinos in matter can be found in the book by Mohapatra and Pal (2004). We shall give here only the essential elements.

We can parameterise the average interaction of neutrinos with the particles of the medium as an effective potential. All neutrinos interact with electrons and quarks by NC weak interactions, independently of their flavour. Only electron neutrinos interact with electrons and quarks also by CC weak interactions. Consequently their potential $V_e(r)$ is different from that of the other flavours $V_{\mu,\tau}(r)$. It can be shown that this difference is

$$\Delta V(r) \equiv V_e(r) - V_{\mu,\tau}(r) = \sqrt{2} G_F N_e(r) \tag{10.34}$$

where $N_e(r)$ is the electron number density at distance r from the centre of the Sun.

In the energy range of the solar neutrinos, muon and tau neutrinos are indistinguishable since their interactions are identical since they are of neutral currents only. If we also assume $\theta_{13} = 0$ in a first approximation, we reduce the problem to that of two neutrino species, v_e and, say, v_a, where the latter is a superposition of v_μ and v_τ that we do not need to define.

Let us start by considering the evolution of the system in a uniform density medium. It is given by

$$i \frac{d}{dt} \begin{pmatrix} v_e(t) \\ v_a(t) \end{pmatrix} = H_m \begin{pmatrix} v_e(t) \\ v_a(t) \end{pmatrix}. \tag{10.35}$$

Without giving the proof (Mohapatra & Pal 2004), we say that the Hamiltonian is the sum of a diagonal term H_{diag}, which we do not need to write down, and of a non-diagonal one, which is the important one. Its expression is

$$H_m = H_{\text{diag}} + \begin{pmatrix} -\dfrac{\delta m^2}{4E} \cos 2\theta_{12} + \sqrt{2} G_F N_e & \dfrac{\delta m^2}{4E} \sin 2\theta_{12} \\ \dfrac{\delta m^2}{4E} \sin 2\theta_{12} & \dfrac{\delta m^2}{4E} \cos 2\theta_{12} \end{pmatrix}. \tag{10.36}$$

Notice that the relevant mixing angle is θ_{12}. Recalling Eq. (10.17) we see that the mixing is determined in matter by an 'effective mixing angle' $\theta_{12,m}$ given by

$$\tan 2\theta_{12,m} = \frac{2H_{m,12}}{H_{m,22} - H_{m,11}} = \frac{\delta m^2 \sin 2\theta_{12}}{\delta m^2 \cos 2\theta_{12} - A} \tag{10.37}$$

where

$$A = 2\sqrt{2}G_F N_e E. \tag{10.38}$$

Notice that A is proportional to the Fermi constant, as anticipated, to the neutrino energy and to the electron density. The effect of matter becomes dramatic for a particular value of the electron density, i.e.

$$N_e = \frac{1}{E} \frac{\delta m^2 \cos 2\theta_{12}}{2\sqrt{2}G_F}. \tag{10.39}$$

At this density, Eq. (10.37) diverges, meaning that the effective mixing angle becomes $\theta_{12,m} = \pi/4$. Consequently, the mixing becomes maximal even if the (vacuum) mixing angle is small. This resonance condition can be reached if the following necessary conditions are satisfied: (1) $\cos 2\theta_{12} > 0$, i.e. θ_{12} is in the first octant; (2) $\delta m^2 > 0$, i.e. $m_2 > m_1$. In particular, there is no resonance for $\theta_{12} = \pi/4$.

To understand the phenomenon we must consider the stationary states of the system in the medium ($\tilde{\nu}_i$) and the corresponding eigenvalues ($\tilde{m}_i$), the effective masses, by diagonalising the Hamiltonian (10.36). Skipping the calculation, we give the result graphically in Fig. 10.9.

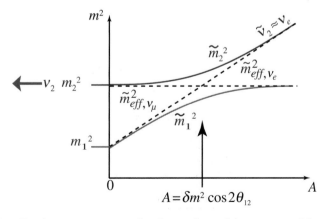

Fig. 10.9. Continuous curves are the eigenvalues of the squares of the masses as functions of A. Dotted lines are the effective squared masses of the flavour states.

We observe that, at high enough densities, electron neutrinos have an effective mass, due to their interaction with the electrons of the medium, larger than the other flavours. The opposite is true at low densities and, in particular, in vacuum. We have a level-crossing phenomenon, a situation also met in other fields of physics.

As observed by Mikheyev and Smirnov (Mikheyev & Smirnov 1985), the crossing of the resonance can induce a change of the neutrino flavour. The eigenstates and their eigenvalues in a non-uniform medium differ from point to point and consequently the description of the propagation of the system is not simple. Actually, the evolution of the system has different characteristics, depending on the various physical quantities of the problem.

However, Nature has chosen the simplest situation in the Sun (but we did not know that at the beginning), and we shall limit our discussion to this. Indeed, for the actual values of the energy, mixing angle and δm^2 the resonance is crossed 'adiabatically', meaning that the electron neutrino state evolves following the upper curve in Fig. 10.9.

First consider the mixing in vacuum, where the mixing angle is θ_{12}. We have

$$\begin{aligned} v_1 &= \cos\theta_{12} \times v_e - \sin\theta_{12} \times v_a \\ v_2 &= \sin\theta_{12} \times v_e + \cos\theta_{12} \times v_a. \end{aligned} \tag{10.40}$$

Then consider the mixing in the high-density core regions where neutrinos are produced. Here $A \gg \delta m^2 \cos 2\theta_{12}$. Correspondingly $\tan 2\theta_m$ is negative and tends to zero, i.e. $\theta_m \to \pi/2$. Consequently we have

$$\begin{pmatrix} \tilde{v}_e \\ \tilde{v}_a \end{pmatrix} = \begin{pmatrix} \cos\theta_{12,m} & \sin\theta_{12,m} \\ -\sin\theta_{12,m} & \cos\theta_{12,m} \end{pmatrix} \begin{pmatrix} \tilde{v}_1 \\ \tilde{v}_2 \end{pmatrix} \approx \begin{pmatrix} 0 & 1 \\ -1 & 0 \end{pmatrix} \begin{pmatrix} \tilde{v}_1 \\ \tilde{v}_2 \end{pmatrix} \Rightarrow \begin{matrix} \tilde{v}_1 \approx \tilde{v}_a \\ \tilde{v}_2 \approx \tilde{v}_e \end{matrix}. \tag{10.41}$$

The important conclusion is that electron neutrinos are produced in a mass eigenstate; to be precise, the one with the larger mass, $\tilde{v}_2$. The eigenstate then propagates toward lower density regions and may encounter a layer in which the resonance conditions are satisfied. If, as we have assumed, adiabaticity is also satisfied, the state follows the upper curve in Fig. 10.9. Finally, when neutrinos reach the surface they leave the Sun still in the mass eigenstate, which is now v_2.

We must now check whether neutrinos do meet the resonance or not. We do this with the (today) known values of δm^2 and θ_{12}. On their journey neutrinos encounter all the electron densities smaller than the central one N_0. Its value is

$$N_0 \approx 6 \times 10^{31} \text{ m}^{-3}. \tag{10.42}$$

Neutrinos will meet the resonance if there is a density smaller than N_0 satisfying Eq. (10.39). Having fixed all the other quantities, this is a condition on the neutrino energy, i.e.

$$E > \frac{\delta m^2 \cos 2\theta_{12}}{2\sqrt{2}G_F N_0} \approx \frac{\delta m^2 \cos 2\theta_{12}}{1.5 \times 10^{-11} \text{ eV}} \approx 2 \text{ MeV}. \tag{10.43}$$

In conclusion, neutrinos emitted at the Sun's surface with energy larger than about 2 MeV are ν_2 and will remain in this state until they propagate in vacuum. A detector on Earth sensitive to ν_e will observe only the component of amplitude $\sin \theta_{12}$, as from Eq. (10.40). The survival probability of electron neutrinos from the production to the detection points is

$$P_{ee} = \sin^2 \theta_{12} \qquad E \gtrsim 2 \text{ MeV}. \tag{10.44}$$

Neutrinos of lower energy do not encounter the resonance and propagate in the Sun as in vacuum. They oscillate with a maximum excursion

$$A(\nu_e \to \nu_a) = \sin^2(2\theta_{12}). \tag{10.45}$$

This factor multiplies the oscillating term. Our detectors take the average value of the oscillation term on times much longer than the oscillation period, which is 1/2. The survival probability is, in conclusion,

$$P_{ee} = 1 - \frac{1}{2}\sin^2 \theta_{12} \qquad E \lesssim 2 \text{ MeV}. \tag{10.46}$$

We now notice that the resonance corresponding to the larger neutrino square-mass difference Δm^2 does not exist in the Sun, because for that to occur neutrino energies should be about 33 times larger than the limit (10.43). Both resonances can exist in supernovae, where the densities are much larger.

10.4 The experiments

The historical process leading to the discovery of the neutrino flavour transitions in the Sun is not due to a single experiment, but rather to a series of experimental and theoretical developments. In 1946 B. Pontecorvo (Pontecorvo 1946) proposed the detection of electron neutrinos by the inverse beta decay reaction

$$\nu_e + {}^{37}\text{Cl} \to e^- + {}^{37}\text{Ar}. \tag{10.47}$$

In 1962 J. Bahcall (Bahcall *et al.* 1963) started the construction of a solar model and the calculation of the expected reaction rate for (10.47). The initial result was discouraging: the rate was too small to be detectable. However, soon afterwards,

Bahcall noticed the presence of a super-allowed transition to an analogue state of ^{37}Ar at 5 MeV. This increased the estimated rate by almost a factor of 20. The experiment was feasible.

R. Davis used 615 t of perchloroethylene (C_2Cl_4) as the target detector medium, in which about one Ar nucleus per day produced by reaction (10.47) was expected. Every few weeks the metastable Ar nuclei within the atoms they had formed were extracted using a helium stream. After suitable chemical processing, the extracted gas was introduced into a counter to detect the ^{37}Ar decays. Since the signal rate is only of a few counts per month, it is mandatory to work deep underground to be shielded from the cosmic rays and to use only materials extremely free of radioactive components. The experiment took place deep underground in the Homestake mine in South Dakota at 1600 m depth (Davis *et al.* 1964).

The energy threshold of the experiment, 814 keV as shown in Fig. 10.8, allows the detection of the highest-energy neutrinos, the beryllium and boron neutrinos. Already in the 1970s the measured flux appeared to be substantially lower than the expected one. It was the beginning of the 'solar neutrino puzzle'. The value of the solar neutrino capture rate resulting from 108 runs between 1970 and 1994 (Cleveland 1998), expressed in Solar Neutrino Units (1 SNU = one capture per 10^{36} atoms per second), is

$$R(\text{Cl}, \text{exp.}) = 2.56 \pm 0.16 \pm 0.16 \, \text{SNU}. \qquad (10.48)$$

This value is about 1/3 of the SSM prediction (Bahcall *et al.* 2005)

$$R(\text{Cl}, \text{SSM}) = 8.1 \pm 1.3 \, \text{SNU}. \qquad (10.49)$$

The first confirmation of the puzzle came from the Kamiokande experiment in 1987, which, like its larger successor Super-Kamiokande, is a water Cherenkov detector. Neutrinos from the Sun were detected through their elastic scattering on electrons

$$v_x + e^- \rightarrow v_x + e^-. \qquad (10.50)$$

All neutrino flavours contribute; however, while electron neutrinos scatter both via NC and CC, the other two flavours scatter only via NC and, consequently, with a cross section about 1/6 of the former. Having a high-energy threshold, the experiment was sensitive to the boron neutrinos only. The rate measured by Kamiokande (Hirata *et al.* 1989) was about one-half of the expected rate, a value that was later confirmed by Super-Kamiokande (Hosaka *et al.* 2005). The measured flux is

$$\Phi_{\text{exp}} = (2.35 \pm 0.02 \pm 0.08) \times 10^{10} \, \text{m}^{-2} \, \text{s}^{-1}. \qquad (10.51)$$

While the theoretical one is

$$\Phi_{SSM} = (5.69 \pm 0.91) \times 10^{10} \text{ m}^{-2} \text{ s}^{-1}. \tag{10.52}$$

By measuring the direction of the electron hit by the neutrino in the elastic scattering, Super-Kamiokande also established that neutrinos were coming from the Sun (the first neutrino telescope). At this point the existence of a problem was well established. However, was the problem due to some flaw in the solar model or to anomalous behaviour of the neutrinos? Indeed, as anticipated, the high-energy neutrino flux is very sensitive to the values of the parameters of the model. For example it depends on the temperature of the core as T^{18}.

The answer could come only from the measurement of the *pp* neutrino flux, which can be calculated from solar luminosity with a 2% uncertainty. Two radiochemical experiments were built for this purpose, GALLEX in the Gran Sasso National Laboratory in Italy and SAGE in the Baksan Laboratory in Russia, both underground. Both employ gallium as the target, 30 t and 60 t respectively. They are sensitive to electron neutrinos via the inverse beta decay reaction

$$\nu_e + {}^{71}\text{Ga} \rightarrow e^- + {}^{71}\text{Ge} \tag{10.53}$$

with 233 keV energy threshold. GALLEX published the first results in 1992 (Anselmann *et al.* 1992): the ratio between measured and expected rates was about 60%. SAGE soon confirmed this value. GALLEX ended in 1997, becoming, in an improved version, GNO, which ended in 2003. SAGE is still running. The measured values of the electron neutrino capture rates (Altman *et al.* 2005, Abdurashitov *et al.* 2002) are

$$R(\text{Ga}, \text{GALLEX} + \text{GNO}) = 69.3 \pm 4.1 \pm 3.6 \text{ SNU}$$
$$R(\text{Ga}, \text{SAGE}) = 70.8^{+5.3+3.7}_{-5.2-3.2} \text{ SNU}. \tag{10.54}$$

These values, in mutual agreement, are again much smaller than the SSM prediction (Bahcall *et al.* 2005)

$$R(\text{Ga}, \text{SSM}) = 126 \pm 10 \text{ SNU}. \tag{10.55}$$

In fact, by 1995 GALLEX (Anselmann *et al.* 1995) had reached such a precision that they could exclude the 'solar solution' of the puzzle by the following argument. The rate measured by GALLEX is the sum of three main contributions: from *pp*, from boron and from beryllium. However, the sum of the first, as evaluated from the solar luminosity, and the second, as measured by Super-Kamiokande, was already larger than the rate measured by GALLEX. Consequently, no space is left in the budget for beryllium neutrinos that must be present, because the boron, which exists as observed by Super-Kamiokande, is a daughter of beryllium. Also this result was later confirmed by SAGE. In conclusion, the neutrino deficit cannot be explained by any modification of the solar model.

The same conclusion is reached by observing that the chlorine experiment measures the sum of the beryllium and boron neutrinos, Super-Kamiokande only the boron neutrinos. However, the deficit of the former is larger than the deficit of the latter.

Further controls, calibrations and independent measurements of the relevant nuclear cross section led in 1997 to the conclusion that the solution of the puzzle was in the anomalous behaviour of neutrinos. The experiments were sensitive only (or almost so in the case of the Cherenkov experiments) to electron neutrinos. Apparently, electron neutrinos were disappearing by a large fraction on their way from the solar centre to the Earth. The most probable hypothesis was the flavour conversion we have described in the previous sections.

The final proof came from an appearance experiment in 2002. The Sudbury Neutrino Observatory (SNO) is a heavy-water Cherenkov detector with 1000 t of D_2O, located in a mine 2000 m deep in Canada. The observations started in 1999. The detector is flavour sensitive, detecting electron neutrinos through the charged-current reaction

$$\nu_e + \text{d} \rightarrow p + p + e^- \tag{10.56}$$

and all flavours through the neutral-current reaction

$$\nu_x + \text{d} \rightarrow p + n + \nu_x. \tag{10.57}$$

Since the NC cross section is independent of the flavour, the rate of (10.57) measures the total neutrino rate, while the rate of (10.56) gives the contribution of electron neutrinos. SNO also measured the elastic cross section, with results in agreement with Super-Kamiokande, but with much poorer statistics. The experiment is sensitive in the higher-energy part of the spectrum, namely to boron neutrinos.

We now extract from the measured rates the fluxes of electron neutrinos and of muon or tau neutrinos (which are indistinguishable) with the help of Fig. 10.10. The CC (10.56) event rate directly gives the electron neutrino flux, independent of the muon/tau neutrino flux. It is the vertical band in the figure. The NC rate (10.57) gives the sum of the three fluxes, the three cross sections being equal. It is the band along the second diagonal of the figure. The elastic scattering (ES) rate

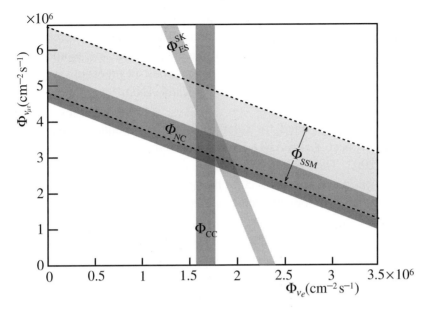

Fig. 10.10.   Fluxes of ^{8}B neutrinos $\Phi(v_e)$ and $\Phi(v_{\mu\tau})$ deduced from SNO CC and NC results and from Super-Kamiokande ES results. Bands represent 1σ uncertainties. SSM predictions are also shown. (Yao *et al.* 2006 by permission of Particle Data Group and the Institute of Physics)

measured by Super-Kamiokande gives the almost vertical band, taking into account that $\sigma\left(v_{\mu,\tau}+e\right)\approx 0.16\sigma(v_e + e)$. Having three relationships between two unknown quantities, we can check for consistency. This is proven by the fact that the three bands cross in the same area. The values of the fluxes (Aharmin *et al.* 2005) are

$$\Phi_{CC}(v_e) = \left(1.68 \pm 0.06^{+0.08}_{-0.09}\right) \times 10^8 \text{ m}^{-2} \text{ s}^{-1} \tag{10.58}$$

and

$$\Phi_{NC}(v_x) = \left(4.94 \pm 0.21^{+0.38}_{-0.34}\right) \times 10^8 \text{ m}^{-2} \text{ s}^{-1}. \tag{10.59}$$

The figure also shows the total neutrino flux as predicted by the SSM, which is in perfect agreement with the measured flux. This proves that the missing electron neutrinos have indeed transformed into muon and/or tau neutrinos.

Solar neutrino data are sensitive to two parameters of the neutrino system, i. e. δm^2 and θ_{12}. We now briefly consider how they are determined. The experiments on solar neutrinos have measured the neutrino flux integrated over different energy intervals. To each value, and to the associated uncertainty, corresponds a region in the parameter plane, δm^2 vs. θ_{12}, different for each experiment. To be compatible with all the measurements, the solution must lie in the intersection of these regions. In practice a best-fit procedure is performed, obtaining, with the further contribution of KamLAND, the values reported in Section 10.2.

The oscillation discovered for solar neutrinos was confirmed by the disappearance experiment KamLAND (Kamioka Liquid Scintillator Anti-Neutrino Detector) in 2002. The experiment used the electron antineutrinos produced by Japanese nuclear power plants. Each of them has a different power and is located at a different distance from the detector. The flux-weighted average distance is about 180 km.

KamLAND is a one-kiloton ultra-pure scintillator detector, located in the old Kamiokande site in Japan. The recoil electrons from neutrino elastic scatterings are detected and their energy, which is strongly correlated to neutrino energy, is measured. Once the backgrounds have been subtracted, the experiment measures the electron antineutrino flux and energy spectrum. This is shown in Fig. 10.11. Comparison with the expected rate and spectrum shows that neutrinos have disappeared. Data are then fitted using the oscillation hypothesis with δm^2 and θ_{12} as free parameters (Eguchi *et al.* 2003, Araki *et al.* 2005).

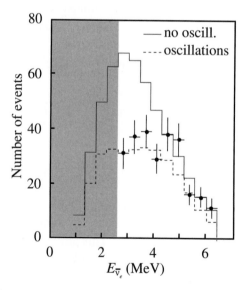

Fig. 10.11.   Electron antineutrino spectrum measured by KamLAND compared with expectations in the absence and in the presence of oscillations. The grey region is not accessible due to background. (Araki *et al.* 2005)

10.5 Limits on neutrino mass

One physical quantity, or more for redundancy, independent of oscillations must be measured to find the neutrino mass spectrum. There are three experimental or observational possibilities:

- beta decay experiments that probe the weighted average $m_{\nu_e}^2 \equiv \sum_{i=1}^{3} |U_{ei}|^2 \, m_i^2$;
- cosmological observations that probe the sum of neutrino masses $\sum_{i=1}^{3} m_i$;
- experiments on neutrinoless double-beta decay, a process that can exist if neutrinos are Majorana particles. They probe the quantity $|m_{ee}| \equiv \left| \sum_{i=1}^{3} U_{ei}^2 m_i \right|$. Notice that the addenda are complex numbers.

We now discuss the first two measurements, which have given only upper limits up to now. We shall give a hint on double-beta decay in the next section.

Cosmology has made tremendous progress in the last several years both in the modelling and in the quantity and, more importantly, the quality of the observational data. A standard model of cosmology has been developed, which consistently explains all the observational data relative to widely different epochs. The basic parameters have been determined with better than 10% accuracy. In this frame, cosmology provides a sensitive, albeit indirect, method to measure or limit neutrino mass.

All the structures present in the Universe were seeded by quantum fluctuations that took place when the Universe was extremely small. These initial fluctuations grew during the evolution of the Universe. Indeed, since the gravitational interaction is only attractive, the regions of higher density attracted more and more mass into them. However, neutrinos, due to their very small masses, may have speeds larger than the escape velocity from the smaller structures. Consequently they can 'free stream' out of those structures diminishing their total mass. As a net result, neutrinos tend to erase the structures at scales smaller than a certain value D_F called the free streaming distance. This is roughly the distance travelled by the neutrinos during a significant fraction of the formation time of the structures. The smaller the sum of the neutrino masses, the larger is D_F and the smaller is the effect. In practice for $\sum m_i$ of $0.5-1$ eV

$$D_F \approx 10-50 \, \text{Mpc} \tag{10.60}$$

where Mpc means megaparsec.

The relevant observable is the mass spectrum, roughly speaking the probability of finding a structure of a given mass as a function of the mass. The scale of the structures we are considering is enormous, from Mpc to Gpc. A galaxy, for comparison, is, in order of magnitude, tens of kpc across.

The cosmological model predicts the shape of the mass spectrum in terms of a small number of parameters. These, as we have said, have been determined with good accuracy. The observed mass spectrum is equal to the predicted one assuming that neutrinos are massless. By studying the effect of increasing the neutrino masses one then obtains the upper limit

$$\sum_{i=1}^{3} m_i \leq 600 \, \text{meV}. \tag{10.61}$$

From neutrino oscillations we know that the mass differences are much smaller than this limit. Consequently, we can say that each mass must be

$$m_i \leq 200 \, \text{meV} \tag{10.62}$$

as in Appendix 3. This is the lowest upper limit on neutrino mass. However, it is indirect and depends on several assumptions. On the other hand, the present rapid progress of cosmology might lead in a few years to the detection of neutrino mass.

We now consider the beta decay of a nucleus. Non-zero neutrino mass can be detected by observing a distortion in the electron energy spectrum, just before its end-point. Clearly, the sensitivity is higher if the end-point energy is lower. The most sensitive choice is the tritium decay

$$^{3}\text{H} \rightarrow {}^{3}\text{He} + e^{-} + \nu_e \tag{10.63}$$

due to its very small Q-value, $Q = m_{^3\text{H}} - m_{^3\text{He}} = 18.6 \, \text{keV}$. Let E_e, p_e and E_ν, p_ν be the energy and the momentum of, respectively, the electron and the neutrino. The electron energy spectrum was calculated by Fermi in his effective four-fermion interaction. We shall give only the result here, which, if neutrinos are massless, is

$$\frac{dN_e}{dE_e} \approx F(Z, E_e) p_e^2 E_\nu p_\nu = F(Z, E_e) p_e^2 (Q - E_e)^2 \tag{10.64}$$

where in the last member we have set $p_\nu = E_\nu$. F is a function of the electron energy characteristic of the nucleus (called the Fermi function). It may be considered a constant in the very small energy range near to the end-point that we are considering. If we plot the quantity $K(E_e) \equiv \sqrt{dN_e/dE_e}/p_e$ versus E_e we obtain a straight line crossing the energy axis at Q. This diagram is called the Kurie plot (Kurie *et al.* 1936) and is shown in Fig. 10.12 as a dotted line.

Let us now suppose that neutrinos have a single mass m_ν. The factor $E_\nu p_\nu$ in (10.64) becomes $(Q - E_e)\sqrt{(Q - E_e)^2 - m_\nu^2}$. In the Kurie plot the end-point

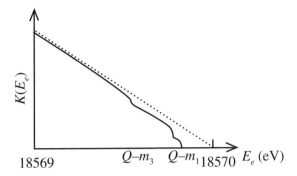

Fig. 10.12. Tritium Kurie plot with three neutrino types.

moves to the left to $Q - m_\nu$ and the slope of the spectrum in the end-point becomes perpendicular to the energy axis.

In the actual situation, with three neutrino types, the spectrum is given by

$$\frac{dN_e}{dE_e} \approx p_e^2 (Q - E_e) \sum_{i=1}^{3} |U_{ei}|^2 \sqrt{(Q - E_e)^2 - m_i^2}. \qquad (10.65)$$

There are now three steps at $Q - m_i$, corresponding to the three eigenstates. Their 'heights' are proportional to $|U_{ei}|^2$. Actually, since $|U_{e3}|^2 = \sin^2 \theta_{13} \leq 0.024$ one step is very small. In Fig. 10.12 we have drawn the qualitative behaviour of (10.64) as a continuous curve for hypothetical values of the masses, assuming $m_3 > m_2$ and $|U_{e3}|^2$ not too small. Notice that the expected effects appear in the last eV of the spectrum.

In practice the energy differences between the steps are so small that they cannot be resolved and the measured, or limited, observable is the weighted average

$$m_{\nu_e}^2 \equiv \sum_{i=1}^{3} |U_{ei}|^2 m_i^2. \qquad (10.66)$$

The experiment is extremely difficult. Firstly, a very intense and pure tritium source is needed. Secondly, the spectrometer must be able to reject the largest part of the spectrum and to provide a superior energy resolution.

The best limit, obtained by the MAINZ experiment in Germany (Kraus *et al.* 2004), is

$$m_{\nu_e} \leq 2.3\,\text{eV}. \qquad (10.67)$$

An only slightly worse limit, 2.5 eV, has been obtained by the TROITSK experiment in Russia (Lobashev *et al.* 2001).

A new experiment, KATRIN, is under construction in Germany aiming to reach a sensitivity of $m_{\nu_e} \leq 200\,\text{meV}$. Notice that, even if this value is close to the present upper limit given by cosmology, KATRIN will provide a direct measurement, while the present limit is obtained by a long chain of inferences.

10.6 Challenges

Neutrino oscillations and flavour changing are not the only phenomena observed beyond the Standard Model. We shall limit our discussion to the experimental challenges at the frontier of contemporary research, naming a few of them here.

- **Majorana vs. Dirac neutrinos**. In the Standard Model neutrinos are assumed to be Dirac particles, however there is no experimental proof of this being true. As we saw in Section 2.7, in 1937 E. Majorana proposed a theory for neutral spinors alternative to the '*simple extension of the Dirac equation to neutral particles*', '*in spite of the fact*', he added, '*that it is probably not yet possible to ask experience to decide*' (Majorana 1937). After 70 years we still do not have an answer to this fundamental question, but we have the means of actively asking experience to provide this *decision*. This can be done by searching for the neutrinoless double-beta decay of some nuclides, a process that is possible only in the Majorana case. Even if it exists, this decay is extremely rare, with lifetimes of 10^{26}–10^{27} years at least. Consequently, the experiments must be done in underground laboratories with the strictest possible control of even the smallest traces of radioactive nuclides in every detector component.

- **SUSY**. The search for supersymmetry (SUSY) is motivated by theoretical arguments. The main argument is the so-called 'hierarchy' problem of the Standard Model. We have seen, for example in Section 9.7, that the gauge boson masses are modified by fermionic and bosonic loops such as those in Fig. 9.23(b). The same happens for the scalar mesons, in particular for the Higgs boson. Here the situation is particularly intriguing. Indeed, the loop corrections would tend 'naturally' to drive the Higgs mass to the enormous energy scale of the 'grand unification', i.e. 10^{16} GeV. A cancellation of 14 orders of magnitude appears to be necessary, something that should be explained. Considering that the problem does not exist for fermions, we can imagine a symmetry that includes integer and half-integer spin particles in the same multiplet. If the symmetry is not broken, the particles of a multiplet have the same mass. This type of symmetry has a rather different mathematics from those we have met and is called supersymmetry. Now suppose that we lodge the Higgs boson and a spin 1/2 partner, called the Higgsino, in the same multiplet. Given that the Higgsino mass is stable against the corrections, because it is a fermion, the Higgs mass, which is equal to the latter, becomes stable too. However, SUSY requires the existence of a partner for each particle, not only for the Higgs boson. Since none of these particles has ever been observed, if they do exist they must have rather large masses and consequently SUSY cannot be exact. It is reasonable to assume that the SUSY particles have masses of the order of a few hundred GeV, because otherwise the hierarchy problem would reappear. If this is indeed the case the experiments at LHC will be able to detect them.

- **Dark matter**. Cosmology has made enormous progress over the last few years. We now have a 'standard cosmological model' that describes all the observational data in a single picture. The components of the mass–energy budget of the Universe have been determined with an accuracy of 10% or better. The mass–energy density is equal to the critical density within 1.5%. However, the contribution of matter is only about 30% of the total. The largest fraction of matter is 'dark'. It neither emits nor absorbs electromagnetic radiation and so it has been inferred indirectly through its gravitational effects. Moreover, the relative abundance of light nuclei, which were produced in the initial phases of the Universe, indicates that 'normal' matter is only a small fraction, about 4%, of the total. Astronomical observations clearly show that the contribution of neutrinos to the budget is also very small, not larger than 1%. Consequently, dark matter is presumably composed of a new class of particles. They presumably have weak interactions and are called WIMPs for Weakly Interacting Massive Particles. We think they have rather large masses, maybe tens or hundreds of GeV, otherwise they would have been discovered, for example at LEP. The supersymmetric theory suggests a candidate, the 'neutralino'. It is stable because it is the lightest SUSY particle. As such, neutralinos produced in the initial phases of the history of the Universe should still be here, around us. The search for WIMPs is very challenging, being carried out with complementary experiments at LHC, in underground laboratories and in space.

- **Dark energy**. The remaining 70% of the Universe is a density of negative pressure, which forces the expansion rate to accelerate, in our cosmological epoch, called 'dark energy'. The direct observational evidence is the following. If the Universe contained only matter, its expansion should have slowed down in all the past epochs under the influence of gravity. Supernovae of type 1a, which are visible even at large distances, can be used as standard candles because their absolute luminosity is known. If expansion were slowing, distant supernovae should appear brighter and closer than their high redshifts might otherwise suggest. On the contrary, the most distant supernovae are dimmer and farther away than expected. This can only be explained if the expansion rate of the Universe is accelerating. From a formal point of view, dark energy can be explained as an effect of the cosmological constant that Einstein introduced in his equations for different reasons. However, from a physical point of view, we do not really know what dark energy is. NASA, the US Space Agency, and other scientific institutions worldwide are developing programmes for this purpose. Telescopes in the visible and in the infrared in orbit will try to map the historical development of the Universe by measuring thousands of type 1a supernovae and the mass distribution at large scales with a number of techniques.

- **Antimatter**. The Standard Model is unable to explain why the Universe is overwhelmingly made of matter, as it appears to be. To check this point, the presence of antimatter will be searched for with particle spectrometers in space, such as PAMELA, which is already in orbit, and AMS that will be launched in the next few years. Experiments on the matter–antimatter asymmetry in the neutrino sector will shed light on the mechanism that is responsible for the asymmetry.

- **Gravitational interaction**. More generally, the Standard Model does not include one of the fundamental interactions, i.e. gravity, for which only a macroscopic theory exists, general relativity. We already know that this theory is only a macroscopic approximation of the true theory, just as Maxwell's equations are the macroscopic approximation of the Standard Model. We know that the structures of the Universe at all scales, superclusters and clusters of galaxies, galaxies, stars and their planets, had their origin in and evolved from the primordial quantum fluctuations that took place when the Universe was very small. However, if general relativity were correct, none of this would exist, including us. The construction of the theory of gravitation needs experimental and observational input. This is extremely challenging, considering how many orders of magnitude separate the scale of our present knowledge from the Planck scale, 10^{19} GeV. The Universe became transparent to electromagnetic waves when electrons and protons combined into atoms, when it was 'only' about 1000 times smaller than now. Gravitational waves appear to be the only messengers reaching us from previous epochs, because they propagate freely everywhere. The LISA interferometer will be deployed in space not very far in the future by NASA and ESA jointly. Being sensitive to gravitational waves of sub-millihertz frequency, LISA might well reveal to us new features of gravity and, moreover, probe the scalar field responsible for cosmic inflation, the inflaton.

We end here our partial and superficial review of open problems. We can guess from the past history of science that unexpected phenomena will open completely new windows to our understanding of Nature. Indeed, we know a lot, but we do not know much more.

Further reading

Bahcall, J. N. (1989); *Neutrino Astrophysics*. Cambridge University Press

Bahcall, J. N. (2002); *Solar models: an historical overview*. arXiv:Astro-ph/0209080 v2

Bettini, A. (2005); The Homi Bhabha Lecture: *Neutrino Physics and Astrophysics*. Proceedings of the 29th International Cosmic Ray Conference. Pune. **10** 73–96

Davis, R. Jr. (2002); Nobel Lecture, *A Half-Century with Solar Neutrinos* http://nobelprize.org/nobel_prizes/physics/laureates/2002/davis-lecture.pdf

Kajita, T. & Totsuka, Y. (2001); *Rev. Mod. Phys.* **73** 85

Kilian, W. (2003); *Electroweak Symmetry Breaking*. Springer Tracts in Modern Physics

Kirsten, T. A. (1999); *Rev. Mod. Phys.* **71** 1213

Koshiba, M. (2002); Nobel Lecture, *Birth of Neutrino Astronomy* http://nobelprize.org/nobel_prizes/physics/laureates/2002/koshiba-lecture.pdf

Mohapatra, R. N. & Pal, P. B. (2004); *Massive Neutrinos in Physics and Astrophysics*. World Scientific. 3rd edn.

Rubin, V. C. (1983); *Dark matter in spiral galaxies. Sci. Am.* June 96

Various authors (2007); *The Large Hadron Collider. Nature* **448** 269–312

Veltman, M. J. G. (1986); *The Higgs boson. Sci. Am.* November 76

Appendix 1

Greek alphabet

alpha	α	A
beta	β	B
gamma	γ	Γ
delta	δ	Δ
epsilon	ε	E
zeta	ζ	Z
eta	η	H
theta	θ, ϑ	Θ
iota	ι	I
kappa	κ	K
lambda	λ	Λ
mu	μ	M
nu	ν	N
xi	ξ	Ξ
omicron	o	O
pi	π	Π
rho	ρ	P
sigma	σ, ς	Σ
tau	τ	T
upsilon	υ	Y, Υ
phi	ϕ, φ	Φ
chi	χ	X
psi	ψ	Ψ
omega	ω	Ω

Appendix 2

Fundamental constants

Quantity	Symbol	Value	Uncertainty
Speed of light in vacuum	c	$299\,792\,458\,\mathrm{m\,s^{-1}}$	exact
Planck constant	h	$6.626\,0693(11)\times 10^{-34}\,\mathrm{J\,s}$	170 ppb
Planck constant, reduced	$\hbar$	$1.054\,571\,68(18)\times 10^{-34}\,\mathrm{J\,s}$	170 ppb
		$6.582\,119\,15(56)\times 10^{-22}\,\mathrm{MeV\,s}$	85 ppb
Conversion constant	$\hbar c$	$197.326\,968(7)\,\mathrm{MeV\,fm}$	85 ppb
Conversion constant	$(\hbar c)^2$	$389.379\,323(67)\,\mathrm{GeV^2\,\mu barn}$	170 ppb
Elementary charge	q_e	$1.602\,176\,53(14)\times 10^{-19}\,\mathrm{C}$	85 ppb
Electron mass	m_e	$9.109\,3826(16)\times 10^{-31}\,\mathrm{kg}$	170 ppb
Proton mass	m_p	$1.672\,621\,71(29)\times 10^{-27}\,\mathrm{kg}$	170 ppb
Bohr magneton	$\mu_{\mathrm{B}} = \dfrac{q_e h}{2m_e}$	$5.788\,381\,804(39)\times 10^{-11}\,\mathrm{MeV\,T^{-1}}$	6.7 ppb
Nuclear magneton	$\mu_N = \dfrac{q_e h}{2m_p}$	$3.152\,451\,259(21)\times 10^{-14}\,\mathrm{MeV\,T^{-1}}$	6.7 ppb
Bohr radius	$a = \dfrac{4\pi\epsilon_0 h^2}{m_e q_e^2}$	$0.529\,177\,2108(18)\times 10^{-10}\,\mathrm{m}$	85 ppb
1/fine structure constant	$a^{-1}(0)$	$137.035\,999\,710(96)$	0.7 ppb
Newton constant	G_{N}	$6.6742(10)\times 10^{-11}\,\mathrm{m^3\,kg^{-1}\,s^{-2}}$	150 ppm
Fermi constant	$G_{\mathrm{F}}/(hc)^3$	$1.166\,37(1)\times 10^{-5}\,\mathrm{GeV^{-2}}$	9 ppm
Weak mixing angle	$\sin^2\theta_W(M_Z)$	$0.231\,22(15)$	650 ppm
Strong coupling constant	$a_s(M_Z)$	$0.1176(20)$	1.7%
Avogadro number	N_{A}	$6.022\,1415(10)\times 10^{23}\,\mathrm{mole^{-1}}$	170 ppb
Boltzmann constant	k_{B}	$1.380\,6505(24)\times 10^{-23}\,\mathrm{J\,K^{-1}}$	1.8 ppm

Values are mainly from CODATA (Committee on Data for Science and Technology) http://physics. nist.gov/cuu/Constants/index.html and Mohr, P. J. and Taylor, B. N., *Rev. Mod. Phys.* **77** (2005). Fine structure constant is from Gabrielse *et al.* (2006). Recent measurement of the Newton constant by Schlamminger *et al.* (2006) gives $6.674\,252(109)(54)\times 10^{-11}\,\mathrm{m^3\,kg^{-1}\,s^{-2}}$, i.e. 16 ppm statistic and 8 ppm systematic uncertainties. Fermi and strong coupling constants and weak mixing angle are from 'Particle Data Group' *Journal of Physics G* **33** (2006); http://pdg.lbl.gov/2006/. The figures in parentheses after the values give the one standard-deviation uncertainties in the last digits.

Appendix 3

Properties of elementary particles

Gauge bosons

		Mass	Width	Main decays
photon	γ	$<6 \times 10^{-17}$ eV	stable	
gluon	g	0 (assumed)	stable	
weak boson	Z^0	91.1876 ± 0.0021 GeV	2.4952 ± 0.0023 GeV	$l^+ l^-$, $\nu \bar{\nu}$, $q\bar{q}$
weak boson	$W^\pm$	80.403 ± 0.029 GeV	2.141 ± 0.041 GeV	$l\bar{\nu}_l$, $q\bar{q}$

Data are from 'Particle Data Group' *Journal of Physics G* **33** (2006); http://pdg.lbl.gov/2006/

Gauge boson couplings. Colour, electric charge (in elementary charge units), weak isospin and its third component and weak hypercharge of the gauge bosons

	Colour	Q	I_W	I_{W_z}	Y_W
g	octet	0	0	0	0
γ	0	0	—	—	0
Z^0	0	0	—	—	0
W^+	0	+1	1	+1	0
W^-	0	−1	1	−1	0

Leptons

	Mass		Lifetime	Main decays
e	0.510 998 92(4) MeV		$>4.6 \times 10^{26}$ yr	
μ	105.658 369(9) MeV		2.197 03(4) μs	$e^- \bar{\nu}_e \nu_\mu$
τ	1776.99 ± 0.28 MeV		290.6 ± 1 fs	$\mu^- \bar{\nu}_\mu \nu_\tau,\ e^- \bar{\nu}_e \nu_\tau,\ h^- \nu_\tau$
ν_1	$0 \le m_1 < 200$ meV	if $m_3 > m_2$	stable	
	50 meV $< m_1 < 200$ meV	if $m_3 < m_2$		
ν_2	9 meV $< m_2 < 200$ meV	if $m_3 > m_2$	stable	
	50 meV $< m_2 < 200$ meV	if $m_3 < m_2$		
ν_3	50 meV $< m_3 < 200$ meV	if $m_3 > m_2$	stable	
	$0 \le m_3 < 200$ meV	if $m_3 < m_2$		

Charged lepton data are from 'Particle Data Group' *Journal of Physics G* **33** (2006); http://pdg.lbl. gov/2006/

Upper limits on neutrino masses come from cosmology, lower limits from oscillations and conversion in matter.

Quarks

	Q	I	I_z	S	C	B	T	$\mathcal{B}$	Y	Mass
d	$-1/3$	$1/2$	$-1/2$	0	0	0	0	$1/3$	$1/3$	3–7 MeV
u	$+2/3$	$1/2$	$+1/2$	0	0	0	0	$1/3$	$1/3$	1.5–3.0 MeV
s	$-1/3$	0	0	-1	0	0	0	$1/3$	$-2/3$	95 ± 25 MeV
c	$+2/3$	0	0	0	$+1$	0	0	$1/3$	$4/3$	1.25 ± 0.09 GeV
b	$-1/3$	0	0	0	0	-1	0	$1/3$	$-2/3$	4.20 ± 0.07 GeV
t	$+2/3$	0	0	0	0	0	$+1$	$1/3$	$4/3$	173 ± 3 GeV

Electric charge Q (in units of elementary charge), strong isospin I and its third component I_z, strangeness S, charm C, beauty B, top T, baryonic number $\mathcal{B}$ and strong hypercharge Y of the quarks. Each quark can have red, blue or green colour.

Quark masses are from 'Particle Data Group' *Journal of Physics G* **33** (2006); http://pdg.lbl.gov/ 2006/

Weak couplings of the fermions

	I_W	I_{W_z}	Q	Y_W	c_Z
ν_{lL}	1/2	+1/2	0	−1	1/2
l_L^-	1/2	−1/2	−1	−1	$-1/2 + s^2$
l_R^-	0	0	−1	−2	s^2
u_L	1/2	+1/2	2/3	1/3	$1/2 - (2/3)s^2$
d_L'	1/2	−1/2	−1/3	1/3	$-1/2 + (1/3)s^2$
u_R	0	0	2/3	4/3	$-(2/3)s^2$
d_R'	0	0	−1/3	−2/3	$(1/3)s^2$
$\bar{\nu}_{lR}$	1/2	−1/2	0	1	$-1/2$
l_R^+	1/2	+1/2	+1	1	$1/2 - s^2$
l_L^+	0	0	+1	2	$-s^2$
$\bar{u}_R$	1/2	−1/2	−2/3	−1/3	$-1/2 + (2/3)s^2$
$\bar{d}_R'$	1/2	+1/2	1/3	−1/3	$1/2 - (1/3)s^2$
$\bar{u}_L$	0	0	−2/3	−4/3	$(2/3)s^2$
$\bar{d}_L'$	0	0	1/3	2/3	$-(1/3)s^2$

Weak isospin, hypercharge, electric charge and Z-charge factor $c_Z = I_{W_z} - s^2 Q$ of the fundamental fermions ($s^2 = \sin^2\theta_W$). The values are identical for every colour.

Quark-gluon colour factors

	R		G	B
$\bar{R}$	$\dfrac{1}{\sqrt{2}} \times g_7$	$\dfrac{1}{\sqrt{6}} \times g_8$	$1 \times g_3$	$1 \times g_5$
$\bar{G}$	$1 \times g_1$		$-\dfrac{1}{\sqrt{2}} \times g_7 \quad \dfrac{1}{\sqrt{6}} \times g_8$	$1 \times g_6$
$\bar{B}$	$1 \times g_2$		$1 \times g_4$	$-\dfrac{2}{\sqrt{6}} \times g_8$

Mesons (lowest levels)

Symbol	$q\bar{q}$	J^P	I^G	Mass (MeV)	Lifetime/Width	Main decay modes
$\pi^\pm$	$u\bar{d}, d\bar{u}$	0^-	1^-	139.57018(35)	26.033(5) ns	$\mu^+\nu_\mu$
π^0	$u\bar{u}, d\bar{d}$	0^-	1^-	134.9766(6)	84 ± 6 as	2γ
η	$u\bar{u}, d\bar{d}, s\bar{s}$	0^-	0^+	547.51 ± 0.18	1.30 ± 0.07 keV	$2\gamma, 3\pi^0, \pi^+\pi^-\pi^0$
ρ	$u\bar{d}/u\bar{u}, d\bar{d}/d\bar{u}$	1^-	1^+	775.5 ± 0.4	149.4 ± 1.0 MeV	2π
ω	$u\bar{u}, d\bar{d}$	1^-	0^-	782.65 ± 0.12	8.49 ± 0.08 MeV	$\pi^+\pi^-\pi^0, \pi^0\gamma$
η'	$u\bar{u}, d\bar{d}, s\bar{s}$	0^-	0^+	957.78 ± 0.14	203 ± 16 keV	$\pi^+\pi^-\eta, \rho\gamma, 2\pi^0\eta$
ϕ	$s\bar{s}$	1^-	0^-	1019.460 ± 0.019	4.26 ± 0.05 MeV	$K\bar{K}, \pi^+\pi^-\pi^0$
K^+, K^-	$u\bar{s}, s\bar{u}$	0^-	$1/2$	493.677 ± 0.016	12.39 ± 0.02 ns	$\mu^+\nu_\mu, \pi^+\pi^0, 3\pi$
K_S^0		0^-		497.648 ± 0.022	89.53 ± 0.05 ps	$2\pi^0, \pi^+\pi^-$
K_L^0		0^-		497.648 ± 0.022	51.14 ± 0.21 ns	$\pi^\pm l^\mp \nu_l, 3\pi$
K^{*+}, K^{*-}	$u\bar{s}, s\bar{u}$	1^-	$1/2$	891.66 ± 0.26	50.8 ± 0.9 MeV	$K\pi$
$K^{*0}, \bar{K}^{*0}$	$d\bar{s}, s\bar{d}$	1^-	$1/2$	896.00 ± 0.25	50.3 ± 0.6 MeV	$K\pi$
D^+, D^-	$c\bar{d}, d\bar{c}$	0^-	$1/2$	1869.3 ± 0.4	1.040 ± 0.007 ps	$K + \cdots$
$D^0, \bar{D}^0$	$c\bar{u}, u\bar{c}$	0^-	$1/2$	1864.5 ± 0.4	0.410 ± 0.002 ps	$K + \cdots$
D_s^+, D_s^-	$c\bar{s}, s\bar{c}$	0^-	0	1968.2 ± 0.5	0.500 ± 0.007 ps	$K + \cdots$
B^+, B^-	$u\bar{b}, b\bar{u}$	0^-?	$1/2$	5279.0 ± 0.5	1.638 ± 0.011 ps	$D + \cdots$
$B^0, \bar{B}^0$	$d\bar{b}, b\bar{d}$	0^-?	$1/2$	5279.4 ± 0.5	1.530 ± 0.009 ps	$D + \cdots$
$B_s^0, \bar{B}_s^0$	$s\bar{b}, b\bar{s}$	0^-?	0	5367.5 ± 1.8	1.466 ± 0.059 ps	$D_s^\pm + \cdots$
B_c^+, B_c^-	$c\bar{b}, b\bar{c}$	0^-?	0	6286 ± 5	0.46 ± 17 ps	$J/\psi l^\mp \nu_l + \cdots$
$\eta_c(1S)$	$c\bar{c}$	0^-	0	2980.4 ± 1.2	25.5 ± 3.4 MeV	$\eta\pi\pi, \eta'\pi\pi, K\pi\pi$
$J/\psi(1S)$	$c\bar{c}$	1^-	0	3096.92 ± 0.011	93.4 ± 2.1 keV	hadrons, $e^+e^-, \mu^+\mu^-$
$\chi_{c0}(1P)$	$c\bar{c}$	0^+	0	3414.76 ± 0.35	10.4 ± 0.7 MeV	hadrons
$\chi_{c1}(1P)$	$c\bar{c}$	1^+	0	3510.66 ± 0.07	0.89 ± 0.05 MeV	hadrons
$\chi_{c2}(1P)$	$c\bar{c}$	2^+	0	3556.20 ± 0.09	2.06 ± 0.12 MeV	hadrons
$\psi(2S)$	$c\bar{c}$	1^-	0	3686.09 ± 0.03	337 ± 13 keV	hadrons
$\psi(3S)$	$c\bar{c}$	1^-	0	3771.1 ± 2.4	23.0 ± 2.7 MeV	$D\bar{D}$
$\Upsilon(1S)$	$b\bar{b}$	1^-	0	9460.3 ± 0.3	54.0 ± 1.3 keV	l^+l^-
$\Upsilon(2S)$	$b\bar{b}$	1^-	0	10023.3 ± 0.3	31.98 ± 2.63 keV	$\Upsilon(1S)2\pi$
$\Upsilon(3S)$	$b\bar{b}$	1^-	0	10352.2 ± 0.5	20.32 ± 1.85 keV	$\Upsilon(2s) + \cdots$
$\Upsilon(4S)$	$b\bar{b}$	1^-	0	10579.4 ± 1.2	20.5 ± 2.5 MeV	$B^+B^-, B^0\bar{B}^0$
$\Upsilon(5S)$	$b\bar{b}$	1^-	0	10865 ± 8	110 ± 13 MeV	$D_s + \cdots$

Data from 'Particle Data Group' *Journal of Physics G* **33** (2006); http://pdg.lbl.gov/2006/

391

Baryons (lowest levels)

Symbol	qqq	J^P	I	Mass (MeV)	Lifetime/Width	Main decay modes
p	uud	$1/2^+$	$1/2$	$938.272\,03(8)$	$>10^{33}$ yr	$pe^-\,\bar{\nu}_e$
n	udd	$1/2^+$	$1/2$	$939.565\,36(8)$	885.7 ± 0.8 s	$p\pi,\ n\pi$
$\Delta(1232)$	uuu	$3/2^+$	$3/2$	1232 ± 2	118 ± 2 MeV	$p\pi^0,\ n\pi^+$
Λ	uds	$1/2^+$	0	1115.683 ± 0.006	263 ± 2 ps	$p\pi^-,\ n\pi^0$
Σ^+	uus	$1/2^+$	1	1189.37 ± 0.07	80.2 ± 0.3 ps	$\Lambda\gamma$
Σ^0	uds	$1/2^+$	1	1192.64 ± 0.02	$(7.4 \pm 0.7) \times 10^{-20}$ s	$n\pi^-$
Σ^-	dds	$1/2^+$	1	1197.45 ± 0.03	148 ± 1 ps	$\Lambda\pi,\ \Sigma\pi$
$\Sigma^+(1385)$	uus	$3/2^+$	1	1382.8 ± 0.4	36 ± 2 MeV	$\Lambda\pi,\ \Sigma\pi$
$\Sigma^0(1385)$	uds	$3/2^+$	1	1383.7 ± 1.0	35.8 ± 0.8 MeV	$\Lambda\pi,\ \Sigma\pi$
$\Sigma^-(1385)$	dds	$3/2^+$	1	1387.2 ± 0.5	39.4 ± 2.1 MeV	$\Lambda\pi^0$
Ξ^0	uss	$1/2^+$	$1/2$	1314.83 ± 0.20	290 ± 9 ps	$\Lambda\pi^-$
Ξ^-	dss	$1/2^+$	$1/2$	1321.31 ± 0.13	163.9 ± 1.5 ps	$\Xi\pi$
$\Xi^0(1530)$	uss	$3/2^+$	$1/2$	1531.80 ± 0.32	9.1 ± 0.5 MeV	$\Xi\pi$
$\Xi^-(1530)$	dss	$3/2^+$	$1/2$	1535.0 ± 0.6	9.9 ± 1.8 MeV	$\Lambda K^+,\ \Xi\pi$
Ω^-	sss	$3/2^+$	0	1672.45 ± 0.29	82.1 ± 1.1 ps	hadrons $S=-1$
Λ_c^+	udc	$1/2^+?$	0	2286.46 ± 0.14	200 ± 6 fs	$\Lambda_c^+\pi^+$
Σ_c^{++}	uuc	$1/2^+?$	1	2452.02 ± 0.18	2.2 ± 0.3 MeV	$\Lambda_c^+\pi^0$
Σ_c^+	udc	$1/2^+?$	1	2452.9 ± 0.4	<4.6 MeV	$\Lambda_c^+\pi^-$
Σ_c^0	ddc	$1/2^+?$	1	2453.76 ± 0.18	2.2 ± 0.4 MeV	hadrons $S=-2$
Ξ_c^+	usc	$1/2^+?$	$1/2$	2467.9 ± 0.4	442 ± 26 fs	hadrons $S=-2$
Ξ_c^0	dsc	$1/2^+?$	$1/2$	2471.0 ± 0.4	112 ± 12 fs	hadrons $S=-3$
Ω_c^0	ssc	$1/2^+?$	0	2697.5 ± 2.6	69 ± 12 fs	$\Lambda_c^+ + \cdots$
Λ_b^0	udb	$1/2^+?$	0	5624 ± 9	1.230 ± 0.074 ps	

Data from 'Particle Data Group' *Journal of Physics G* **33** (2006); http://pdg.lbl.gov/2006/

Three more metastable hyperons have been recently observed, $\Sigma_b^+ = uub$, $\Sigma_b^- = ddb$ and $\Xi_b^- = dsb$. Not yet observed are $\Sigma_b^0 = udb$, $\Xi_b^0 = usb$, $\Omega_b^- = ssb$ and the double and triple beauties $\Xi_{bb}^0 = ubb$, $\Xi_{bb}^- = dbb$, $\Omega_{bb}^- = sbb$ and $\Omega_{bbb}^- = bbb$.

Appendix 4

Clebsch–Gordan coefficients

$$\langle J_1, J_{z1}; J_2, J_{z2}|J, J_z; J_1, J_2 \rangle = (-1)^{J-J_1-J_2}\langle J_2, J_{z2}; J_1, J_{z1}|J, J_z; J_2, J_1 \rangle$$

1/2⊗1/2

		J, M			
m_1	m_2	1, +1	1, 0	0, 0	1, −1
+ 1/2	+ 1/2	1			
+1/2	−1/2		$\sqrt{1/2}$	$\sqrt{1/2}$	
− 1/2	+1/2		$\sqrt{1/2}$	$-\sqrt{1/2}$	
− 1/2	− 1/2				1

1⊗1/2

		J, M					
m_1	m_2	3/2, + 3/2	3/2, + 1/2	1/2, + 1/2	3/2, − 1/2	1/2, − 1/2	3/2, − 3/2
+ 1	+1/2	1					
+1	− 1/2		$\sqrt{1/3}$	$\sqrt{2/3}$			
0	+1/2		$\sqrt{2/3}$	$-\sqrt{1/3}$			
0	− 1/2				$\sqrt{2/3}$	$\sqrt{1/3}$	
− 1	+1/2				$\sqrt{1/3}$	$-\sqrt{2/3}$	
− 1	− 1/2						1

1⊗1

							J, M				
m_1	m_2	2, +2	2, +1	1, +1	2, 0	1, 0	0, 0	2, −1	1, −1	2, −2
+1	+1	1								
+1	0		$\sqrt{1/2}$	$\sqrt{1/2}$						
0	+1		$\sqrt{1/2}$	$-\sqrt{1/2}$						
+1	−1				$\sqrt{1/6}$	$\sqrt{1/2}$	$\sqrt{1/3}$			
0	0				$\sqrt{2/3}$	0	$-\sqrt{1/3}$			
−1	+1				$\sqrt{1/6}$	$-\sqrt{1/2}$	$\sqrt{1/3}$			
0	−1							$\sqrt{1/2}$	$\sqrt{1/2}$	
−1	0							$\sqrt{1/2}$	$-\sqrt{1/2}$	
−1	−1									1

Appendix 5

Spherical harmonics and d-functions

Spherical harmonics

$$Y_l^m(\theta, \phi) = \sqrt{\frac{(2l+1)(l-m)!}{4\pi(l+m)!}} P_l^m(\cos\theta) e^{im\phi}.$$

$$P_l^{-m}(\cos\theta) = (-1)^m \frac{(l-m)!}{(l+m)!} P_l^m(\cos\theta).$$

$$Y_l^{-m}(\theta, \phi) = (-1)^m Y_l^{m*}(\theta, \phi).$$

$$Y_0^0 = \sqrt{\frac{1}{4\pi}}.$$

$$Y_1^0 = \sqrt{\frac{3}{4\pi}} \cos\theta \qquad Y_1^1 = -\sqrt{\frac{3}{8\pi}} \sin\theta\, e^{i\phi}.$$

$$Y_2^0 = \sqrt{\frac{5}{4\pi}} \left(\frac{3}{2}\cos^2\theta - \frac{1}{2}\right) \qquad Y_2^1 = -\sqrt{\frac{15}{8\pi}} \sin\theta \cos\theta\, e^{i\phi}.$$

$$Y_2^2 = \frac{1}{4}\sqrt{\frac{15}{2\pi}} \sin^2\theta\, e^{i2\phi}.$$

d-functions

$$d_{m,m'}^j(\theta, \phi) = (-1)^{m-m'} d_{m,m'}^j(\theta, \phi) = d_{-m,-m'}^j(\theta, \phi).$$

$$d_{0,0}^1 = \cos\theta.$$

$$d_{1,1}^1 = \frac{1+\cos\theta}{2} \qquad d_{1,0}^1 = -\frac{\sin\theta}{\sqrt{2}} \qquad d_{1,-1}^1 = \frac{1-\cos\theta}{2}.$$

$$d_{\frac{1}{2},\frac{1}{2}}^1 = \cos\frac{\theta}{2} \qquad d_{\frac{1}{2},-\frac{1}{2}}^1 = -\sin\frac{\theta}{2}.$$

395

Appendix 6

Experimental and theoretical discoveries in particle physics

This table gives some indication of the historical development of particle physics. However, the discoveries are rarely due to a single person and never happen instantaneously. The dates indicate the year of the most relevant publication(s), the names are those of the main contributors.

1896	H. Bequerel	Discovery of particle radiation (radioactivity)
1897	J. J. Thomson	Discovery of the electron
1912	V. Hess	Discovery of the cosmic rays
	C. T. R. Wilson	Cloud chamber
1928	P. A. M. Dirac	Relativistic wave equation for the electron
	H. Geiger	Geiger counter
1929	E. Hubble	Expansion of the Universe
1930	W. Pauli	Neutrino hypothesis
	E. Lawrence	Cyclotron
1932	J. Chadwick	Discovery of the neutron
	C. Anderson, P. Blackett and G. Occhialini	Discovery of the positron
1933	F. Zwicky	Discovery of dark matter in the Universe
1934	E. Fermi	Theory of beta decay
1935	H. Yukawa	Theory of strong nuclear forces
	P. Cherenkov	Cherenkov effect
1937	J. Street and E. Stevenson, C. Anderson and S. Neddermeyer	Penetrating component of cosmic rays (muon)
	E. Majorana	Wave equation for completely neutral fermions
1944/45	V. Veksler, E. McMillan	Principle of phase stability in accelerators
1947	W. Lamb	Lamb shift
	P. Kusch	Measurement of electron magnetic moment
	M. Conversi, E. Pancini, O. Piccioni	Leptonic character of the muon

	G. Occhialini,	Discovery of the pion
	C. Powell *et al.*	
	G. Rochester and	Discovery of V^0 particles
	C. Butler	
1948	S. Tomonaga,	Quantum electrodynamics
	R. Feynman,	
	J. Schwinger *et al.*	
1952	*Cosmotron operational at BNL at 3 GeV*	
	D. Glaser	Bubble chamber
	E. Fermi *et al.*	Discovery of the baryon resonance $\Delta(1236)$
1953	Cosmic ray experiments	$\theta-\tau$ puzzle
	M. Gell-Mann,	Strangeness hypothesis
	K. Nishijima	
1954	*Bevatron operational at Berkeley at 7 GeV*	
1955	O. Chamberlain *et al.*	Discovery of the antiproton
	M. Conversi, A. Gozzini	Spark chamber
	M. Gell-Mann, A. Pais	K^0 oscillation proposal
1956	F. Reines *et al.*	Discovery of electron antineutrino
	T. D. Lee and C. N. Yang	Hypothesis of parity violation
1957	C. S. Wu *et al.*	Discovery of parity violation
	B. Pontecorvo	Neutrino oscillations hypothesis
	Synchro-phasatron operational at Dubna at 10 GeV	
1959	*Proton synchrotrons PS at CERN, AGS at BNL operational at 30 GeV*	
1960	B. Touschek	Proposal of e^+e^- storage ring (ADA)
1961	L. Alvarez and others	Discovery of meson resonances
	S. Glashow	First work on electroweak gauge theory
1962	M. Schwartz,	Discovery of muon neutrino
	L. Lederman,	
	J. Steinberger *et al.*	
1963	N. Cabibbo	Flavour mixing
1964	V. Fitch and	Discovery of $\mathcal{CP}$ violation
	J. Cronin *et al.*	
	G. Zweig, M. Gell-Mann	Quark model
	Bubble chamber	Discovery of the Ω^-
	experiment at BNL	
	P. W. Higgs, P. Brout,	Theoretical mechanism for mass generation
	F. Engrelt	
	A. Salam	Electroweak model
1967	S. Weinberg	Electroweak model
	J. Friedman, H. Kendall	Quark structure of the proton
	and R. Taylor *et al.*	
	Proton synchrotron operational at Serpukhov at 76 GeV	
	Electron linear accelerator operational at SLAC at 20 GeV	
1968	C. Charpak *et al.*	Multi-wire proportional chamber
1970	R. Davis *et al.*, J. Bahcall	Solar neutrino puzzle
	S. Glashow, I. Iliopoulos,	Fourth quark hypothesis
	L. Maiani	
1971	K. Niu *et al.*	Discovery of charm
	A. H. Walenta *et al.*	Drift chamber
	Intersecting proton Storage Rings (ISR) operational at CERN $(30+30 GeV)$	

1972	*Fermilab proton synchrotron operational at 200 GeV, later at 500 GeV*	
	SPEAR e^+e^- (4 + 4 GeV) storage ring operational at Stanford	
	G. 't Hooft and	Renormalisability of electroweak theory
	M. Veltman	
1973	Gargamelle bubble	Discovery of weak neutral currents
	chamber	
	D. Gross, D. Politzer,	Quantum chromodynamics
	F. Wilczek, H. Fritzsch,	
	M. Gell-Mann,	
	G. 't Hooft *et al.*	
	M. Kobaiashi,	Quark mixing for three families
	K. Maskawa	
1974	B. Richter *et al.*,	Discovery of J/ψ hidden charm particle
	S. Ting *et al.*	
1975	M. Perl *et al.*	Discovery of the τ lepton
1976	L. Lederman *et al.*	Discovery of Υ hidden beauty particles
	Super Proton Synchrotron (SPS) operational at CERN at 400 GeV	
1979	PETRA experiments	Observation of gluon jets
	at DESY	
1981	*First collisions in the SPS $p\bar{p}$ storage ring at CERN (270 + 270 GeV)*	
1983	C. Rubbia *et al.*	Discovery of the W and Z particles
1986	*TRISTAN e^+e^- (15+15 GeV) storage ring operational at KEK at Tsukuba*	
1987	M. Koshiba *et al.*	Observation of neutrinos from a supernova
1989	*Stanford Linear Collider e^+e^- (50 + 50 GeV) operational*	
	LEP e^+e^- storage ring operational at CERN (50 + 50 GeV). Later	
	105+105 GeV	
	LEP experiments	Three neutrino types
	Kamiokande experiment	Confirmation of solar neutrino deficit
1990	T. Berners-Lee,	World Wide Web proposal
	R. Cailliau (CERN)	
1991	*HERA ep collider operational at DESY (30 + 820 GeV). Later 30*	
	+ 920 GeV	
1992	GALLEX experiment	Solar neutrino deficit at low energy
1995	CDF experiment	Discovery of the top-quark
1997	LEP experiments	W bosons self-coupling
1998	Solar and atmospheric ν	Discovery of neutrino oscillations
	experiments	
1999	*'Beauty factories' operational, KEKB at Tsukuba and PEP2 at Stanford*	
2001	K. Niwa *et al.*	Discovery of the tau neutrino
2008	*Commissioning of the LHC proton collider at CERN (7 + 7 TeV)*	

Solutions

1.2. $s = (3E)^2 - 0 = 9E^2 = 9(p^2 + m^2) = 88.9$ GeV^2; $m = \sqrt{s} = 9.43$ GeV.

1.3. $\Gamma_{\pi^\pm} = h/\tau_{\pi^\pm} = \left(6.6 \times 10^{-16} \text{ eVs}\right)/\left(2.6 \times 10^{-8} \text{ s}\right) = 25$ neV,
$\Gamma_K = 54$ neV, $\Gamma_\Lambda = 2.5$ μeV.

1.6. Our reaction is $p + p \to p + p + m$. In the CM frame the total momentum is zero. The lowest energy configuration of the system is when all particles in the final state are at rest.

    a. Let us write down the equality between the expressions of s in the CM and L frames, i.e.

$$s = \left(E_p + m_p\right)^2 - p_p^2 = \left(2m_p + m\right)^2.$$

Recalling that $E_p^2 = m_p^2 + p_p^2$, we have $E_p = \dfrac{\left(2m_p + m\right)^2 - 2m_p^2}{2m_p} = m_p + 2m + \dfrac{m^2}{2m_p}$.

    b. The two momenta are equal and opposite because the two particles have the same mass, hence we are in the CM frame. The threshold energy E_p^* is given by $s = \left(2E_p^*\right)^2 = \left(2m_p + m\right)^2$ which gives $E_p^* = m_p + m/2$.

    c. $E_p = 1.218$ GeV; $p_p = 0.78$ GeV; $T_p = 280$ MeV; $E_p^* = 1.007$ GeV; $p_p^* = 0.36$ GeV.

1.7. a. $s = \left(E_\gamma + m_p\right)^2 - p_\gamma^2 = \left(E_\gamma + m_p\right)^2 - E_\gamma^2 = \left(m_p + m_\pi\right)^2 = 1.16$ GeV^2,
hence we have $E_\gamma = 149$ MeV.

    b. $s = \left(E_\gamma + E_p\right)^2 - \left(\mathbf{p}_\gamma + \mathbf{p}_p\right)^2 = m_p^2 + 2E_\gamma E_p - 2\mathbf{p}_\gamma \cdot \mathbf{p}_p$. For a given proton energy, s reaches a maximum for a head-on collision. Consequently, $\mathbf{p}_\gamma \cdot \mathbf{p}_p = -E_\gamma p_p$ and, taking into account that the energies are very large, $s = m_p^2 + 2E_\gamma \left(E_p + p_p\right) \approx m_p^2 + 4E_\gamma E_p$. In conclusion

$$E_p = \frac{s - m_p^2}{4E_\gamma} = \frac{(1.16 - 0.88) \times 10^{18} \text{ eV}^2}{4 \times 10^{-3} \text{ eV}} = 7 \times 10^{19} \text{ eV} = 70 \text{ EeV}.$$

c. The attenuation length is $\lambda = 1/(\sigma\rho) = 1.7 \times 10^{25}$ m $\simeq 500$ Mpc (1 Mpc $= 3.1 \times 10^{22}$ m). This is a short distance on the cosmological scale. The cosmic ray spectrum (Fig. 1.10) should not go beyond the above computed energy. This is called the Greisen, Zatzepin and Kusmin (GZK) bound. The Auger Observatory is now exploring this extreme energy region.

1.11. We must consider the reaction

$$M \rightarrow m_1 + m_2.$$

$$m_2, \mathbf{p}_{2f}^*, E_2^* \quad \longleftarrow \quad M \quad \longrightarrow \quad m_1, \mathbf{p}_{1f}^*, E_1^*$$

The figure defines the CM variables.
We can use equations (P1.5) and (P1.6) with $\sqrt{s} = M$, obtaining

$$E_{2f}^* = \frac{M^2 + m_2^2 - m_1^2}{2M} \qquad E_{1f}^* = \frac{M^2 + m_1^2 - m_2^2}{2M}.$$

The corresponding momenta are

$$\mathbf{p}_f^* \equiv \mathbf{p}_{1f}^* = -\mathbf{p}_{2f}^* = \sqrt{E_{1f}^{*2} - m_1^2} = \sqrt{E_{2f}^{*2} - m_2^2}.$$

1.14. Let us call x a coordinate along the beam. The velocity of the pions in L should not be larger than the velocity of the muon in the CM, i.e. $\beta_\pi \leq \beta_{\mu x}^* \leq \beta_\mu^*$. Let us use the formulae found in Problem 1.11 to calculate the Lorentz parameters for the CM–L transformation:

$$\beta_\mu^* = \frac{p^*}{E_\mu^*} = \frac{m_\pi^2 - m_\mu^2}{m_\pi^2 + m_\mu^2} \qquad \gamma_\mu^* = \frac{E_\mu^*}{m_\mu} = \frac{m_\pi^2 + m_\mu^2}{2m_\mu m_\pi} \Rightarrow \beta_\mu^* \gamma_\mu^* = \frac{m_\pi^2 - m_\mu^2}{2m_\mu m_\pi}.$$

The condition $\beta_\pi < \beta_\mu^*$ gives $p_\pi = \beta_\pi \gamma_\pi m_\pi < \beta_\mu^* \gamma_\pi^* m_\pi = \dfrac{m_\pi^2 - m_\mu^2}{2m_\mu} = 39.35\,\text{MeV}.$

1.15. When dealing with a Lorentz transformation problem, the first step is the accurate drawing of the momenta in the two frames and the definition of the kinematic variables.

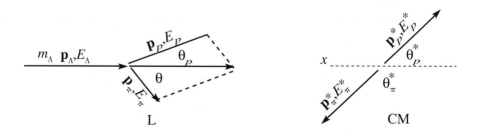

Using the expressions we found in the introduction we have:

a. $E_\pi^* = \dfrac{m_A^2 - m_p^2 + m_\pi^2}{2m_A} = 0.17\,\text{GeV}; \qquad E_p^* = 0.95\,\text{GeV};$

$p_\pi^* = p_p^* = \sqrt{E_\pi^{*2} - m_\pi^2} = 0.096\,\text{GeV}.$

b. We calculate the Lorentz factors for the transformation:

$$E_A = \sqrt{p_A^2 + m_A^2} = 2.29\,\text{GeV}; \; \beta_A = \frac{p_A}{E_A} = 0.87; \; \gamma_A = \frac{E_A}{m_A} = 2.05.$$

c. We do the transformation and calculate the requested quantities

$p_\pi \sin\theta_\pi = p_\pi^* \sin\theta_\pi^* = 0.096 \times \sin 210° = -0.048\,\text{GeV}.$

$p_\pi \cos\theta_\pi = \gamma_A\left(p_\pi^* \cos\theta_\pi^* + \beta_A E_\pi^*\right)$
$= 2.05(0.096 \times \cos 210° + 0.87 \times 0.17) = 0.133\,\text{GeV}.$

$\tan\theta_\pi = \dfrac{-0.048}{0.133} = -0.36, \quad \theta_\pi = -20°, \quad p_\pi = \sqrt{\left(p_\pi \sin\theta_\pi\right)^2 + \left(p_\pi \cos\theta_\pi\right)^2}$
$= 0.141\,\text{GeV}.$

$p_p \sin\theta_p = p_p^* \sin\theta_p^* = 0.048\,\text{GeV}.$

$p_p \cos\theta_p = \gamma_A\left(p_p^* \cos\theta_p^* + \beta_A E_p^*\right) = 2.05(0.096 \times \cos 30° + 0.87 \times 0.95)$
$= 1.86\,\text{GeV}.$

$\tan\theta_p = \dfrac{0.048}{1.86} = 0.026, \quad \theta_p = 1.5°.$

$p_p = \sqrt{\left(p_p \sin\theta_p\right)^2 + \left(p_p \cos\theta_p\right)^2} = 1.9\,\text{GeV}, \quad \theta = \theta_p - \theta_\pi = 21.5°.$

1.17. We continue to refer to the figure of Solution 1.15. We shall solve our problem in two ways: by performing a Lorentz transformation and by using the Lorentz invariants.

We start with the first method. We calculate the Lorentz factors. The energy of the incident proton is $E_1 = \sqrt{p_1^2 + m_p^2} = 3.143\,\text{GeV}$. Firstly, let us calculate the CM energy squared of the two-proton system (i.e. its mass squared).

$p_{pp} = p_1 = 3\,\text{GeV}; \; E_{pp} = E_1 + m_p = 4.081\,\text{GeV}.$
Hence $s = 2m_p^2 + 2E_1 m_p = 7.656\,\text{GeV}^2.$

The Lorentz factors are $\beta_{pp} = p_{pp}/E_{pp} = 0.735$ and $\gamma_{pp} = E_{pp}/\sqrt{s_{pp}} = 1.47.$

Since all the particles are equal, we have $E_1^* = E_2^* = E_3^* = E_4^* = \sqrt{s}/2$
$= 1.385$ GeV; $p_1^* = p_2^* = p_3^* = p_4^* = \sqrt{E_1^{*2} - m_p^2} = 1.019$ GeV.

We now perform the transformation. To calculate the angle we must calculate firstly the components of the momenta

$$p_3 \sin \theta_{13} = p_3^* \sin \theta_{13}^* = 1.019 \times \sin 10° = 0.177 \text{ GeV.}$$

$$\begin{aligned} p_3 \cos \theta_{13} &= \gamma\left(p_3^* \cos \theta_{13}^* + \beta E_3^*\right) \\ &= 1.473 \times (1.019 \times \cos 10° + 0.735 \times 1.385) = 2.978 \text{ GeV.} \end{aligned}$$

$$\tan \theta_{13} = \frac{0.177}{2.978} = 0.0594, \qquad \theta_{13} = 3°.$$

$$-p_4 \sin \theta_{14} = -p_4^* \sin \theta_{14}^* = -1.019 \times \sin 170° = -0.1769 \text{ GeV.}$$

$$\begin{aligned} p_4 \cos \theta_{14} &= \gamma\left(p_4^* \cos \theta_{14}^* + \beta E_4^*\right) \\ &= 1.473 \times (1.019 \times \cos 170° + 0.735 \times 1.385) = 0.0213 \text{ GeV.} \end{aligned}$$

$$\tan \theta_{14} = -0.1769/0.0213 = -8.305,$$

$$\theta_{14} = -83° \quad \Rightarrow \quad \theta_{34} = \theta_{13} - \theta_{14} = 86°.$$

In relativistic conditions the angle between the final momenta in a collision between two equal particles is always, as in this example, smaller than 90°.

We now solve the problem using the invariants and the expressions in the introduction. We want the angle between the final particles in L. We then write down the expression of s in L in the initial state, which we have already calculated, i.e.

$$s = (E_3 + E_4)^2 - (\mathbf{p}_3 + \mathbf{p}_4)^2 = m_3^2 + m_4^2 + 2E_3E_4 - 2\mathbf{p}_3 \cdot \mathbf{p}_4$$

that gives $\mathbf{p}_3 \cdot \mathbf{p}_4 = m_p^2 + E_3E_4 - s/2$ and hence

$$\cos \theta_{34} = \frac{m_p^2 + E_3E_4 - s/2}{p_3 p_4}.$$

We need E_3 and E_4 (and their momenta); we can use (P1.13) if we have t. With the data of the problem we can calculate t in the CM:

$$\begin{aligned} t &= 2m_p^2 + 2p_i^{*2} \cos \theta_{13}^* - 2E_i^{*2} = 2p_i^{*2}\left(\cos \theta_{13}^* - 1\right) \\ &= 2 \times 1.019^2 (\cos 10° - 1) = -0.0316 \text{ GeV}^2. \end{aligned}$$

We then obtain

$$E_3 = \frac{s + t - 2m_p^2}{2m_p} = \frac{7.656 - 0.0316 - 2 \times 0.938^2}{2 \times 0.938} = 3.126 \text{ GeV,}$$

$$p_3 = 2.982 \text{ GeV.}$$

From energy conservation we have

$$E_4 = E_1 + m_p - E_3 = 3.143 + 0.938 - 3.126 = 0.955\,\text{GeV},$$

$$p_4 = 0.179\,\text{GeV}.$$

Finally we obtain

$$\cos\theta_{34} = \frac{0.938^2 + 3.126 \times 0.955 - 7.656/2}{2.982 \times 0.179} = 0.0696 \quad \Rightarrow \theta_{34} = 86°.$$

1.20. In this case the reference frames L and CM coincide. We have

$$\mathbf{p}_{\pi^0} + \mathbf{p}_n = 0 \quad \Rightarrow \quad p_{\pi^0} = p_n = p^*.$$

The total energy is $E = E_{\pi^0} + E_n = m_{\pi^-} + m_n = 1079\,\text{MeV}$.

Subtracting the members of the two relationships $E_n^2 = p^{*2} + m_n^2$ and $E_{\pi^0}^2 = p^{*2} + m_{\pi^0}^2$ we obtain $E_n^2 - E_{\pi^0}^2 = m_n^2 - m_{\pi^0}^2$.

From $E_n = E - E_{\pi^0}$ we have $E_n^2 = E^2 + E_{\pi^0}^2 - 2EE_{\pi^0}$; and finally

$$E_{\pi^0} = \frac{E^2 + E_{\pi^0}^2 - E_n^2}{2E} = \frac{E^2 + m_{\pi^0}^2 - m_n^2}{2E} = 138.8\,\text{MeV},$$

$T_n = E - E_{\pi^0} - m_n = 0.6\,\text{MeV}$.

The Lorentz factors are $\gamma_{\pi^0} = E_{\pi^0}/m_{\pi^0} = 1.028$ and

$$\beta_{\pi^0} = \sqrt{1 - 1/\gamma_{\pi^0}^2} = 0.23.$$

The distance travelled in a lifetime is then

$$l = \gamma_{\pi^0}\tau_{\pi^0}\beta_{\pi^0}c = 1.028 \times 8.4 \times 10^{-17} \times 0.23 \times 3 \times 10^8 = 6\,\text{nm}.$$

1.21. The equation of motion is $q\mathbf{v} \times \mathbf{B} = d\mathbf{p}/dt$. Since in this case the Lorentz factor γ is constant, we can write $q\mathbf{v} \times \mathbf{B} = \gamma m \dfrac{d\mathbf{v}}{dt}$. The centripetal acceleration is then: $\left|\dfrac{d\mathbf{v}}{dt}\right| = \dfrac{qvB}{\gamma m} = \dfrac{v^2}{\rho}$. Simplifying we obtain $p = qB\rho$. We now want pc in GeV, B in tesla and ρ in metres. Starting from $pc = qcB\rho$ we have

$$pc[\text{GeV}] \times 1.6 \times 10^{-10}[\text{J / GeV}]$$
$$= 1.6 \times 10^{-19}[\text{C}] \times 3 \times 10^8[\text{m/s}] \times B[\text{T}] \times \rho[\text{m}].$$

Finally in NU: $p[\text{GeV}] = 0.3 \times B[\text{T}] \times \rho[\text{m}]$.

1.23. The Lorentz factor of the antiproton is $\gamma = \sqrt{p^2 + m^2}/m = 1.62$ and its velocity $\beta = \sqrt{1 - \gamma^{-2}} = 0.787$. The condition in order to have the antiproton above the Cherenkov threshold is that the index is $n \geq 1/\beta = 1.27$.

If the index is $n = 1.5$, the Cherenkov angle is given by $\cos\theta = 1/n\beta = 0.85$. Hence $\theta = 32°$.

1.24. The speed of a particle of momentum $p = m\gamma\beta$ is

$$\beta = \left(1 + \frac{m^2}{p^2}\right)^{-1/2} \approx 1 - \frac{m^2}{2p^2},$$ which is a good approximation for speeds

close to c. The difference between the flight times is $\Delta t = L\dfrac{m_2^2 - m_1^2}{2p^2}$ in

NU. In order to have $\Delta t > 600\,\text{ps}$, we need a base-length $L > 26\,\text{m}$.

1.26. Superman saw the light blue-shifted due to the Doppler effect. Taking for the wavelengths $\lambda_R = 650\,\text{nm}$ and $\lambda_G = 520\,\text{nm}$, we have $\nu_G/\nu_R = 1.25$.

Solving for β the Doppler shift expression $\nu_G = \nu_R\sqrt{\frac{1+\beta}{1-\beta}}$, we obtain $\beta = 0.22$.

2.3. The second photon moves backwards. The total energy is $E = E_1 + E_2$; the total momentum is $P = p_1 - p_2 = E_1 - E_2$. The square of the mass of the two-photon system is equal to the square of the pion mass:
$m_{\pi^0}^2 = (E_1 + E_2)^2 - (E_1 - E_2)^2 = 4E_1E_2$, from which we obtain

$$E_2 = \frac{m_{\pi^0}^2}{4E_1} = \frac{135^2}{4\times150} = 30.4\,\text{MeV}. \quad \text{The speed of the } \pi^0 \text{ is}$$

$$\beta = \frac{P}{E} = \frac{E_1 - E_2}{E_1 + E_2} = 0.662.$$

2.4. The Lorentz factor for $E_\mu = 5\,\text{GeV}$ is $\gamma = E_\mu/m_\mu = 47$. In its rest frame the distance to the Earth surface is $l_0 = l/\gamma = 630\,\text{m}$. For $E_\mu = 5\,\text{TeV}$, the distance to the Earth is $l_0 = l/\gamma = 0.63\,\text{m}$. The first muon travels in a lifetime $\gamma\beta c\tau \approx \gamma c\tau = 28\,\text{km}$, the second would travel $28\,000\,\text{km}$ if it did not hit the surface first.

2.8. Since the decay is isotropic, the probability of observing a photon is a constant $P(\cos\theta^*, \phi^*) = K$. We determine K by imposing that the probability of observing a photon at any angle is 2, i.e. the number of photons.

We have $2 = \int K\sin\theta^*\,d\theta^*\,d\phi = \int_0^{2\pi} d\phi \int_0^\pi Kd(\cos\theta^*) = K4\pi$. Hence $K = 1/2\pi$ and $P(\cos\theta^*, \phi^*) = 1/2\pi$.

The distribution is isotropic in azimuth in L too. To have the dependence

of θ that is given by $P(\cos\theta) \equiv \dfrac{dN}{d\cos\theta} = \dfrac{dN}{d\cos\theta^*}\dfrac{d\cos\theta^*}{d\cos\theta}$, we must cal-

culate the 'Jacobian' $J = \dfrac{d\cos\theta^*}{d\cos\theta}$.

Calling β and γ the factors of the transformation and taking into account that $p^* = E^*$, we have

$$p\cos\theta = \gamma(p^*\cos\theta^* + \beta E^*) = \gamma p^*(\cos\theta^* + \beta)$$
$$E = p = \gamma(E^* + \beta p^*\cos\theta^*) = \gamma p^*(1 + \beta\cos\theta^*).$$

We differentiate the first and third members of these relationships, taking into account that p^* is a constant. We obtain

$$dp \times \cos\theta + p \times d(\cos\theta) = \gamma p^* d(\cos\theta^*) \quad \Rightarrow \quad \frac{dp}{d\cos\theta^*}\cos\theta + p\frac{d\cos\theta}{d\cos\theta^*} = \gamma p^*$$

$$dp = \gamma\beta p^* d(\cos\theta^*) \quad \Rightarrow \quad \frac{dp}{d\cos\theta^*} = \gamma\beta p^*$$

and $\quad J^{-1} = \dfrac{d\cos\theta}{d\cos\theta^*} = \gamma\dfrac{p^*}{p}(1 - \beta\cos\theta).$

The inverse transformation is $\quad E^* = \gamma(E - \beta p\cos\theta),\quad$ i.e.

$$p^* = \gamma p(1 - \beta\cos\theta), \text{ giving } J^{-1} = \frac{d\cos\theta}{d\cos\theta^*} = \gamma^2(1 - \beta\cos\theta)^2.$$

Finally we obtain $\quad P(\cos\theta) \equiv \dfrac{dN}{d\cos\theta} = \dfrac{1}{2\pi}\gamma^{-2}(1 - \beta\cos\theta)^{-2}.$

2.11. The energy needed to produce an antiproton is minimum when the Fermi motion is opposite to the beam direction. If E_f is the total energy of the target proton and p_f its momentum, the threshold condition is $(E_p + E_f)^2 - (p_p - p_f)^2 = (4m_p)^2$. From this we have $E_p E_f + p_p p_f = 7m_p^2$. We simplify by setting $p_p \approx E_p$ obtaining

$$E_p = \frac{7m_p^2}{E_f + p_f} \approx \frac{7m_p^2}{m_p + p_f} \approx 7m_p\left(1 - \frac{p_f}{m_p}\right) = 5.5\,\text{GeV}.$$

This value should be compared to $E_p = 6.6\,\text{GeV}$ on free protons.

2.13. Considering the beam energy and the event topology, the event is probably an associate production of a K^0 and a Λ. Consequently the V^0 may be one of these two particles. The negative track is in both cases a π, while the positive track may be a π or a proton. We need to measure the mass of the V. With the given data we start by calculating the Cartesian components of the momenta:

$$p_x^- = 121 \times \sin(-18.2°)\cos 15° = -36.5\,\text{MeV};$$
$$p_y^- = 121 \times \sin(-18.2°)\sin 15° = -9.8\,\text{MeV};$$
$$p_z^- = 121 \times \cos(-18.2°) = 115\,\text{MeV}.$$
$$p_x^+ = 1900 \times \sin(20.2°)\cos(-15°) = 633.7\,\text{MeV};$$
$$p_y^+ = 1900 \times \sin(20.2°)\sin(-15°) = -169.8\,\text{MeV};$$
$$p_z^+ = 1900 \times \cos(20.2°) = 1783.1\,\text{MeV}.$$

Summing the components, we obtain the momentum of the V, i.e. $p = 1998\,\text{MeV}$.

The energy of the negative pion is $E^- = \sqrt{(p^-)^2 + m_\pi^2} = 185\,\text{MeV}$. If the positive track is a π its energy is $E_\pi^+ = \sqrt{(p^+)^2 + m_\pi^2} = 1905\,\text{MeV}$, while if it is a proton its energy is $E_p^+ = 2119\,\text{MeV}$.

The energy of the V is $E_\pi^V = 2090\,\mathrm{MeV}$ in the first case, $E_p^V = 2304\,\mathrm{MeV}$ in the second case. The mass of the V is consequently $m_\pi^V = \sqrt{E_\pi^{V2} - p^2} = 620\,\mathrm{MeV}$ in the first hypothesis, $m_p^V = 1150\,\mathrm{MeV}$ in the second. Within the $\pm 4\%$ uncertainty, the first hypothesis is incompatible with any known particle, while the second is compatible with the particle being a Λ.

3.2. Strangeness conservation requires that a K^+ or a K^0 be produced together with the K^-. The third component of the isospin in the initial state is $-1/2$. Let us check if it is conserved in the two reactions. The answer is yes for $\pi^- + p \to K^- + K^+ + n$ because in the final state we have $I_z = -1/2 + 1/2 + 1/2 = +1/2$, and yes also for $\pi^- + p \to K^- + K^0 + p$ because in the final state we have $I_z = -1/2 - 1/2 + 1/2 = -1/2$. The threshold of the first reaction is just a little smaller than that of the second reaction because $m_n + m_{K^+} < m_p + m_{K^0}$ (1433 MeV < 1436 MeV). For the former we have

$$E_\pi = \frac{(2m_K + m_n)^2 - m_\pi^2 - m_p^2}{2m_p} = 1.5\,\mathrm{GeV}.$$

3.4. (1) OK, S; (2) OK, W; (3) Violates $\mathcal{L}_\mu$; (4) OK, EM; (5) Violates $\mathcal{C}$; (6) Cannot conserve both energy and momentum; (7) Violates $\mathcal{B}$ and S; (8) Violates $\mathcal{B}$ and S; (9) Violates J and $\mathcal{L}_e$; (10) Violates energy conservation.

3.8. (a) NO for J and $\mathcal{L}$; (b) NO for J and $\mathcal{L}$; (c) YES; (d) NO for $\mathcal{L}$; (e) YES; (f) NO for $\mathcal{L}_e$ and $\mathcal{L}_\mu$; (g) NO for $\mathcal{L}$; (h) YES.

3.9.

$$|\pi^- p\rangle = |1, -1\rangle\left|\frac{1}{2}, +\frac{1}{2}\right\rangle = \sqrt{\frac{1}{3}}\left|\frac{3}{2}, -\frac{1}{2}\right\rangle - \sqrt{\frac{2}{3}}\left|\frac{1}{2}, -\frac{1}{2}\right\rangle$$

$$|\pi^+ p\rangle = |1, +1\rangle\left|\frac{1}{2}, +\frac{1}{2}\right\rangle = \left|\frac{3}{2}, +\frac{3}{2}\right\rangle$$

$$|\Sigma^0 K^0\rangle = |1, 0\rangle\left|\frac{1}{2}, -\frac{1}{2}\right\rangle = \sqrt{\frac{2}{3}}\left|\frac{3}{2}, -\frac{1}{2}\right\rangle + \sqrt{\frac{1}{3}}\left|\frac{1}{2}, -\frac{1}{2}\right\rangle$$

$$|\Sigma^- K^+\rangle = |1, -1\rangle\left|\frac{1}{2}, +\frac{1}{2}\right\rangle = \sqrt{\frac{1}{3}}\left|\frac{3}{2}, -\frac{1}{2}\right\rangle - \sqrt{\frac{2}{3}}\left|\frac{1}{2}, -\frac{1}{2}\right\rangle$$

$$|\Sigma^+ K^+\rangle = |1, +1\rangle\left|\frac{1}{2}, +\frac{1}{2}\right\rangle = \left|\frac{3}{2}, +\frac{3}{2}\right\rangle$$

$$\langle K^+\Sigma^+|\pi^+ p\rangle = A_{3/2}; \quad \langle \Sigma^- K^+|\pi^- p\rangle = \sqrt{\frac{1}{3}}\sqrt{\frac{1}{3}}A_{3/2} = \frac{1}{3}A_{3/2};$$

$$\langle \Sigma^0 K^0|\pi^- p\rangle = \sqrt{\frac{2}{3}}\sqrt{\frac{1}{3}}A_{3/2} = \frac{\sqrt{2}}{3}A_{3/2}.$$

Hence:

$$\sigma(\pi^+ p \to \Sigma^+ K^+) : \sigma(\pi^- p \to \Sigma^- K^+) : \sigma(\pi^- p \to \Sigma^0 K^0) = 9 : 1 : 2.$$

3.12. We can proceed as in the previous solutions or also as follows:

$$|p, d\rangle = \left|\frac{1}{2}, \frac{1}{2}\right\rangle |0, 0\rangle = \left|\frac{1}{2}, +\frac{1}{2}\right\rangle$$

$$\left|\frac{1}{2}, +\frac{1}{2}\right\rangle = -\sqrt{\frac{1}{3}}|1, 0\rangle \left|\frac{1}{2}, +\frac{1}{2}\right\rangle + \sqrt{\frac{2}{3}}|1, 1\rangle \left|\frac{1}{2}, -\frac{1}{2}\right\rangle$$

$$= -\sqrt{\frac{1}{3}}|\pi^0\rangle |^3\text{He}\rangle + \sqrt{\frac{2}{3}}|\pi^+\rangle |^3\text{H}\rangle.$$

$$\sigma(p + d \to {}^3\text{He} + \pi^0)/\sigma(p + d \to {}^3\text{H} + \pi^+) = 1/2.$$

3.15. $\sigma(1) : \sigma(2) : \sigma(3) = \left|-\frac{1}{\sqrt{6}}A_0 + \frac{1}{2}A_1\right|^2 : \left|\frac{1}{\sqrt{6}}A_0\right|^2 : \left|\frac{1}{\sqrt{6}}A_0 + \frac{1}{2}A_1\right|^2.$

3.17. (a) The initial parity is $P_i = P(\pi)P(d)(-1)^{l_i} = (-)(+)(+) = -$ and the final one is $P_f = P(n)P(n)(-1)^{l_f} = (-1)^{l_f}$. Parity conservation requires $l_f = 1, 3, 5 \ldots$ Angular momentum conservation requires that $l_f < 3$. Only $l_f = 1$ remains. The two-neutron wave function must be completely antisymmetric. Since the spatial part is antisymmetric, the spin part must be symmetric. In conclusion the state is ${}^3S_1$, with total spin $S = 1$.

(b) Since $P_i = +$, l_f is even. The spin function is antisymmetric. Hence the state is ${}^1S_0$ and its total spin is $S = 0$.

3.19. 1. $C(\bar{p}p) = (-1)^{l+s} = C(n\pi^0) = +$. Then $l + s = $ even. The possible states are ${}^1S_0, {}^3P_1, {}^3P_2, {}^3P_3, {}^1D_2$.

2. The orbital momentum is even, because the wave function of the $2\pi^0$ state must be symmetric. Since the total angular momentum is just orbital momentum, only the states ${}^1S_0, {}^3P_2, {}^1D_2$ are left. Parity conservation gives $P(2\pi^0) = + = P(\bar{p}p) = (-1)^{l+1}$. Hence, $l = $ odd, leaving only ${}^3P_2$.

3.21. It is convenient to prepare a table with the possible values of the initial J^{PC} and of the final l^{PC} with $l = J$ to satisfy angular momentum conservation. Only the cases with the same parity and charge conjugation are allowed. Recall that $P(\bar{p}p) = (-1)^{l+1}$ and $C(\bar{p}p) = (-1)^{l+s}$.

	${}^1S_0$	${}^3S_1$	${}^1P_1$	${}^3P_0$	${}^3P_1$	${}^3P_2$	${}^1D_2$	${}^3D_1$	${}^3D_2$	${}^3D_3$
J^{PC}	0^{-+}	1^{--}	1^{+-}	0^{++}	1^{++}	2^{++}	2^{-+}	1^{--}	2^{--}	3^{--}
l^{PC}	0^{++}	1^{--}	1^{--}	0^{++}	1^{--}	2^{++}	2^{++}	1^{--}	2^{++}	3^{--}
		Y		Y		Y		Y		Y

In conclusion: (1) $^1S_0$; (2) $^3S_1$, $^3D_1$; and (3) $^3P_2$.

3.23. Λ_b is neutral. (a) Violates charm and beauty, (b) and (c) are allowed, (d) violates beauty, (e) violates baryon number.

4.3. The ρ decays strongly into 2π, hence $G=+$. The possible values of its isospin are 0, 1 and 2. In the three cases the Clebsch–Gordan coefficients are $\langle 1,0|1,0;1,0\rangle = 0$, $\langle 0,0|1,0;1,0\rangle \neq 0$ and $\langle 2,0|1,0;1,0\rangle \neq 0$. Hence $I=1$. Since $I=1$, the isospin wave function is antisymmetric. The spatial wave function must consequently be antisymmetric, i.e. the orbital momentum of the two π must be $l=$ odd. The ρ spin is equal to l. $C=(-1)^l=-1$. $P=(-1)^l=-1$.

4.6. (a) The decay is strong. (b) The initial strangeness in the reaction $K^- + p \to \pi^- + \Sigma^+(1385)$ is $S=-1$. The strangeness of the $\Sigma(1385)$ is $S=-1$. Since the isospin is conserved in the strong decay, the isospin of the $\Sigma(1385)$ is equal to the isospin of the $\pi^+\Lambda$ system, i.e. is 1.

4.7. (1) Two equal bosons cannot be in an antisymmetric state; (2) $C(2\pi^0)=+1$; (3) the Clebsch–Gordan coefficient $\langle 1,0;1,0|1,0\rangle = 0$.

4.11. It is useful to prepare a table with the quantum numbers of the relevant states

	$\bar{p}p\,^3S_1$	$\bar{p}p\,^3S_1$	$\bar{p}p\,^1S_0$	$\bar{p}p\,^1S_0$	$\bar{p}n\,^3S_1$	$\bar{p}n\,^1S_0$
J^P	1^-	1^-	0^-	0^-	1^-	0^-
C	$-$	$-$	$+$	$+$	X	X
I	0	1	0	1	1	1
G	$-$	$+$	$+$	$-$	$+$	$-$

$\bar{p}n \to \pi^-\pi^-\pi^+$. Since $G=-1$ in the final state, there is only one possible initial state, i.e. $^1S_0$.

$$|\bar{p},n\rangle = |1,-1\rangle = \frac{1}{\sqrt{2}}|1,0;1,-1\rangle - \frac{1}{\sqrt{2}}|1,-1;1,0\rangle$$
$$= \frac{1}{\sqrt{2}}|\rho^0;\pi^-\rangle - \frac{1}{\sqrt{2}}|\rho^-;\pi^0\rangle$$

hence $R(\bar{p}n \to \rho^0\pi^-)/R(\bar{p}n \to \rho^-\pi^0) = 1$.

$$|\bar{p},p\rangle = |1,0\rangle = \frac{1}{\sqrt{2}}|\rho^-;\pi^+\rangle + 0\frac{1}{\sqrt{2}}|\rho^0;\pi^0\rangle - \frac{1}{\sqrt{2}}|\rho^+;\pi^-\rangle$$

hence $R(\bar{p}p(I=1) \to \rho^+\pi^-):R(\bar{p}p(I=1) \to \rho^0\pi^0):R(\bar{p}p(I=1) \to \rho^-\pi^+)$ $=1:0:1$.

$$|\bar{p},p\rangle = |0,0\rangle = \frac{1}{\sqrt{3}}|\rho^-;\pi^+\rangle - \frac{1}{\sqrt{3}}|\rho^0;\pi^0\rangle + \frac{1}{\sqrt{3}}|\rho^+;\pi^-\rangle$$

hence $R(\bar{p}p(I=0) \to \rho^+\pi^-) : R(\bar{p}p(I=0) \to \rho^0\pi^0) : R(\bar{p}p(I=0) \to \rho^-\pi^+)$
$= 1:1:1$.

4.13. The matrix element $\mathcal{M}$ must be symmetric under the exchange of each pair of pions. Consequently:

1. If $J^P = 0^-$, $\mathcal{M} = $ constant. There are no zeros.
2. If $J^P = 1^-$, $\mathcal{M} \propto \mathbf{q}(E_1 - E_2)(E_2 - E_3)(E_3 - E_1)$; zeros on the diagonals and on the border.
3. If $J^P = 1^+$, $\mathcal{M} \propto \mathbf{p}_1 E_1 + \mathbf{p}_2 E_2 + \mathbf{p}_3 E_3$; zero in the centre, where $E_1 = E_2 = E_3$; zero at $T_3 = 0$, where $\mathbf{p}_3 = 0$, $\mathbf{p}_2 = -\mathbf{p}_1$; $E_2 = E_1$.

4.15. A baryon can contain between 0 and 3 c valence quarks; therefore the charm of a baryon can be $C = 0, 1, 2, 3$. Since the charge of c is equal to 2/3, the baryons with $Q = +1$ can have charm $C = 2$ (*ccd*, *ccs*, *ccb*), $C = 1$ (e.g. *cud*) or $C = 0$ (e.g. *uud*). If $Q = 0$, one c can be present, as in *cdd*, or none as in *udd*. Hence $C = 1$ or $C = 0$.

4.17. *sss*, *uuc*, *usc*, *ssc*, *udb*.

4.23. $\gamma = E/m = 20/1.86 = 10.7$. The condition $I = I_0 e^{-\frac{t}{\tau}} > 0.9 I_0$ gives $t < \gamma\tau\ln\left(\frac{1}{0.9}\right)$. We need to resolve distances $d = ct < 139\,\mu\text{m}$.

Possible instruments: bubble chambers, emulsions, silicon microstrips.

4.24. We start from $\sigma(E) = \dfrac{3\pi}{E^2}\dfrac{\Gamma_e\Gamma_f}{(E - M_R)^2 + (\Gamma/2)^2} = \dfrac{12\pi\Gamma_e\Gamma_f}{\Gamma^2}\dfrac{1}{E^2}$
$\times \dfrac{1}{[2(E - M_R)/\Gamma]^2 + 1}$.

In the neighbourhood of the resonance peak the factor $1/E^2$ varies only slowly, compared to the resonant factor, and we can approximate it with the constant $1/M_R^2$, i.e.

$$\int_{-\infty}^{+\infty} \sigma(E)dE = \frac{12\pi\Gamma_e\Gamma_f}{\Gamma^2}\int_{-\infty}^{+\infty}\frac{1}{E^2}\frac{1}{[2(E - M_R)/\Gamma]^2 + 1}dE$$

$$\approx \frac{12\pi\Gamma_e\Gamma_f}{\Gamma^2 M_R^2}\int_{-\infty}^{+\infty}\frac{1}{[2(E - M_R)/\Gamma]^2 + 1}dE.$$

Setting $\tan\theta = \dfrac{2(E - M_R)}{\Gamma}$, we have

$$\int_{-\infty}^{+\infty} \sigma(E)dE = \frac{12\pi\Gamma_e\Gamma_f}{\Gamma^2 M_R^2}\int_{-\infty}^{+\infty}\frac{1}{\tan^2\theta + 1}dE = \frac{12\pi\Gamma_e\Gamma_f}{\Gamma^2 M_R^2}\int_{-\infty}^{+\infty}\cos^2\theta\,dE$$

$$= \frac{12\pi\Gamma_e\Gamma_f}{\Gamma^2 M_R^2}\int_{-\pi/2}^{+\pi/2}\cos^2\theta\frac{dE}{d\theta}\,d\theta.$$

We find that $\dfrac{dE}{d\theta} = \dfrac{dE}{d\tan\theta}\dfrac{d\tan\theta}{d\theta} = \dfrac{\Gamma}{2}\dfrac{1}{\cos^2\theta}$, obtaining

$$\int\limits_{-\infty}^{+\infty}\sigma(E)dE = \frac{6\pi\Gamma_e\Gamma_f}{\Gamma M_R^2}\int\limits_{-\pi/2}^{+\pi/2}d\theta = \frac{6\pi^2\Gamma_e\Gamma_f}{\Gamma M_R^2}.$$

5.2. Since the speeds are small enough, we can use non-relativistic concepts and expressions. The electron potential energy, which is negative, becomes smaller with its distance r from the proton as $-1/r$. The closer the electron is to the proton, the better is its position defined and consequently the larger is the uncertainty of its momentum p. Actually, the larger the uncertainty of p, the larger is its average value and, with it, the electron kinetic energy. The radius of the atom is the distance at which the sum of potential and kinetic energies is minimum.

Due to its large mass, we consider the proton to be immobile. At the distance r the energy of the electron is $E = \dfrac{p^2}{2m_e} - \dfrac{1}{4\pi\varepsilon_0}\dfrac{q_e^2}{r}$.

The uncertainty principle dictates $pr = \hbar$ and we have $E = \dfrac{\hbar^2}{2m_e r^2} - \dfrac{1}{4\pi\varepsilon_0}\dfrac{q_e^2}{r}$.

To find the minimal radius a we set $\left(\dfrac{dE}{dr}\right)_a = 0 = -\dfrac{\hbar^2}{m_e a^3} + \dfrac{1}{4\pi\varepsilon_0}\dfrac{q_e^2}{a^2}$, obtaining $a = \dfrac{4\pi\varepsilon_0\hbar^2}{m_e q_e^2} = 52.8\,\text{pm}$, which is the Bohr radius.

5.6. At the next to the tree-level order in the t channel there are the eight diagrams in the figure

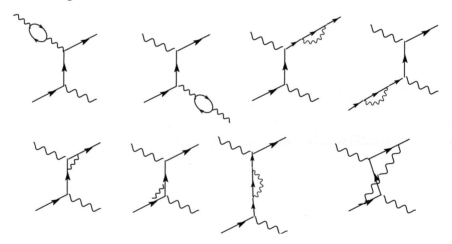

There are as many diagrams in the *s* channel. The last one is

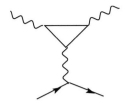

5.8.
$$\sigma_{m_\psi}(e^+e^- \to \mu^+\mu^-) = \frac{12\pi}{m_\psi^2}\frac{\Gamma_e^2}{\Gamma^2} = \frac{12\pi}{(3.097)^2}\left(5.9\times10^{-2}\right)^2$$
$$\times 389\left(\mu\text{b/GeV}^{-2}\right) = 5.3\,\mu\text{b}.$$

$$\sigma_{m_\psi}(e^+e^- \to \text{hadrons}) = \sigma_{m_\psi}(e^+e^- \to \mu^+\mu^-)\frac{87.7\%}{5.9\%} = 84\,\mu\text{b}.$$

6.2. The first case is below charm threshold, hence $R(u, d, s) = 2$; the second case is above the charm threshold and below the beauty one, hence $R(u, d, s, c) = 10/3 = 3.3$.

6.5. Having the alpha particle charge $z = 2$, the cross section is

$$\frac{d\sigma}{d\Omega} = \frac{Z^2\alpha^2}{4E_k^2\sin^4(\theta/2)} = \frac{Z^2\alpha^2}{E_k^2}\frac{1}{(1-\cos\theta)^2}.$$

Integrating on the angles we have

$$\int_0^{2\pi}d\phi\int_{\theta_1}^{\theta_2}d\cos\theta\frac{d\sigma}{d\Omega} = \frac{Z^2\alpha^2}{E^2}2\pi\int_{\theta_1}^{\theta_2}\frac{1}{(1-\cos\theta)^2}d\cos\theta$$
$$= \left(\frac{Z^2\alpha^2}{E^2}2\pi\right)\frac{\theta_2}{\theta_1}\bigg|\frac{1}{\cos\theta-1}.$$

Hence $\left(\dfrac{d\sigma}{d\Omega}\right)_{\theta>90°}\bigg/\left(\dfrac{d\sigma}{d\Omega}\right)_{\theta>10°} = 0.0074.$

6.8. $\cos\theta = 1 - \dfrac{E/E'-1}{E/M} = 1 - \dfrac{2.5-1}{20} = 0.925;\ \theta = 22°.$

6.11. From (6.27) with $W = m_p$, $2m_pv = Q^2$ follows and then from (6.31) we have $x = 1$.

6.12. From Solution 6.11 we have for elastic scattering $2m_pv = Q^2$. Using (6.26) we obtain $2m_pv = Q^2 = 2EE'(1 - \cos\theta)$, and then (6.11) because $v = E - E'$.

6.15. For every x, the momentum transfer Q^2 varies from a minimum to a maximum value when the electron scattering angle varies from $0°$ to $180°$. From Eqs. (6.26) and (6.29) that are valid in the L frame and (6.31) we obtain

$$Q^2 = \frac{2E^2\left(1 - \cos\theta_f\right)}{1 + \left(E/xm_p\right)\left(1 - \cos\theta_f\right)}.$$

Clearly, we have $Q^2 = 0$ in the forward direction ($\theta = 0$). The maximum momentum transfer is for background scattering ($\theta = 180°$), i.e.

$$Q^2_{\text{max}} = \frac{4E^2}{1 + 2E/xm_p} \approx 2Exm_p.$$

For $E = 100\,\text{GeV}$, $x = 0.2$ we have $Q^2_{\text{max}} = 37.5\,\text{GeV}^2$, corresponding to a resolving power of 32 am.

6.17. We write (5.37) with $\mu^2 = m_Z^2$ and $a^{-1}(m_Z^2) = 129$, as $a^{-1}(Q^2) = 129 - 0.71 \times \ln\left(\frac{Q^2}{m_Z^2}\right)$.

Hence, $a^{-1}(10^2) = 132$, $a^{-1}(100^2) = 129$.
Equation (6.69) with $n_f = 5$ gives: $a_s^{-1}(Q^2) = \frac{33 - 10}{12\pi}\ln\left(\frac{|Q|^2}{\lambda_{\text{QCD}}^2}\right) = 0.61$
$\times \ln\left(\frac{|Q|^2}{0.04}\right)$.

Hence $a_s^{-1}(10^2) = 4.8$ and $a_s^{-1}(100^2) = 7.6$.
The ratios are $a_s(10^2)/a(10^2) = 27.5$ and $a_s(100^2)/a(100^2) = 16.9$.

6.21. The colour wave function is $\frac{1}{\sqrt{6}}[RGB - RBG + GBR - GRB + BRG - BGR]$, which is completely antisymmetric. Since the space wave function is symmetric, the product of the spin and isospin wave functions must be completely symmetric for any two-quark exchange. The system *uud* is obviously symmetric in the exchange within the *u* pair. Consider the *ud* exchange. The totally symmetric combination *uud + udu + duu* has isospin 3/2 and is not the proton.

The isospin 1/2 wave function contains terms that are antisymmetric under the exchange of the second and third quarks, such as *uud–udu*. We obtain symmetry by multiplying by a term with the same antisymmetry in spin, namely ($\uparrow\uparrow\downarrow - \uparrow\downarrow\uparrow$). We thus obtain a term symmetric under the exchange of the second and third quarks:

$$(u\uparrow)(u\uparrow)(d\downarrow) - (u\uparrow)(d\uparrow)(u\downarrow) - (u\uparrow)(u\downarrow)(d\uparrow) + (u\uparrow)(d\downarrow)(u\uparrow).$$

Similarly for the first two quarks we have

$$(u\uparrow)(u\uparrow)(d\downarrow) - (d\uparrow)(u\uparrow)(u\downarrow) - (u\downarrow)(u\uparrow)(d\uparrow) + (d\downarrow)(u\uparrow)(u\uparrow),$$

and for the first and third

$$(d\downarrow)(u\uparrow)(u\uparrow) - (u\downarrow)(d\uparrow)(u\uparrow) - (d\uparrow)(u\downarrow)(u\uparrow) + (u\uparrow)(d\downarrow)(u\uparrow).$$

In total we have 12 terms. We take their sum and normalise, obtaining

$$\frac{1}{\sqrt{12}}[2(u\uparrow)(u\uparrow)(d\downarrow) + 2(d\downarrow)(u\uparrow)(u\uparrow) + 2(u\uparrow)(d\downarrow)(u\uparrow)$$
$$- (u\uparrow)(d\uparrow)(u\downarrow) - (u\uparrow)(u\downarrow)(d\uparrow) - (d\uparrow)(u\uparrow)(u\downarrow)$$
$$- (u\downarrow)(u\uparrow)(d\uparrow) - (u\downarrow)(d\uparrow)(u\uparrow) - (d\uparrow)(u\downarrow)(u\uparrow)]$$

that is, as required, completely antisymmetric for the exchange of any pair.

7.1. $K^{*+} \rightarrow K^0 + \pi^+$. We start by writing the valence quark compositions of all the particles, i.e. $(u\bar{s}) \rightarrow (d\bar{s}) + (u\bar{d})$ and then draw diagram (a). Since it is a strong process we do not draw any gauge bosons.

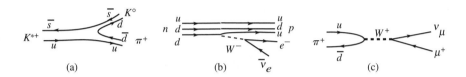

$n \rightarrow p + e^- + \bar{\nu}_e$. It is a weak process. In order to draw the diagram (b) we consider two steps: the emission of a W, $(udd) \rightarrow (udu) + W^-$ and its decay $W^- \rightarrow e^- + \bar{\nu}_e$.

$\pi^+ \rightarrow \mu^+ + \nu_\mu$. We have $u\bar{d} \rightarrow W^+$ followed by $W^+ \rightarrow \mu^+ + \nu_\mu$; diagram (c).

7.5. $\Gamma(\tau \rightarrow e\nu_e\bar{\nu}_\tau)/\Gamma(\mu \rightarrow e\nu_e\bar{\nu}_\mu) = m_\tau^5/m_\mu^5 = 1.35 \times 10^6$, and

$$\tau_\tau = \frac{2.2 \times 10^{-6} \times 0.16}{1.35 \times 10^6} = 2.6 \times 10^{-13} \text{ s}.$$

7.8. (a) $\nu_\mu + p \rightarrow \mu^- + p + \pi^+$; (b) $\nu_\mu + n \rightarrow \mu^- + n + \pi^+$ and $\nu_\mu + n \rightarrow \mu^- + p + \pi^0$. Both $\mu^- \rightarrow e^+ + \gamma$ and $\mu^+ \rightarrow e^+ + e^+ + e^-$ violate lepton and muon flavour. They do not exist.

7.9. The quantity $\mathbf{p}_\Lambda \cdot \boldsymbol{\sigma}_\Lambda$ is a pseudoscalar. It must be zero if parity is conserved, therefore the polarisation must be perpendicular to $\mathbf{p}_\Lambda$.

7.13. The neutrino flux through a generic normal surface S is $\Phi = N_\nu/S$. The corresponding target is a cylinder of section S and length $2R$. Its mass is $M = \rho S 2R$, containing $N_b = M N_A 10^3 = \rho S 2R N_A 10^3$ nucleons. Therefore the number of interactions is $\Phi N_b \sigma = N_\nu \rho 2R N_A 10^3 \sigma = 25.2$.

7.16. In order to have a rate $R = 1/84\,600$, we need $N_{71} = \frac{R}{\Phi \times \sigma \times \varepsilon} = 10^{29}$ ^{71}Ga nuclei, corresponding to $N_{\text{mol}} = N_{71}/N_A = 1.7 \times 10^5$ moles. The ^{71}Ga mass is $M_{71} = N_{\text{mol}} \times 10^{-3} \times 71\,\text{kg} = 12\text{t}$ and the total Ga mass is $M = M_{71}/a = 30\text{t}$.

7.18. The decay $c \rightarrow d + e^+ + \nu_e$ is disfavoured because its amplitude is proportional to $\sin\theta_C$. The decay $c \rightarrow s + e^+ + \nu_e$ is favoured because its amplitude is proportional to $\cos\theta_C$. We write down the valence quark compositions: $D^+ = c\bar{d}, K^+ = u\bar{s}, K^- = s\bar{u}, \bar{K}^0 = s\bar{d}$. Consequently the decays of D^+

in final states containing a K^- or $\bar{K}^0$ are favoured. For example, $D^+ \to K^- + \pi^+ + e^+ + v_e$, $D^+ \to \bar{K}^0 + e^+ + v_e$, $D^+ \to \bar{K}^{*0} + e^+ + v_e$ are favoured. $D^+ \to \pi^- + \pi^+ + e^+ + v_e$, $D^+ \to \pi^0 + \pi^+$ and $D^+ \to \rho^0 + e^+ + v_e$ are disfavoured.

7.20. Having V_{tb} very close to 1, the dominant decay is $t \to bW$. There are seven diagrams:

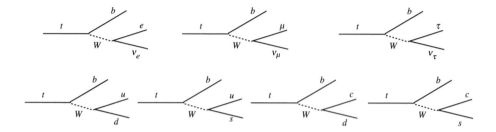

7.22. We approximate the semileptonic decay of the D^+ hadron with the decay of the quark $c \to se^+ v_e$. Then we have $\Gamma(c \to s + e^+ + v_e)/\Gamma(\mu^+ \to e^+ v_e \bar{v}_\mu) = (m_c/m_\mu)^5 \cos^2\theta_C = 2.1 \times 10^5$, with an uncertainty of at least 35% due to the 7% uncertainty on m_c. The experimental value is 1.6×10^5.

7.23. We start by writing the valence quark contents of the hadrons, we then identify the decay at the quark level and which quark acts as a spectator.

1. At the hadron level we have $c\bar{d} \to s\bar{d} + u\bar{d}$ and at the quark level $c \to su\bar{d}$ with a spectator $\bar{d}$. The decay rate is proportional to $|V_{cs}|^2|V_{ud}|^2 \approx \cos^4\theta_C$.

2. At hadron level it is $c\bar{d} \to u\bar{s} + s\bar{d}$ and at the quark level it is $c \to su\bar{s}$ with a spectator $\bar{d}$. The decay probability is proportional to $|V_{cs}|^2|V_{us}|^2 \approx \sin^2\theta_C \cos^2\theta_C$.

3. The π^0 has a $u\bar{u}$ and a $d\bar{d}$ component. The decay picks up the latter. At hadron level it is $c\bar{d} \to u\bar{s} + d\bar{d}$ and at the quark level it is $c \to du\bar{s}$ with a spectator $\bar{d}$. The decay probability is proportional to $|V_{cd}|^2|V_{us}|^2 \approx \sin^4\theta_C$.

7.24. Let us consider the valence quarks of the hadrons. The decay $\Sigma^-(dds) \to n(ddu) + e^- + \bar{v}_e$ corresponds to $s \to u + e^- + \bar{v}_e$ at the quark level, with two ds as 'spectators'. The decay $\Sigma^+(uus) \to n(ddu) + e^+ + v_e$ does not correspond to $u \to d + e^+ + v_e$ because an initial s should transform into a final d. It is a violation of the $\Delta S = \Delta Q$ rule.

7.28. There are $n_{Fe} = \rho \times N_A 10^3/A = 10^{17} \times 6 \times 10^{23} \times 10^3/56 = 1.1 \times 10^{42}\,\mathrm{m}^{-3}$ nucleons per unit volume. Consequently the mean free path is

$$\lambda_v = \frac{1}{n_{Fe}\sigma} = \frac{1}{1.1 \times 10^{42} \times 3 \times 10^{-46}} = 3\,\text{km}. \text{ This distance is smaller than the}$$

radius of the supernova core.

8.1. From isospin conservation we have $|\phi\rangle = |0,0\rangle = \frac{1}{\sqrt{2}}|K^+\rangle|K^-\rangle - \frac{1}{\sqrt{2}}|K^0\rangle|\bar{K}^0\rangle$.

Therefore $\phi \to \bar{K}^0 K^0 / \phi \to K^+ K^- = 1$. The ratio of the phase space volumes is

$$\frac{p_{\pm}^*}{p_0^*} = \frac{127}{110} = 1.15.$$

The two mesons are in the P wave, i.e. in a spatially antisymmetric state. Consequently they cannot be identical. The only possibility is $K_1^0 K_2^0$.

8.4. $\gamma = E/m_K = 4.05$ and $\beta = 0.97$; $t = \gamma\tau \ln 10 = 1.1 \times 10^{-7}\,\text{s}$ and $d = \beta c t = 32\,\text{m}$.

8.7. The first B lived $n = \dfrac{l}{\beta\gamma\tau_B c} = \dfrac{120\,\mu\text{m}}{257\,\mu\text{m}} = 0.47$ lifetimes, the second one 1.9

lifetimes. The μ^+ comes from a $\bar{b}$ decay (as opposed to a b decay) therefore the hadron is a B^0. The sister B has negative beauty at production but can have any beauty at decay, due to the oscillation. Therefore the second μ can have both signs.

8.9. We invert the system of Eq. (8.32), i.e. $\sqrt{2}|K_L^0\rangle = (1+\varepsilon)|K^0\rangle + (1-\varepsilon)|\bar{K}^0\rangle$ and the corresponding one for K_S^0, i.e. $\sqrt{2}|K_S^0\rangle = (1+\varepsilon)|K^0\rangle - (1-\varepsilon)|\bar{K}^0\rangle$, taking into account that $|\varepsilon|$ is small.

We obtain $\sqrt{2}|K^0\rangle = (1-\varepsilon)|K_S^0\rangle + (1-\varepsilon)|K_L^0\rangle$, $\sqrt{2}|K^0\rangle = (1+\varepsilon)|K_S^0\rangle - (1+\varepsilon)|K_L^0\rangle$. The decay amplitudes can be written as $\sqrt{2}A(\bar{K}^0 \to \pi^+\pi^-) = (1+\varepsilon)[A(K_S^0 \to \pi^+\pi^-) - A(K_L^0 \to \pi^+\pi^-)] = A(K_S^0 \to \pi^+\pi^-)(1+\varepsilon)(1-\eta_{+-})$ and similarly $\sqrt{2}A(K^0 \to \pi^+\pi^-) = A(K_S^0 \to \pi^+\pi^-)(1-\varepsilon)(1+\eta_{+-})$. We finally

obtain $\left|\dfrac{A(\bar{K}^0 \to \pi^+\pi^-)}{A(K^0 \to \pi^+\pi^-)}\right| = \left|\dfrac{(1+\varepsilon)-(1+\varepsilon)\eta_{+-}}{(1-\varepsilon)+(1-\varepsilon)\eta_{+-}}\right| = \left|\dfrac{(1+\varepsilon)-(1+\varepsilon)(\varepsilon+\varepsilon')}{(1-\varepsilon)+(1-\varepsilon)(\varepsilon+\varepsilon')}\right| \approx$

$1 - 2\text{Re}\varepsilon'$.

9.1. The requested interaction rate is $R = 1/(4 \times \Delta t) = 42/\text{s}$. On the other hand we have $R = \Phi\sigma N_e$. The number of electrons is half of the number of nucleons. Therefore their number in the fiducial mass M is $N_e = \frac{1}{2}M N_A 10^3 = 1.64 \times 10^{32}$. The cross section at $E_v = 24\,\text{GeV}$ is $\sigma = 4.1 \times 10^{-44}\,\text{m}^2 = 0.41$ fb. The requested flux is $\Phi = R/(\sigma N_e) = 1.5 \times 10^{13}\,\text{s}^{-1}\,\text{m}^{-2}$ and the beam intensity $I = \Phi \times A = 1.6 \times 10^{14}\,\text{s}^{-1}$. The duty cycle is $2\Delta t/T = 0.8 \times 10^{-3}$.

9.3. In the rest frame of the pion the neutrino energy is $E_v^* = (m_\pi^2 - m_\mu^2)/(2m_\pi) = 30\,\text{MeV}$. The Lorentz factors are $\gamma = E_\pi/m_\pi = 1429$ and $1 - \beta = 1 - \sqrt{1-\gamma^{-2}} \approx \frac{1}{2}\gamma^{-2} = 2.4 \times 10^{-7}$. The neutrino energy in the L frame is $E_\nu = \gamma(E_\nu^* + \beta p^* \cos\theta^*) = \gamma E_\nu^*(1 + \beta\cos\theta^*)$. Its maximum for $\theta^* = 0$ is $E_\nu^{\max} = \gamma E_\nu^*(1+1) = 1429 \times 30 \times 10^{-3} (\text{GeV})2 = 85.7\,\text{GeV}$. Its minimum for $\theta^* = \phi$ is $E_\nu^{\min} = \gamma E_\nu^*(1-\beta) = 10\,\text{keV}$.

We use the Lorentz transformations of the components of the neutrino momentum to find the relationship between the angle θ^* in CM and θ in L.

$p_\nu \sin\theta = p^* \sin\theta^*$; $p_\nu \cos\theta = \gamma(p^*\cos\theta^* + \beta E_\nu^*) \approx \gamma p^*(\cos\theta^* + 1)$,

which gives $\tan\theta = \dfrac{\sin\theta^*}{\gamma(\cos\theta^* + 1)} \approx \dfrac{0.05}{1429 \times 2} = 22 \times 10^{-6} \quad \Rightarrow \theta = 22\,\mu\text{rad}$.

9.7. For each reaction we check whether charge Q and hypercharge Y are conserved. We write the hypercharge values explicitly.

For $W^- \to d_L + \bar{u}_L$ we have $0 \to 1/3 + 0$. It violates Y.

For $W^- \to u_L + \bar{u}_R$ we have $0 \to 1/3 - 1/3$. It conserves Y, but violates Q.

For $Z \to W^- + W^+$ we have $0 \to 0 + 0$. OK.

For $W^+ \to e_R^+ + \bar{\nu}_R$ we have $0 \to 1 - 1$. OK.

9.9. $\Gamma_\nu = \dfrac{G_F M_Z^3}{3\sqrt{2}\pi}\left(\dfrac{1}{2}\right)^2 \approx 660 \times 1/4\,\text{MeV} = 165\,\text{MeV}$.

$\Gamma_l = \dfrac{G_F M_Z^3}{3\sqrt{2}\pi}\left[\left(-\dfrac{1}{2} + s^2\right)^2 + s^4\right] \approx 660 \times 0.148 \approx 98\,\text{MeV}$.

$\Gamma_u = \Gamma_c = 3\dfrac{G_F M_Z^3}{3\sqrt{2}\pi}\left[\left(\dfrac{1}{2} - \dfrac{2}{3}s^2\right)^2 + \left(-\dfrac{2}{3}s^2\right)^2\right] \approx 3 \times 660 \times 0.173 \approx 342\,\text{MeV}$.

$\Gamma_d = \Gamma_s = \Gamma_b = 3\dfrac{G_F M_Z^3}{3\sqrt{2}\pi}\left[\left(-\dfrac{1}{2} + \dfrac{1}{3}s^2\right)^2 + \left(\dfrac{1}{3}s^2\right)^2\right] \approx 3 \times 660 \times 0.207$

$\approx 410\,\text{MeV}$.

$\Gamma_Z = 3 \times 165 + 3 \times 98 + 2 \times 342 + 3 \times 410 = 2.7\,\text{GeV}$.

$\Gamma_h = 2 \times 342 + 3 \times 410 = 1910\,\text{MeV}$; $\Gamma_\mu/\Gamma_h = \dfrac{98}{1910} = 5.1\%$.

9.12. $\dfrac{g_{Zee}^2}{g_W^2} = \dfrac{(-1/2 + s^2)^2 + s^4}{1/2} = 0.25$ and $\dfrac{\Gamma(Z \to e^+e^-)}{\Gamma(W \to e^+\nu_e)} = \dfrac{g_{Zee}^2}{g_W^2}\dfrac{M_Z^3}{M_W^3} = 0.25$

$\times 1.45 = 0.36$.

9.15. $\Gamma_Z = 3\Gamma_l + 2\Gamma_u + 3\Gamma_d + N\Gamma_\nu = 3 \times 83 + 2 \times 280 + 3 \times 360 + N \times 166$

$= 1889 + N \times 166$.

$\dfrac{\Gamma_\mu}{\Gamma_Z}(3) = 3.48\%$, $\dfrac{\Gamma_\mu}{\Gamma_Z}(4) = 3.25\%$ and $\dfrac{\Gamma_\mu}{\Gamma_Z}(5) = 3.05\%$.

$\sigma_0^h(3) : \sigma_0^h(4) : \sigma_0^h(5) = \Gamma_Z^{-2}(3) : \Gamma_Z^{-2}(4) : \Gamma_Z^{-2}(5) = 1 : 0.87 : 0.77$.

9.17. The energy of a Z of momentum $p_Z = 140\,\text{GeV}$ is $E_Z = 167\,\text{GeV}$ and the Lorentz parameters are $\gamma_Z = p_Z/M_Z = 1.54$ and $\beta_Z = p_Z/E_Z = 0.84$. In the CM frame the components of the momenta of the electrons perpendicular to the beams are $p_n^{*+} = -p_n^{*-} = 45\,\text{GeV}$, while their longitudinal components are zero. Also $E^{*+} = E^{*-} = 45$ GeV. In the L frame, $p_n^+ = p_n^{*+} = 45$ GeV, $p_n^- = p_n^{*-} = 45$ GeV. The longitudinal momentum and the energy of both the electron and the positron are $p_L = 0 + \gamma_Z \beta_Z E^* = 58$ GeV and $E_L = \sqrt{p_L^2 + p_n^2 + m_Z^2} = 117$ GeV. Their angles at the two sides of the beams are $\theta_L = \tan^{-1}(p_n/p_L) = 38°$.

9.18. $m^2 = 4E_1 E_2 \sin^2 \theta/2 \quad \Rightarrow m = 92$ GeV.

$$\frac{\sigma_{M_Z}}{M_Z} = \frac{1}{2}\sqrt{\left(\frac{\sigma(E_1)}{E_1}\right)^2 + \left(\frac{\sigma(E_2)}{E_2}\right)^2 + \left(\frac{\sigma(\theta)}{\tan \theta/2}\right)^2}$$

$$= \frac{1}{2} 10^{-2}\sqrt{2.4^2 + 2^2 + 0.6^2} = 1.6\%.$$

9.21. The energy squared in the quark–antiquark CM frame is $\hat{s} = x_q x_{\bar{q}} s$. Assuming, for the sake of our evaluation, $x_q = x_{\bar{q}}$, we have $x_q = x_{\bar{q}} = \sqrt{\hat{s}/s} = M_Z/\sqrt{s} = 0.045$.

The sea quark structure functions are about $x\bar{d}(0.045) \approx x\bar{u}(0.045) \approx xd(0.045) \approx 0.5xu(0.045)$. The momentum fraction of the Z with longitudinal momentum $P_Z = 100$ GeV is $x_Z = x_q - x_{\bar{q}} = p_Z/p_{beam} = 0.1$. By substitution into $m_Z^2 = x_q x_{\bar{q}} s$ we obtain $m_Z^2 = x_q(x_q - 0.1)s$ and $x_q^2 - 0.1 x_q - m_Z^2/s = 0$ or, numerically, $x_q^2 - 0.1 x_q - 0.002 = 0$. Its solution is $x_q = 0.1 \pm \sqrt{0.1^2 + 4 \times 0.002} = 0.234$. The other solution is negative and therefore not physical.

9.26. The target contains the same number of up and down quarks. In the charged-current case neutrinos interact as $\nu_{\mu L} + d_L \rightarrow \mu_L^- + u_L$, antineutrinos as $\bar{\nu}_{\mu R} + u_L \rightarrow \mu_R^+ + d_L$. As we saw in Problem 9.4, in the latter case (LR) there is a factor $1/3$ relative to the former (LL). Therefore the ratio is $\sigma_{CC}(\nu_L)/\sigma_{CC}(\bar{\nu}_R) = 3$.

All the target quarks u_R, u_L, d_R and d_L contribute to the neutral-current interactions each proportionally to its Z-charge factor squared c_Z^2. We sum their contributions taking into account the $1/3$ factor for the LR and RL contributions relative to the LL and RR contributions.

We have $\dfrac{\sigma_{NC}(\nu_L)}{\sigma_{NC}(\bar{\nu}_R)} = \dfrac{[c_Z^2(u_L) + c_Z^2(d_L)] + \frac{1}{3}[c_Z^2(u_R) + c_Z^2(d_R)]}{\frac{1}{3}[c_Z^2(u_L) + c_Z^2(d_L)] + [c_Z^2(u_R) + c_Z^2(d_R)]}$,

giving

$$\frac{\sigma_{NC}(\nu)}{\sigma_{NC}(\bar{\nu})} = \frac{[(\frac{1}{2} - \frac{2}{3}s^2)^2 + (-\frac{1}{2} + \frac{1}{3}s^2)^2] + \frac{1}{3}[(-\frac{2}{3}s^2)^2 + (\frac{1}{3}s^2)^2]}{\frac{1}{3}[(\frac{1}{2} - \frac{2}{3}s^2)^2 + (-\frac{1}{2} + \frac{1}{3}s^2)^2] + [(-\frac{2}{3}s^2)^2 + (\frac{1}{3}s^2)^2]}$$

$$= \frac{\frac{1}{2} - s^2 + \frac{20}{27}s^4}{\frac{1}{3}(\frac{1}{2} - s^2 + \frac{20}{9}s^4)} = 2.3.$$

References

Abbiendi, G. *et al.* (2004); *Euro. Phys. J.* **C33** 173

Abbiendi, G. *et al.* (2006); *Euro. Phys. J.* **C45** 1

Abdurashitov, J. N. *et al.* (2002); *JETP* **95** 181

Abe, F. *et al.* (1995); *Phys. Rev. Lett.* **74** 2626

Abe, K. *et al.* (2005); *Phys. Rev.* **D71** 072003

Abrams, G. S. *et al.* (1974); *Phys. Rev. Lett.* **33** 1453

Abulencia, *et al.* (2006); *Phys. Rev. Lett.* **97** 242003

Achard, P. *et al.* (2005) (L3 Collaboration); *Phys. Lett.* **B623** 26

Ageno, M. *et al.* (1950); *Phys. Rev.* **79** 720

Aharmin, B. *et al.* (2005); nucl-ex/0502021/

Ahn, M. H. *et al.* (2003); *Phys. Rev. Lett.* **90** 041801; *ibid.* **94** (2005) 081802

Aihara, H. *et al.* (1988); *Phys. Rev. Lett.* **61** 1263

Alavi-Harati, A. *et al.* (2003); *Phys. Rev.* **D67** 012005; *ibid.* **D70** 079904 (erratum)

Albajar, C. *et al.* (1987); *Z. Phys.* **C36** 33

Albajar, C. *et al.* (1989); *Z. Phys.* **C44** 15

Alff, C. *et al.* (1962); *Phys. Rev. Lett.* **9** 325

Allison, W. W. M. *et al.* (1974a); *Proposal* CERN/SPSC 74–45; CERN/SPSC 75–15 and *Phys. Lett.* **B93** (1980)

Allison, W. W. M. *et al.* (1974b); *Nucl. Instrum. Methods* **119** 499

Alston, M. *et al.* (1961); *Phys. Rev. Lett.* **6** 300

Altarelli, G. & Parisi, G. (1977); *Nucl. Phys.* **B126** 298

Altman, M. *et al.* (2005); *Phys. Lett.* **B616** 174

Alvarez, L. W. *et al.* (1963); *Phys. Rev. Lett.* **10** 184; also M. Alston *et al.* (1960); *Phys. Rev. Lett.* **5** 520

Alvarez, L. W. (1972); In *Nobel Lectures, Physics 1963–1970*, Elsevier Publishing Company

Amaldi, U. *et al.* (1987); *Phys. Rev.* **D36** 1385

Ambrosino, F. *et al.* (2006); *Phys. Lett.* **B636** 173

Anderson, C. D. & Neddermeyer, S. H. (1937); *Phys. Rev.* **51** 884; *Phys. Rev.* **54** (1938) 88

Anderson, C. D. (1933); *Phys. Rev.* **43** 491

Anderson, H. L. *et al.* (1952); *Phys. Rev.* **85** 936; also *ibid.* p. 934 and p. 935

Andrews, A. *et al.* (1980); *Phys. Rev. Lett.* **44** 1108

Anselmann, P. *et al.* (1992); *Phys. Lett.* **B285** 375 and 390

Anselmann, P. *et al.* (1995); *Phys. Lett.* **B357** 237

Anthony, P. L. *et al.* (2004); *Phys. Rev. Lett.* **92** 181602

Apollonio, M. *et al.* (1999); *Phys. Lett.* **B466** 415
Araki, T. *et al.* (2005); *Phys. Rev. Lett.* **94** 081801
Arnison, G. *et al.* (1983a); *Phys. Lett.* **B122** 103
Arnison, G. *et al.* (1983b); *Phys. Lett.* **B126** 398
Ashie, Y. *et al.* (2005); *Phys. Rev.* **D71** 112005
Asner, D. *et al.* (2007); HFAG group, http://www.physics.uc.edu/~schwartz/hfag/
    results_mixing.html
Aubert, B. *et al.* (2004); *Phys. Rev. Lett.* **93** 131801
Aubert, B. *et al.* (2006); *Proc. 33rd International Conference on High-Energy Physics,*
    ICHEP06
Aubert, B. *et al.* (2007); *Phys. Rev. Lett.* **98** 211802
Aubert, J.J. *et al.* (1974); *Phys. Rev. Lett.* **33** 1404
Augustin, J.E. *et al.* (1974); *Phys. Rev. Lett.* **33** 1406
Bacci, C. *et al.* (1974); *Phys. Rev. Lett.* **33** 1408
Bagnaia, P. *et al.* (1983); *Phys. Lett.* **B129** 130
Bahcall, J.N. *et al.* (1963); *Astr. J.* **137** 334
Bahcall, J.N. *et al.* (2005); *Astrophys. J.* **621** L85
Banner, M. *et al.* (1982); *Phys. Lett.* **B118** 203
Banner, M. *et al.* (1983); *Phys. Lett.* **B122** 476
Barnes, V.E. *et al.* (1964); *Phys. Rev. Lett.* **12** 204
Barnett, R.M. *et al.* (1996); *Phys. Rev.* **D54** 1
Barr, G.D. *et al.* (1993); *Phys. Lett.* **B317** 233
Batlay, J.R. *et al.* (2002); *Phys. Lett.* **B544** 97
Beherend, H.J. *et al.* (1987); *Phys. Lett.* **B183** 400
Bellgardt, U. *et al.* (1988); *Nucl. Phys.* **B299** 1
Bernardini, M. *et al.* (1967); '*A proposal to search for leptonic quarks and heavy leptons
    produced at ADONE*' INFN/AE-67/3. V. Alles-Borelli *et al.* (1970); *Lett. Nuovo
    Cimento* **4** 1156. M. Bernardini *et al.* (1973); *Nuovo Cimento* **17A** 383
Bethe, H.A. (1930); *Ann. Phys.* **5** 321
Bethe, H.A. (1947); *Phys. Rev.* **72** 339
Bjorken, J.D. (1969); *Phys. Rev.* **179** 1547
Blackett, P.M.S. & Occhialini, G.P.S. (1933); *Proc. R. Soc. London* **A139** 699
Bona, M. *et al.* (2006); hep-ph/0606167/; also http://utfit.roma1.it
Boyarski, A.M. (1975); *Phys. Rev. Lett.* **34** 1357
Brandelik, R. *et al.* (1980); *Phys. Lett.* **B97** 453
Breidenbach, M. *et al.* (1969); *Phys. Rev. Lett.* **23** 935; also Wu, S.L. (1984); *Phys. Rep.*
    107 59
Brooks, M.L. *et al.* (1999); *Phys. Rev. Lett.* **83** 1521
Broser, I. & Kallmann, H. (1947); *Z. Naturforsch* **2a** 439, 642
Brown, R.H. *et al.* (1949); *Nature* **163** 47
Burfening, J. *et al.* (1951); *Proc. Phys. Soc.* **A64** 175
Cabibbo, N. (1963); *Phys. Rev. Lett.* **10** 531
Cartwright, W.F. *et al.* (1953); *Phys. Rev.* **91** 677
Ceccucci, A. *et al.* (2006) in Yao *et al.* (2006) p. 138
Chamberlain, O. *et al.* (1950); *Phys. Rev.* **79** 394
Chamberlain, O. *et al.* (1955); *Phys. Rev.* **100** 947
Chao, Y. *et al.* (2004); *Phys. Rev. Lett.* **93** 191802
Charles, J. *et al.* (2006) *International Conference on High Energy Physics* ICHEP06; also
    http://www.slac.stanford.edu/xorg/ckmfitter/
Charpak, G. *et al.* (1968); *Nucl. Instrum. Methods* **62**, 262
Chekanov, S. *et al.* (2001); *Eur. Phys. J.* **C21** 443

Cherenkov, P. A. (1934); *C. R. Acad. Sci. USSR* **8** 451

Christenson, J. *et al.* (1964); *Phys. Rev. Lett.* **13** 138

Clark, D. L. *et al.* (1951); *Phys. Rev.* **83** 649

Cleveland, B. T. (1998); *Astrophys. J.* **496** 505

CNGS (1998); CERN 98–02, INFN/AE-98/05 and CERN-SL/99–034(DI), INFN/AE-99/05

Connoly, P. L. *et al.* (1963); *Phys. Rev. Lett.* **10** 114

Conversi, M. *et al.* (1947); *Phys. Rev.* **71** 209 (L)

Conversi, M. & Gozzini, A. (1955); *Nuovo Cimento* **2** 189

Costa, G. *et al.* (1988); *Nucl. Phys.* **B297** 244

Courant, E. D. & Snyder, H. S. (1958); *Ann. Phys.* **3** 1

Cowan, C. L. *et al.* (1956); *Science* **124** 103

Dalitz, R. H. (1956); *Proceedings of the Rochester Conference*. See also R. H. Dalitz
    (1953); *Philos. Mag.* **44** 1068 and *ibid.* **94** (1954) 1046

Danby, G. *et al.* (1962); *Phys. Rev. Lett.* **9** 36

Davies, J. H. *et al.* (1955); *Nuovo Cimento* **2** 1063

Davis, R. *et al.* (1964); *Phys. Rev. Lett.* **12** 302. See also *ibid.* **20** (1968) 1205

Day, T. B. *et al.* (1960); *Phys. Rev. Lett.* **3** 61

Dirac, P. A. M. (1931); *Proc. R. Soc. London* **A133** 60

Dokshitzer, Yu. L. (1977); *Sov. Phys. JETP* **46** 641

Durbin, R. *et al.* (1951); *Phys. Rev.* **83** 646

Eguchi, K. *et al.* (2003); *Phys. Rev. Lett.* **90** 021802

Englert, F. & Brout, R. (1964); *Phys. Rev. Lett.* **13** 321

Erwin, A. R. *et al.* (1961); *Phys. Rev. Lett.* **6** 628

Fermi, E. (1934); *Nuovo Cimento* **11** 1; *Z. Phys.* **88** 161 [transl. into English by
    F. L. Wilson; *Am. J. Phys.* **36** (1968) 1150]

Fermi, E. (1949); *Phys. Rev.* **75** 1169

Feynman, R. P. (1948); *Phys. Rev.* **74** 1430; *ibid.* **76** (1949) 749 and 769

Feynman, R. P. (1969); *Phys. Rev. Lett.* **23** 1415

Fogli, G. L. *et al.* (2006); hep-ph/0608060/

Frank, I. M. & Tamm, I. E. (1937); *C. R. Acad. Sci. USSR* **14** 107

Fukuda, Y. *et al.* (1998); *Phys. Rev. Lett.* **81** 1562

Fukui, S. & Myamoto, S. (1959); *Nuovo Cimento* **11** 113

Gabrielse, G. *et al.* (2006); *Phys. Rev. Lett.* **97** 030802

Geiger, H. & Mueller, W. (1928); *Naturwissenschaften* **16** 617

Geiregat, D. *et al.* (1991); *Phys. Lett.* **B259** 499

Gell-Mann, M. (1953); *Phys. Rev.* **92** 833

Gell-Mann, M. (1964); *Phys. Lett.* **8** 214

Gell-Mann, M. & Pais, A. (1955); *Phys. Rev.* **97** 1387

Gibbons, L. K. *et al.* (1993); *Phys. Rev. Lett.* **70** 1203

Gjesdal, S. *et al.* (1974); *Phys. Lett.* **B52** 113

Glaser, D. A. (1952); *Phys. Rev.* **87** 665, *ibid.* **91** (1953) 496

Glashow, S. L. (1961); *Nucl. Phys.* **22** 579

Glashow, S. L. *et al.* (1970); *Phys. Rev.* **D2** 1285

Goldberg, M. *et al.* (1964); *Phys. Rev. Lett.* **12** 546

Goldhaber, M. *et al.* (1958); *Phys. Rev.* **109** 1015

Goldhaber, G. *et al.* (1976); *Phys. Rev. Lett.* **37** 255

Gribov, V. N. & Lipatov, L. N. (1972); *Sov. J. Nucl. Phys.* **15** 438 and Lipatov, L. N.
    (1975); *ibid.* **20** 95

Grodznis, L. (1958); *Phys. Rev.* **109** 1015

Gross, D. J. & Wilczek, F. (1973); *Phys. Rev. Lett.* **30** 1343

Grossman, Y. *et al.* (2005); *Phys. Rev.* **D72** 031501

Hasert, F. J. *et al.* (1973); *Phys. Lett.* **B46** 121 and 138; *Nucl. Phys.* **B73** (1974) 1

Heisenberg, W. (1932); *Z. Phys.* **120** 513, 673

Herb, S. W. *et al.* (1977); *Phys. Rev. Lett.* **39** 252

Hess, V. F. (1912); *Physik. Z.* **13** 1084

Higgs, P. W. (1964); *Phys. Lett.* **12** 132; *Phys. Rev. Lett.* **13** 508

Hirata, K. S. *et al.* (1989); *Phys. Rev. Lett.* **63** 16

Hori, M. *et al.* (2003); *Phys. Rev. Lett.* **91** 123401

Hosaka, J. *et al.* (2005); hep-ex/0508053/

Kalbfleish, G. R. *et al.* (1964); *Phys. Rev. Lett.* **12** 527

Kallmann, H. (1950); *Phys. Rev.* **78** 621

Katayama, Y. *et al.* (1962); *Prog. Theor. Phys.* **28** 675

Kobaiashi, M. & Maskawa, K. (1973); *Prog. Theor. Phys.* **49** 652

Kodama, K. *et al.* (2001); *Phys. Lett.* **B504** 218

Koks, F. & van Klinken, J. (1976); *Nucl. Phys.* **A272** 61

Kraus, Ch. *et al.* (2004); *Eur. Phys. J.* **C40** 447

Kusch, P. & Foley, H. M. (1947); *Phys. Rev.* **72** 1256; *ibid.* **73** (1948) 412; *ibid.* **74** (1948) 250

Kurie, F. N. D. *et al.* (1936); *Phys. Rev.* **49** 368

Lamb, W. E. Jr. & Retherford, R. C. (1947); *Phys. Rev.* **72** 241

Landau, L. D (1957); *Zh. Eksp. Teor. Fiz.* **32** 405 [*JETP* **5** 1297]

Lattes, C. M. G. *et al.* (1947); *Nature* **159** 694; *ibid.* **160** 453 and 486

Lee, T. D. & Yang, C. N. (1956); *Phys. Rev.* **104** 254; **105** (1957) 1671

LEP (2006); hep-ex/0612034/

LEP & SLD (2006); *Phys. Rep.* **427** 257

Leprince-Ringuet, L. & l'Héritier, M. (1944); *Compt. Rend.* **219** 618

Lobashev, V. M. *et al.* (2001); *Nucl. Phys. B Proc. Suppl.* **91** 280

Lorentz, H. A. (1904); *Versl. Kon. Akad. v. Wet.*, Amsterdam, Dl. **12** 986 [English transl. *Proc. Acad. Sci. Amsterdam*, **6** (1904) 809]

Maglic, B. *et al.* (1961); *Phys. Rev. Lett.* **7** 178

Majorana, E. (1937); *Nuovo Cimento* **5** 171

Maki, Z. *et al.* (1962); *Prog. Theor. Phys.* **28** 870

McDonough, J. M. *et al.* (1988); *Phys. Rev.* **D38** 2121

McMillan, E. (1945); *Phys. Rev.* **68** 143

Mele, S. (2005); *International Conference on High-Energy Physics* HEP2005

Michael, D. G. *et al.* (2006); *Phys. Rev. Lett.* **97** 191801

Mikheyev, S. P. & Smirnov, A. Yu. (1985); *Yad. Fiz.* **42** 1441 [*Sov. J. Nucl. Phys.* **42** (1985) 913]

Mohapatra, R. & Pal, P. (2004); *Physics of Massive Neutrinos*, World Scientific

Muller, F. *et al.* (1960); *Phys. Rev. Lett.* **4** 418

Nakato, T. & Nishijima, K. (1953); *Prog. Theor. Phys.* **10** 581

Naroska, B. (1987); *Phys. Rep.* **148** 67

Nefkens, B. M. K. *et al.* (2005); *Phys. Rev.* **C72** 035212

Niu, K. *et al.* (1971); *Prog. Theor. Phys.* **46** 1644. For a story of the charm discovery in Japan see: K. Niu, *Proc. 1st Int. Workshop on Nuclear Emulsion Techniques*, Nagoya 1998, preprint DPNU-98–39

Noecker, M. C. *et al.* (1988); *Phys. Rev. Lett.* **61** 310

Nygren, D. R. (1981); *Phys. Scr.* **23** 584

OPERA (2000); *Proposal.* LNGS P25/2000. July 10, 2000; CERN/SPSC 2000–028; SPSC/P318

422   *References*

Orear, J. *et al.* (1956); *Phys. Rev.* **102** 1676

Pais, A. & Piccioni, O. (1955); *Phys. Rev.* **100** 1487

Panowsky, W. K. H. *et al.* (1951); *Phys. Rev.* **81** 565

Perl, M. L. *et al.* (1975); *Phys. Rev. Lett.* **35** 1489

Perkins, D. H. (2004); *Introduction to High Energy Physics*, Cambridge University Press, 4th edn.

Peruzzi, I. *et al.* (1976); *Phys. Rev. Lett.* **37** 569

Pevsner, A. *et al.* (1961); *Phys. Rev. Lett.* **7** 421

Pjerrou, G. M. *et al.* (1962); *Phys. Rev Lett.* **9** 180

Planck, M. (1906); *Verh. Deutsch. Phys. Ges.* **8** 136

Pocanic, D. *et al.* (2004); *Phys. Rev. Lett.* **93** 181803

Poincaré, H. (1905); *C. Rendues Acad. Sci. Paris* **140** 1504

Politzer, D. (1973); *Phys. Rev. Lett.* **30** 1346

Pontecorvo, B. (1946); Chalk River Lab. PD-205 report

Pontecorvo, B. (1957); *Zh. Eksp. Teor. Fiz.* **33** 549 [*Sov. Phys. JETP* **6** (1957) 429]

Pontecorvo, B. (1959); *Zh. Eksp. Teor. Fiz.* **37** 1751 [*Sov. Phys. JETP* **10** (1960) 1236]

Pontecorvo, B. (1967); *Zh. Eksp. Teor. Fiz.* **53** 1717 [*Sov. Phys. JETP* **26** (1968) 984]

Prescott, C. Y. *et al.* (1978); *Phys. Lett.* **B77** 347; *ibid.* **B84** (1979) 524

Raaf, J. L. (2006); 'Recent proton decay results from Super-Kamiokande', *Neutrino 2006, Journal of Physics: Conference Series*

Reines, F. *et al.* (1996); *Rev. Mod. Phys.* **68** 315; also *Phys. Rev.* **117** (1960) 159

Reynolds, G. T. *et al.* (1950); *Phys. Rev.* **78** 488

Richter, B. (1977); Nobel Lecture. *Rev. Mod. Phys.* **49** (1977) 251

Rochester, G. D. & Butler, C. C. (1947); *Nature* **160** 855

Rohlf, J. W. (1994); *Modern Physics From a to Z^0*. John Wiley & Sons

Rossi, B. (1930); *Nature* **125** 636

Rossi, B. (1933); *Z. Phys.* **82** 151; *Rend. Lincei* **17** (1933) 1073

Rossi, B. (1952); *High-Energy Particles*, Prentice-Hall

Rossi, B. & Nereson, N. (1942); *Phys. Rev.* **62** 17

Rubbia, C. (1985); *Rev. Mod. Phys.* **57** 699

Rubbia, C. *et al.* (1976); *Proc. Int. Neutrino Conf., Aachen*, eds. Faissner, H., Reithler, H. and Zerwas, P., Vieweg, Braunschweig, p. 683 (1977)

Salam, A. & Ward, J. C. (1964); *Phys. Lett.* **13** 168; see also A. Salam in *Elementary Particle Theory*, ed. Svartholm, N., Almquist and Wiksel, Stockholm, p. 367 (1968)

Schlamminger, St. *et al.* (2006); *Phys. Rev.* **D74** 082001

Schlein, P. E. *et al.* (1963); *Phys. Rev. Lett.* **11** 167

Schwartz, M. (1960); *Phys. Rev. Lett.* **4** 306

Schwinger, J. (1948); *Phys. Rev.* **73** 416; *ibid.* **74** (1948) 1439; *ibid.* **75** (1949) 651; *ibid.* **76** (1949) 790

Starich, M. *et al.* (2007); *Phys. Rev. Lett.* **98** 211803

Street, J. C. & Stevenson, E. C. (1937); *Phys. Rev.* **52** 1003 (L)

't Hooft, G. (1971); *Nucl. Phys.* **B35** 167 and 't Hooft & Veltman, M. (1972); *ibid.* **B44** 189

Taylor, R. E. (1991); *Rev. Mod. Phys.* **63** 573

Thomson, J. J. (1897); *Proc. Cambridge Philos. Soc.* **9**; *The Electrician*, May 21, 1897; *Philos. Mag.* **48** (1899) 547

Tomonaga, S. (1946); *Prog. Theor. Phys. (Kyoto)* **27** 1

Tonner, N. (1957); *Phys. Rev.* **107** 1203

Touschek, B. (1960); Laboratori Nazionali di Frascati, *Int. Rep.* **62**; see also L. Bonolis (2005); *Riv. Nuovo Cimento* **28** 11

Vavilov, S. I. (1934); *C. R. Acad. Sci. USSR* **8** 457

Veksler, V. I. (1944); *C. R. Acad. Sci. URSS (Doklady)* **43** 329; *ibid.* **44** 365

Vilain, P. *et al.* (1994); *Phys. Lett.* **B335** 246

Walenta, A. H. *et al.* (1971); *Nucl. Instrum. Methods* **92** 373

Weinberg, S. (1967); *Phys. Rev. Lett.* **19** 1264

Wilson, C. T. R. (1912); *Proc. R. Soc.* **A87** 277

Wilson, C. T. R. (1933); *Proc. R. Soc.* **A142** 88

Wolfenstein, L. (1978); *Phys. Rev.* **D17**, 2369; *ibid.* **D20** (1979) 2634

Wu, C. S. *et al.* (1957); *Phys. Rev.* **105** 1413; *ibid.* **106** (1957) 1361

Wu, T. T. & Yang, C. N. (1964); *Phys. Rev.* **85** 947

Yao, W.-M. *et al.* (2006); *J. Phys. G* **33** 1

Yukawa, H. (1935); *Proc. Phys. Math. Soc. Japan* **17** 48

Zel'dovich, Ya. (1959); *Sov. Phys. JETP* **94** 262

Zemach, C. (1964); *Phys. Rev.* **B133** 1202

Zweig, G. (1964); *CERN report* 8182/Th. 401 Unpublished. See also G. Zweig, Erice Lecture (1964) in *Symmetries in Elementary Particle Physics*, ed. A. Zichichi, Academic Press (1965)

Index